ZnO and Their Hybrid Nano-Structures

Potential Candidates for Diverse Applications

Edited by

Gaurav Sharma

Pooja Dhiman

Amit Kumar

International Research Centre of Nanotechnology for Himalayan Sustainability (IRCNHS), Shoolini University, India

Published by **Materials Research Forum LLC**
Millersville, PA 17551, USA

Published as part of the book series
Materials Research Foundations
Volume 146 (2023)
ISSN 2471-8890 (Print)
ISSN 2471-8904 (Online)

Print ISBN 978-1-64490-238-7
eBook ISBN 978-1-64490-239-4

Distributed worldwide by

Materials Research Forum LLC
105 Springdale Lane
Millersville, PA 17551
USA
https://www.mrforum.com

Manufactured in the United States of America
10 9 8 7 6 5 4 3 2 1

This Book is dedicated to my beloved mother Late Smt. Parkash Sharma. Thanks for all the Love & Sacrifices. Your memories will be a blessing and motivation for life

Gaurav Sharma

Table of Contents

Preface

With the upgradation of technology and commercialisation of nano-materials worldwide, nanotechnology has become hotspot of research fetching great attention of the scientific community. Among nanostructures there exist some novel materials which serve almost in every sector in the field of nanotechnology. Various metal oxides and their hybrid nanostructures are seeking importance because of their diverse applications fulfilling the current technology needs.

ZnO is one such wonder material which has unique optical, physical and chemical properties enabling it and its hybrid nanostructure to find a space in major applications like, optoelectronics, bio-medical, agricultural, and in almost every environmental application. Looking into the past few decades' of data, ZnO and its hybrid nanostructures are being extensively explored for more diverse future applications. Although, ZnO is itself a multi-functional material with several unique properties such as good transparency, high electron mobility, broad band gap, and strong room temperature luminescence.

ZnO material has inherent capacity of exhibiting application in cosmetic industries, solar cell applications, photo-voltaic devices, bio-medical applications, agriculture applications and photo-catalytic degradation of organic pollutants. ZnO being a highly biocompatible material and also easy to fabricate is associated with potential applications like fuel cells, uv filters, lasers, light-emitting diodes, photo-detectors, and sensors such as bio sensors, chemical sensors, gas sensors etc. ZnO is even capable enough to have role in spin-tronic devices and can serve as better host matrix in diluted magnetic semiconductors. A broader view of applications of ZnO suggests its utilization in more advanced devices like nano-generators and piezotronics respectively. Currently, ZnO based nanostructures are being continuously explored for electrochemical sensing applications, and photo-catalytic applications against harmful organic pollutants like dyes, heavy metals, antibiotics.

In summary, ZnO is bestowed with substantial potential to cater to the need of today's technology. It is the necessity of time to accumulate diverse applications of ZnO and ZnO based nanostructures at one spot knowledgeable source.

The focus of this book is concentrated on the basics of ZnO, recent processing techniques of ZnO and their nano hetero-structures, latest trends, advances, and diverse applications where scientists can share their knowledge, experiences, and ideas. Industry people could find ideas of tailored and engineered materials for new era devices with sustainable growth. Finally, this book aims to cover bountiful knowledge of physics, chemistry, engineering, and biology associated with ZnO along with the author's actual experimental experiences.

Editors

Dr. Gaurav Sharma
Dr. Pooja Dhiman
Dr. Amit Kumar

International Research Centre of Nanotechnology for Himalayan Sustainability (IRCNHS), Shoolini University, India

Chapter 1

Nano ZnO: Structure, Synthesis Routes, and Properties

Mojdeh Rahnama Ghahfarokhi [1], Minoo Alizadeh Pirposhte [2#], Debjita Mukherjee [3#], Azadeh Jafarizadeh Dehaghani [4#], Jhaleh Amirian [5,6*], Agnese Brangule [5,6*], Dace Bandere [5,6*]

[1]Department of Materials Engineering, Faculty of Materials Processing and Fabrication, Isfahan University of Technology, Isfahan, Iran

[2]Department of Materials Engineering, Faculty of advanced materials, Isfahan University of Technology, Isfahan, Iran

[3]College of Medical, Veterinary and Life Sciences, University of Glasgow, University Ave, Glasgow G12 8QQ, United Kingdom

[4]Department of Materials Engineering, Faculty of Materials Processing and Fabrication, Ph.D. Student, Isfahan University of Technology, Isfahan, Iran

[5]Department of Pharmaceutical Chemistry, Riga Stradiņš University, Dzirciema 16, LV-1007, Riga, Latvia

[6]Baltic Biomaterials Centre of Excellence, Headquarters at Riga Technical University, Kalku street 1, LV-1658 Riga, Latvia

# these authors contributed equally

* Prof. Dace Bandere (dace.bandere@rsu.lv) Dr. Agnese Brangule (agnese.brangule@rsu.lv), Dr. Jhaleh Amirian (jalehamirian@gmail.com)

Abstract

Nanoscale zinc oxide (ZnO) is one of the most important materials in semiconductor applications today. The ZnO nanoparticles (ZnO-NPs) have received the most interest among the various nanoparticles. The ZnO nanostructures are composed mainly of ZnO and have at least one dimension on the nanometer scale (1-100 nm). ZnO is a wide-bandgap semiconductor with an energy gap of 3.37 eV at room temperature. Different methods have been used to synthesize ZnO NPs, which has led to different physical and chemical properties. The high surface energy of the particles produced in most of these methods tends to accumulate them. Therefore, nanoparticles of ZnO are used in biosensors, gas sensors, solar cells, ceramics, nanogenerators, photodetectors, catalysts, and active fillers in rubber and plastic due to their unique properties. As a UV absorber, it can also be used in cosmetics, photocatalysis, electrical and optoelectronic systems, and as an additive in a wide variety of industrial products.

Keywords

Zinc Oxide (ZnO), Nanoparticles (NPs), Semiconductor, Synthesis, Structure, Properties

Contents

1.   Introduction

1.1   Nano-ZnO

Zinc oxide nanoparticles (ZnO NPs) are nanoparticles of ZnO having a diameter smaller than 100 nanometers. Additionally, they have large surfaces per unit volume, which makes them a highly active catalyst. On the other hand, physical and chemical properties of ZnO NPs are dependent on the method of synthesis and preparation procedure, like those of other nanomaterials [1]. This semiconductor has a large excitation coupling energy of 60 meV at room temperature, high conductivity, and good electrical conductivity. Due to its radiation resistance, ZnO can be used in various spatial applications [2]. The most common use of ZnO NPs is in sunscreens. The reason for using these nanoparticles in sunscreen is that they effectively absorb ultraviolet light, but due to their large bandwidth, they are completely transparent to visible light as well [3, 4]. Furthermore, due to its selective toxicity to bacteria [5], they are being studied for their use in killing harmful microorganisms in packaging and in UV protection materials like textiles [6]. Considering ZnO NPs are a relatively new material, there are concerns about the risks they may pose. These nano-materials can easily penetrate tissue because of their small structure, and they can travel throughout the body, as has been demonstrated in animal studies. This penetration can reach the placenta, the blood-brain barrier, individual cells, and cell nuclei. However, human skin is an effective barrier against ZnO NPs. Up until 2011, no human disease had been caused by nanoparticles [7].

1.2 Structure of ZnO

ZnO crystalizes in three forms namely: cubic, wurtzite, and zinc blende (Figure 1). Most popular and thermodynamically stable wurtzite structure of ZnO consists of two interpenetrating hexagonal packed sub lattices of Zn and O displaced by cation-anion bond (u parameter) in c-axis. The unit cell of hexagonal ZnO is characterized by a, c lattice parameters, where 'a' is approximately. 3.2500 Å and 'c' is 5.2060 Å respectively. The c/a ratio of ZnO structure is 1.60 which is slightly deviated from 1.63 for ideal hexagonal packed structure. ZnO comes from the family of space group C^4_{6v} as per Schoenflies notation and $P_{63}mc$ in the Hermann–Mauguin notation. Figure 2 presents the schematics of wurtzite structure of ZnO.

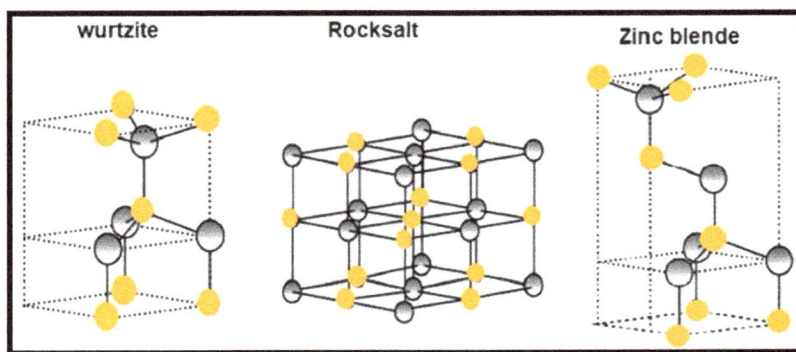

Figure 1: Crystal structure of ZnO in wurtzite , rocksalt and zinc blende form, Copyright © 2022, American Chemical Society with permission from the reference [8].

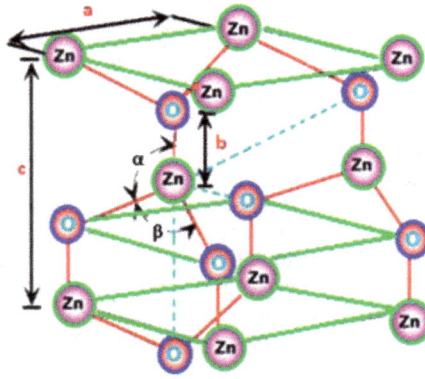

Figure 2: Schematic diagram of wurtzite structure of ZnO, Copyright © 2020, Springer-Verlag GmbH Germany, part of Springer Nature [9].

The structure is also characterized by 'u' parameter. The u parameter refers to the length of the cation-anion bond parallel to the c axis divided by c lattice parameter. It should be noted that there is a high correlation between the c/a ratio and the u parameter, such that when the c/a ratio decreases, the u parameter increases so that the four tetrahedral distances remain nearly constant despite a distortion of tetrahedral angles caused by long-range polar interactions. The slightly varying bond lengths are equal if the following relation holds good.

$$u = \left(\frac{1}{2}\right)\frac{a^2}{c^2} + \frac{1}{4}$$

1

The tetrahedral coordinated crystal structures of diamond, zinc blende, and wurtzite are suited for covalent chemical bonding via sp^3 hybridization. However, ZnO is substantially ionic in nature. That is why one can call ZnO as "centre of solid state physics". Owing to the partial ionic nature of ZnO, The bottom of the conduction band is created mostly from the 4s levels of Zn^{2+}, whereas the top of the valence band is formed from the 2p levels of O2-. At low temperatures, the band gap between the conduction and valence bands is around 3.437 eV.

1.3   Nano ZnO versus bulk ZnO

Small sizes and, therefore, high surface-to-volume ratios are key reasons why nanoparticles have superior properties compared to their bulk counterparts. Furthermore, engineered nanoparticles are more homogeneous in shape, size, and composition than bulk materials [11 ,10]. As a result, they are used in a variety of industrial and commercial applications.

ZnO materials have been used in industry for decades and have recently been undergoing a scientific and technological revival [12].

By 2024, the global ZnO market needs are estimated at $ 5.7 billion. In addition to rubber, which was the largest application of the ZnO market in 2018, this white mineral compound has been widely used in ceramics, cosmetics, pharmaceuticals, chemicals and glass [13] As well as being used as a catalyst or adsorbent, it has found use in drilling fluids, varistors, solar cells, textiles, and even has found its way into agriculture and food production. Nano-sized ZnO counterparts seem to be better alternatives than their bulky counterparts in many business processes, which is a combination of small size and high surface area. Nano-sized ZnO reduces the total amount of ZnO used in various applications, enables the engineering of materials with advanced properties, and allows the development of innovative applications in all specialized fields [13, 14].

The crystalline structure of ZnO determines many of its properties. ZnO is usually formed in a hexagonal crystal structure (wurtzite) with lattice parameters a = 3.25 Å and c = 5.12 [15]. However, it has been found to crystallize in the cubic composition of zinc and rock salt structures. The exact shape of the crystals of a mineral can vary depending on how it is produced.

The X-ray powder diffraction pattern (XRD) of the mass and nanomaterials such as zincite - mineral consisting of ZnO with a hexagonal wurtzite crystal structure reveals high crystallinity shapes. Transmission electron microscopy (TEM) imaging for both ZnO-NP and ZnO bulk provides information on morphology including particle size, shape and arrangement. Some nanoparticles have triangular edges, while others have circular shapes. The particle size and shape of ZnO (ordinary ZnO powder) particles range from square and triangular particles with a size of more than 500 nm to smaller spherical particles and longer rod-like particles.

Besides surface chemistry, nanoparticle size and shape also contribute to toxicity. Numerous studies have shown that rod-shaped nanoparticles produce a wider range of cytotoxic effects than their spherical counterparts. In contrast to their spherical counterparts, bars and wires are thought to penetrate cell walls more easily [16]. In fact, it has been speculated that the genetic toxicity of different nanoparticles may primarily be determined by their shape, rather than their chemical composition [17, 18]. However, the details of how the shape of the nanoparticle may affect its toxicity to cells are still unknown.

2.   Types of nano-ZnO

2.1   Various ZnO nanostructures

ZnO can be classified as a new material due to its diversity of nanometer-sized crystal structures with potential applications in many nanotechnology fields. ZnO can occur as a one-(1D), two-(2D), and three-dimensional (3D) structure. 1D structures is the largest group and include nanorods [19], nanowires [20-22], nanoparticles [23], needles [24], spirals, springs and rings [25], strips [26], nanotubes [27, 28], nanobelts [29], wires [30-

32] and combs [33]. In two-dimensional structures, ZnO can be obtained in two-dimensional structures such as thin films [34], nanosheets / nano-sheets and nanoplate [35]. Three-dimensional structures of ZnO include, flower [35], urchin [35], dandelions, snowflakes, conical hedgehogs and etc. [36-39]. It is known that ZnO has one of the most diverse particle structures among all known materials.

Such nanostructures grow easily at relatively low temperatures, and various growth methods have been reported for the synthesis of ZnO nanostructures, including chemical and physical methods such as thermal evaporation [40, 41], chemical vapor deposition (CVD), and CVD cyclic feeding [42], sol-gel deposition [43], electrochemical deposition [44], hydrothermal and solvothermal growth. As a result of the aforementioned fields, applications, and growth techniques, ZnO will be an important material for future research and applications [19, 40, 45].

In the wake of the discovery of semiconductor oxide nanobelts [26] in 2011, interest has grown in single-dimensional oxide-based nanostructures due to their unique and diverse applications in optics, optoelectronics, catalysts, and piezoelectrics. Semiconductor oxide nanowires are a unique group of quasi-one-dimensional nanomaterials that have been systematically studied for a wide range of materials with specific chemical compositions and crystallographic structures.

By simply evaporating the desired commercial metal oxide powders at high temperatures, nanobelt-like, quasi-one-dimensional structures (called nanobelts) have been synthesized for semiconductive oxides of Zn, Sn, In, Cd, and Ga. As synthesized, the oxide nanobelts are pure, structurally uniform, monocrystalline, and mostly dislocation-free; their cross-section is rectangular in shape with constant dimensions. Belt-like morphology is a characteristic structural property of this family of semiconducting oxides with cations of different valence states and materials of distinct crystallographic structures. The recent synthesis of nanobelts, nanosprings, and nanorings that show piezoelectric properties may make them good candidates for nanoscale sensors, actuators, and transducers.

ZnO has the unique property of exhibiting both semiconducting and piezoelectric activities, unlike other functional oxides like perovskites, rutiles, CaF_2, spinels, and wurtzites. Zinc oxide is a material that has highly diverse structures, which is richer in configurations than any other nanomaterials, including carbon nanotubes. Several ZnO nanostructures have been synthesized by using solid state thermal sublimation and controlling the growth kinetics, the local growth temperature, and the chemical composition of the source materials [40].

Many of the wurtzite family members have important applications such as ZnO, GaN, AlN, ZnS and CdSe, which are widely used in optoelectronics, lasers and piezoelectric devices. Wurtzites have a non-central symmetry, as well as polar surfaces, which are the key characteristics of their structure. A structure of ZnO, for example, is characterized by an arrangement of alternate plates consisting of O^{2-} and Zn^{2+} ions stacked alternately along the c-axis in a quadrilateral arrangement. Ions with opposite charges produce polar surfaces with positive charge (0001) -Zn and negative charge (0001) -O, which will result in a normal dipole moment and spontaneous polarization along the c-axis as well as divergence in surface energy. There are numerous ways in which different structures of ZnO can be produced by regulating the raw materials and introducing various additives [25].

3. Synthesis methods

Recently, the use of nanoparticles with the size range of 1-100 nm has attracted much attention due to their unique and new properties, and extensive research has been conducted on them [46-48]. Due to the numerous applications, studies on the synthesis methods of these materials have been considered. Among the various nanoparticles, ZnO NPs have received the most attention and various approaches have been used to synthesize ZnO NPs. Synthesis methods of these materials include sol-gel method, spray pyrolysis, microemulsion techniques, thermal evaporation, laser ablation, chemical vapor deposition, mechanical mill, photo thermal synthesis, thermal plasma synthesis, flame synthesis, electrowave microwave deposition and synthesis method named hydrothermal [49-59]. However, in most of these techniques, nanoparticles tend to accumulate due to their high surface energy. Among the above techniques, the solution-based approach is the simplest. In this technique, the morphology of nanoparticles can be controlled by optimizing various reaction conditions such as pH, precursor concentration, temperature and reaction time [60].

3.1 Commonly used methods for synthesis

In the previous sections, we discussed the different ways in which ZnO NPs are synthesized according to their intended applications, which we will address here. ZnO can be produced by metallurgical processes, chemical processes such as vapor deposition, mechanochemistry, controlled precipitation, precipitation in water solution, solvothermal and hydrothermal methods, the sol-gel process, precipitation from microemulsions, methods based upon emulsions, microemulsions, etc. These methods enable the production of particles with different shapes, sizes and spatial configurations.

3.1.1 Metallurgical process

In this approach, zinc oxide was obtained based on the roasting of zinc ore. According to ISO 9298, ZnO can be classed as either type A, derived from a direct process (the American process), or type B, derived from an indirect process (the French process). A direct process is characterized by the reduction of zinc ore by the use of coal (such as anthracite), followed by the oxidation of zinc vapor in the same reactor. Samuel Wetherill developed this process, which is conducted in a furnace in which the bottom layer consists of a coal bed that is heated by the heat left from the previous charge. There is another layer that consists of zinc ore and coal on top of this bed. Blast air is fed into the system from below, so it can provide heat to both layers and transport carbon monoxide to the zinc reduction process. ZnO made by type 'A' process contains impurities in the form of metal compounds found in zinc ores. ZnO particles formed by this way are mostly needle-shaped, but they can also be spherical. In order to get a permanent white color, the metal oxides present, such as lead, iron and cadmium, are converted into sulfates. In order to increase the permanence of the color, water-soluble substances must be added to the product, as well as increasing its acidity. The presence of acids in rubber processing technology prolongs the time of prevulcanization and ensures the safe processing of mixtures [61].

The indirect (French) process involves melting metallic zinc in a furnace and vaporizing it at around 910°C. The ZnO particles are formed by the immediate reaction of zinc vapor with oxygen in the air. The ZnO particles are transported via a cooling duct to a bag filter station, where they are collected. LeClaire introduced the indirect process in 1844, and since then, it has been known as the French process. The product's typical particle size between 0.1 and a few microns and these particles are mainly in spherical shapes. French processes are carried out using vertical furnaces, including a vertical charge, a vertical refining column, a vaporizer with an electric arc, and a rotary combustion chamber. The ZnO produced by the indirect (French) process has a higher purity level than that produced by the direct (American) process [62, 63].

3.1.2 Chemical processes

3.1.2.1 Mechano-chemical process

One of the easiest and most cost-effective methods for producing large quantities of nanoparticles is through chemical-mechanical processes. It involves high-energy dry milling at low temperatures, which occurs through ball-powder impacts in a ball mill. In this process, a "thinner" is added to the system, such as NaCl, which acts as a catalyst and separates the nanoparticles being formed. One fundamental problem with this method is the uniform grinding of powder and reduction of grains to the required size, both of which decrease as the time and energy spent milling increases. An extended milling time usually leads to a greater amount of impurities. On the other hand, it offers significant advantages such as low production costs, small particle sizes, a low tendency for particles to aggregate and high homogeneity of crystalline structure and morphology. With a longer milling time, the crystals become smaller, which may be an indication of a "critical moment." At the same time, increasing the calcination temperature will result in a larger crystal. Number of researchers has synthesized ZnO nanostructures by this route [64-66]. It has been observed that particle morphology strongly depends on the milling time of the reactant mixture; an increase in milling time will result in smaller particles. Moreover, by studying the mechanically and chemically synthesized powders by X-ray (XRD) and Raman spectroscopy, we can evaluate the complete long-range order, the net vortex structure, regardless of the milling time. Furthermore, Raman spectroscopy exhibits a different arrangement of oxide powders.

3.1.2.2 Controlled precipitation

In this method, ZnO NPs can be synthesized with reproducible properties, which has made it one of the most widely used methods. This method consists of using a reducing agent to rapidly reduce a solution of zinc salt, to limit particle growth with defined dimensions, followed by precipitation of a ZnO precursor from the solution. Next, this precursor is subjected to thermal treatment, followed by milling to remove the impurities. Due to the formation of agglomerates, calcined powders exhibit a high level of particle aggregation. The precipitation process is controlled by a number of factors, including pH, temperature, and duration. ZnO can also be precipitated from zinc chloride and zinc acetate solutions [67]. Controlled parameters in this process included the concentration of reagents, the rate of addition of substrates and the reaction temperature. A monomodal particle size distribution and increased surface area were found in ZnO produced by this process. Particle growth can be controlled using surfactants in precipitation-based nanopowder synthesis processes. The presence of these compounds affects not only the nucleus and particle growth, but also the coagulation and clotting of particles [67-73]. Figure 1 shows how surfactant can affect particle coagulation and clotting.

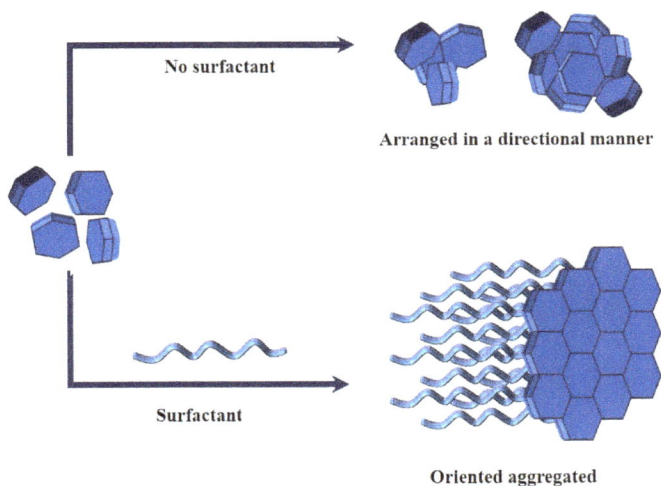

Figure 1: Effect of surfactant on the structure of a ZnO crystal.

3.1.2.3 Sol-gel method

One of the most often utilized techniques is sol-gel; it is mostly employed to create thin films and powdered catalysts. Numerous investigations found that many variations and adaptations of the method have been utilized to create pure thin films or powders in high homogenous concentrations and under stoichiometry control [74-80]. Sol-gel materials are metastable solids that are created from the molecular precursors that serve as the foundation for subsequent materials in kinetically controlled processes [81]. Usually, a metal-organic precursor is hydrolyzed and then condensed, either with or without the aid of a catalyst. A gel is created when these reactions continue and the solution's viscosity rises. A Nano porous structure is produced after the evaporation of chemical byproducts. Processing variables including the kind and quantity of catalyst and/or solvent, temperature, and in certain circumstances, the presence and arrangement of templating molecules, have an impact on the size, distribution, and interconnectivity of pores. These porous gels make up a really intriguing class of nanomaterial's on their own [82].

Due to its simplicity, low cost, repeatability, reliability and relative mild synthesis conditions, the sol-gel method is an attractive method of obtaining ZnO nanopowders that can be surface-modified with selected organic compounds. These properties are modified by some organic compounds, resulting in a wider array of applications [83-87]. Sol-gel nanoparticles provide favorable modified properties, as reflected in a variety of scientific publications based on this methodology. This method is schematically shown in Figure 2. However, by varying the experimental conditions, metal salts, and even the calcination temperature results into varied morphology, dimensions, shape and altered properties.

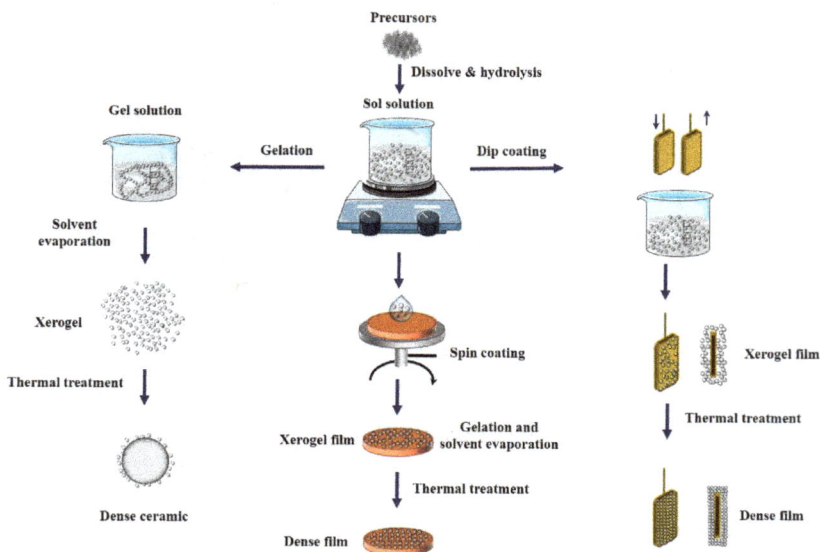

Figure 2: Schematic of the sol-gel method.

Sensitivity of the method can be checked by merely observing results of two recently reported works on ZnO NPS synthesized by sol-gel method. Figure 3 (a-d) shows the SEM images of flowered shaped ZnO NPs synthesized by sol-gel method using Zn acetate as starting raw material [88]. However, Figure 3(e-h) demonstrates the TEM images showing some hexagonal shaped ZnO and Sm doped ZnO NPs [89]. These nanoparticles were synthesized by modified sol-gel route. The results clearly indicate that merely changing the experimental conditions and steps in the same synthesis route may results into different morphology of the nanoparticles [90]. Researchers have noticed the impact of annealing temperature on the morphology, electrical and optical properties of ZnO NPs [91]. Also, the ageing time of the sol during synthesis may also results into varied morphological and associated properties of ZnO nanostructures[92].

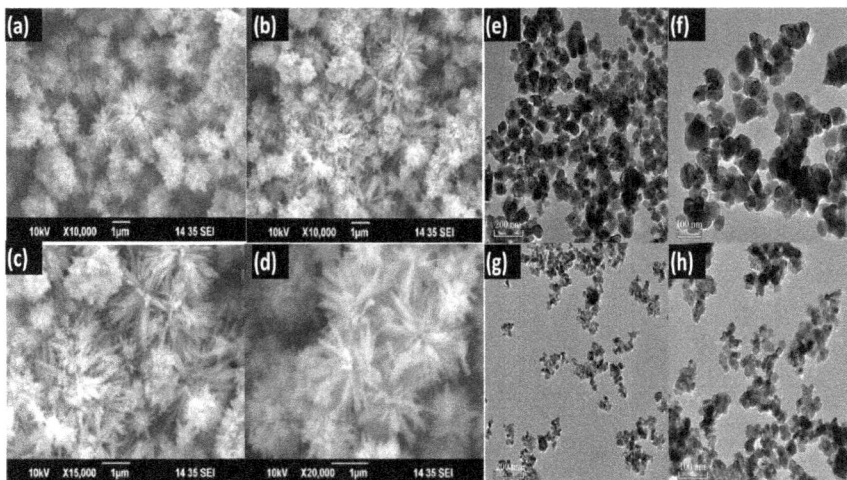

Figure 3: (a-d)SEM Images of ZnO NPs synthesized using sol-gel method (using Zn acetate as raw salt, annealing temperature 300 ℃), Copyright © 2022 Elsevier B.V. with permission from the reference [88], (e-h) TEM images of pure ZnO and Sm doped ZnO NPs (using Zn and Sm nitrate and citric acid as raw salts, annealing temperature 500 ℃), Copyright © 2022 Chinese Society of Rare Earths published by Elsevier B.V. with permission from the reference [89].

3.1.2.4 Thermal and hydrothermal solution method

As hydrothermal methods do not require grinding or calcination, they are known as being simple and environmentally friendly due to the fact that they eliminate the use of organic solvents. As stated above, the synthesis occurs in an autoclave where the substrate mixtures are gradually heated to a temperature of 100–300 °C (low temperature) and allowed to stand for several days. The formation of crystals occurs by heating followed by cooling, which results in the formation of nuclei and subsequent growth of those nuclei. The benefits of this process include the ability to carry out the synthesis at low temperatures, the high degree of crystallinity of the product, and the high purity of the material that results, as well as the possibility of synthesis at low temperatures [93, 94]. Figure 4 schematically illustrates this method. In a recent work (Figure 5), four different kinds of morphology of ZnO was observed by varying the amount of chelating agent, replacing acetates salts with nitrates, and by changing the amount of NaOH etc. in the same hydrothermal route [95]. Such results indicate that hydrothermal is an appropriate, low temperature technique to change the morphology of the ZnO NPs and also their properties.

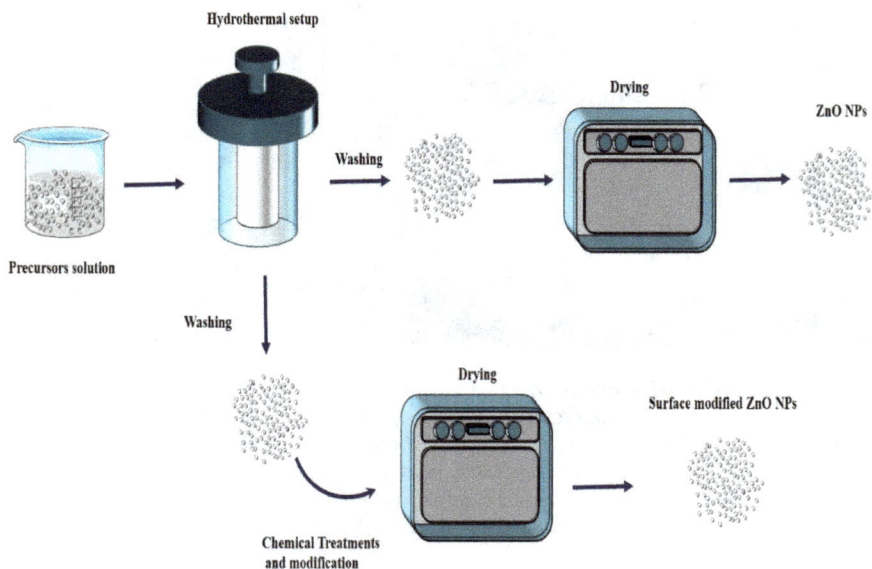

Figure 4: Schematic representation of the hydrothermal method of obtaining ZnO NPs and their chemical modification.

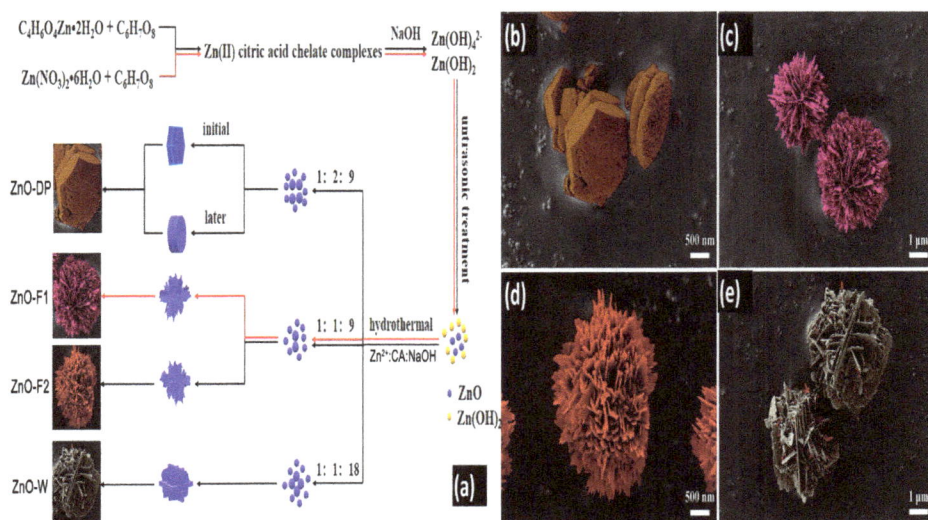

Figure 5: (a) Schematic representation of the hydrothermal method opted with varied experimental conditions for obtaining ZnO NPs and (b) corresponding SEM images, Copyright © 2022 Elsevier B.V. with permission from the reference [95].

3.1.2.5 Emulsion or microemulsion medium

Emulsions are defined as mixtures of two or more liquids that are normally immiscible (unmixable or unblendable) due to liquid-liquid phase separation. Emulsions of oil-in-water (O/W) and water-in-oil (W/O) are commonly known as these two types. These two groups are commonly referred to as oil-in-water (O/W) and water-in-oil (W/O) emulsions. It is very general to use the terms "oil" and "water" when describing this process. Hydrophobic and non-polar liquids are classified as "oil," while highly polar and hydrophilic liquids are classified as "water" [53, 96]. The particle size range of ZnO deposited in this emulsion system is smaller (as compared to ZnO deposited in a traditional system). The particles produced by this method are almost spherical. An illustration of this method is shown schematically in Figure 6.

Materials Research Forum LLC

https://doi.org/10.21741/9781644902394-1

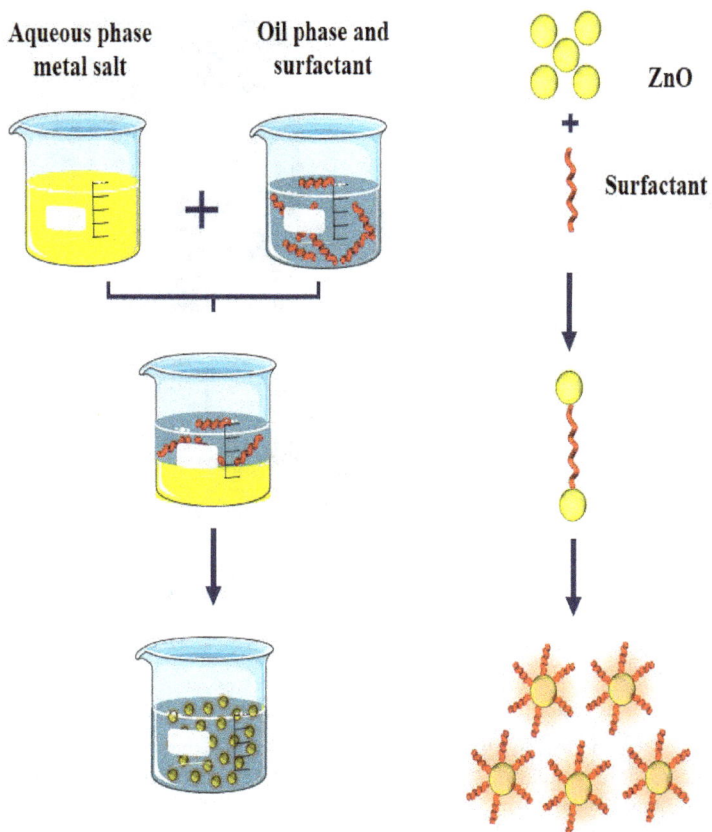

Figure 6: Schematic of the Emulsion method for obtaining ZnO NPs.

Table 2: Summary of the preparation methods, obtained shapes, sizes of nanoparticles, and references for the ZnO NPs.

Method	Precursors	Properties and application	Ref.
Precipitation process	$Zn(CH_3COO)_2$ and KOH as a water solutions	particles diameter: 160–500 nm, BET: 4–16 m^2/g	[67]
	$Zn(NO_3)_2$	wurtzite structure; particles diameter: 50 nm; application: as a gas sensor	[69]
	$Zn(CH_3COO)_2$, $(NH_4)_2CO_3$, PEG10000 as a water solutions	zincite structure; spherical particles (D ~ 30 nm); application: as a photocatlyst in photocatalytic degradation	[97]
	$Zn(NO_3)_2$, NaOH	particles of spherical size of around 40 nm	[70]
	$ZnSO_4$, NH_4HCO_3, ethanol	wurtzite structure; crystallite size 9–20 nm; particle size D: ~12 nm, BET: 30–74 m^2/g	[98]
	$Zn(CH_3COO)_2$, NH_3 aq.	hexagonal structure, shape of rods, flower-like particles: L: 150 nm, D: 200 nm	[72]
	$ZnSO_4$, NH_4OH, NH_4HCO_3	hexagonal structure, flake-like morphology (D: 0.1–1 μm, L: 60 nm)	[99]
	microsized ZnO powder, NH_4HCO_3	hexagonal wurtzite structure; flower-like and rod-like shape (D: 15–25 nm, BET: 50–70 m^2/g)	[100]
	$Zn(CH_3COO)_2$, NaOH	hexagonal structure; flower shape (L: >800 nm); application: antimicrobial activity	[101]
Mechano-chemical process	$ZnCl_2$, Na_2CO_3, NaCl	hexagonal structure; particles diameter: 21–25 nm regular shape of particles; diameter ~27 nm, BET: 47 m^2/g	[102-104]
Sol-gel	$Zn(CH_3COO)_2$, oxalic acid $(C_2H_2O_4)$, ethanol	hexagonal wurtzite structure; uniform, spherically shaped of particles	[84]
	zinc 2-ethylhexanoate, tetramethylammonium (TMAH) $((CH3)4NOH)$, ethanol and 2-propanol	cylinder-shaped crystallites, D: 25–30 nm; L: 35–45 nm	[105]
	$Zn(CH_3COO)_2$, oxalic acid, ethanol and methanol	zincite structure; aggregate particles: ~100 nm; shape of rod; particles L: ~500 nm, D: ~100 nm BET: 53 m^2/g	[106]

		application: decontamination of sarin (neuro-toxic agent)	
	$Zn(CH_3COO)_2$, diethanolamine, ethanol	hexagonal structure; particles: nanotubes of 70 nm	[86]
Solvothermal hydrothermal and microwave techniques	$Zn(CH_3COO)_2$, NaOH, HMTA (hexamethylenetetraamine)	spherical shape; particles diameter: 55–110 nm	[107]
	$Zn(CH_3COO)_2$, NH_3, zinc 2-ethylhexanoate, TMAH, ethanol, 2-propanol	particles with irregular ends and holes; aggregates consist particles of 20–60 nm, BET: 0.49–6.02 m^2/g	[108]
	trimethylamine N-oxide, 4-picoline N-oxide, HCl, toluene, ethylenediamine (EDA), tetramethylethylenediamine (TMEDA)	wurtzite structure; particles morphology: nanorods (40–185 nm), nanoparticles (24–60 nm)	[109]
	zinc acetylacetonate, methoxy-ethoxy- and n-butoxyethanol, zinc oximate	zincite structure; average crystallite size: 9–31 nm; particles diameter: 40–200 nm; BET: 10–70 m^2/g	[110]
Emulsion	$Zn(NO_3)_2$, surfactant (ABS, Tween-80 and 40, $C_{21}H_{38}BrN$)	grain size: cationic surfactants (40–50 nm), non-ionic surfactants (20–50 nm), anionic surfactants (~20 nm)	[69]
	$Zn(C_{17}H_{33}COO)_2$, NaOH, decane, water, ethanol	particles morphology: irregular particles aggregates (2–10 μm); needle-shaped (L: 200–600 nm, T: 90–150 nm); nearly spherical and hexagonal (D: 100–230 nm); spherical and pseudospherical aggregates (D: 150 nm	[111]
Microemulsion	$Zn(NO_3)_2$, NaOH, heptane, hexanol, Triton X-100, PEG400	hexagonal (wurtzite e) structure; particles morphology: needle (L: 150–200 nm, D: ~55 nm), nanocolumns (L: 80–100 nm, D: 50-80 nm), spherical (~45 nm)	[53]
	$Zn(NO_3)_2$, oxalic acid, isooctane, benzene, ethanol, diethyl ether, chloroform, acetone, methanol, Aerosol OT	equivalent spherical diameter: 11.7–12.9 nm, BET: 82–91 m^2/g	[112]
	$Zn(CH_3COO)_2$, Aerosol OT, glycerol, $C_{20}H_{37}NaO_7S$, n-heptane, NaOH, methanol, chloroform	hexagonal wurtzite structure, spherical shape (15–24 nm), rods shape (L: 66–72 nm, D: 21–28 nm)	[113]
	ZnC_{12}, $Zn(CH_3COO)_2$, heptane, BTME (1,2-trimethoxysilyl)ethane, TMOS (tetramethoxysilane), methanol, Aerosol OT, NaOH	hexagonal structure, uniformly dispersed small particles, size of particles ~10 nm	[114]

16

Materials Research Forum LLC
https://doi.org/10.21741/9781644902394-1

A modification of the synthesis method allows different morphologies of ZnO nanostructures to be produced. A study by Yu et al. shows that hexagonal zinc nanotoxide sheaths crystallize from oxygen vapor in silicon by non-catalytic vapor transfer at 630° C [115]. Further, they also reported the methods for synthesizing uniform quadruped-like ZnO nanostructures n bulk by thermal evaporation using zinc metal powder at 850-925 °C [116]. First reports of comb-like ZnO nanostructures and their growth were also made by Wang et al.[117]. Furthermore, Yan et al. developed highly functional microscale composite structures based on a single-crystalline ZnO nanowire array synthesized via chemical vapor transfer and compression of zinc metal powder at 800-900 °C on a silicon substrate[118]. Gao et al. report on the synthesis of polar surface-dominated ZnO NPs arrays using a dual-stage VS mechanism, producing the nanoparticles after first heating the ZnO, SnO_2, and graphite powders at 1200° C for 60 minutes, followed by a subsequent 1300° C thermal treatment at 30° C in polycrystalline Al_2O_3. [119].

4. Properties

4.1 Physical properties

Table 3 shows the main physical properties of ZnO [45]. There is, however, an important note that with the reduction in size of materials, some of their physical properties will change – this phenomenon is referred to as the "quantum size effect". Quantum confinement, for instance, increases the band gap energy of quasi-one-dimensional (Q1D) ZnO [120]. This size dependence is also observed in the bandgap of ZnO NPs [121]. Furthermore, with downsizing of ZnO nanorods, X-ray absorption spectroscopy and scanning photoelectron microscopy reveal enhanced surface states [122]. In addition, surface states can significantly affect the carrier concentration in Q1D systems, according to studies of nanowire chemical sensing [123-125]. A fundamental understanding of physical properties is essential for designing effective devices.

Table 3: Physical properties of wurtzite ZnO.

Properties	Value
Lattice constants (T = 300 K)	
• $a_0 = b_0$	3.2469 nm
• c_0	5.2069 nm
Density	5.606 g/cm^3
Melting point	2248 K
Relative dielectric constant	8.66
Gap Energy	3.4 eV, direct
Intrinsic carrier concentration	<10^6 cm^{-3}
Exciton binding energy	60 meV
Electron effective mass	0.24
Electron mobility (T = 300 K)	200 cm^2/Vs
Hole effective mass	0.59
Hole mobility (T = 300 K)	5–50 cm^2/Vs

4.2   Mechanical properties

The mechanical characteristics of materials include a variety of concepts, including hardness, stiffness, piezoelectric constants, Young's modulus, bulk modulus, and yield strength. The direct measurement of the mechanical behavior of single nanostructures can be quite challenging due to the fact that the traditional measurement methods that apply to bulk materials are not applicable to these materials. However, scientists have opted various micro and nano-identification methods to measure the hardness of ZnO nanostructures. On the (0,0,0,1) surface of the crystal, hardness measurements are often taken with a standard pyramidal or spherical diamond tip, or alternatively, with a sharp triangular indenter. To analyze the mechanical properties of (0001) ZnO nanowires, in situ HRTEM equipped with AFM was utilized and the young's modulus was determined to have value of 249 GPa for nanowires with diameter of 40 nm. In the experiment, thick nanowires were reported as brittle while thin wires were found to have high flexibility[126]. In a report, hardness of the ZnO thin films was found to vary with deposition temperature of films. Figure 7 shows the impact of deposition temperature on hardness of ZnO thin films. According to Hall–Petch relationship, reduction in the grain size leads to hardness of the material [127]. However, there are other reported works also, which are not in agreement with H-P theory. The measurements of depth-sensing indentation offer comprehensive data on the hardness and pressure-induced phase change of semiconductor materials. Through an electric field resonance excitation, TEM can be used to determine the bending modulus of ZnO nanotubes [128]. To achieve this method, an oscillating electric field is applied to a ZnO nanofiber attached to a fixed electrode using a special TEM sample holder. This electric field stimulates the vibration of the nanowire, and by adjusting the frequency of motion, the oscillation is intensified. In terms of nano-intensifiers, ZnO nanofibers will be highly promising.

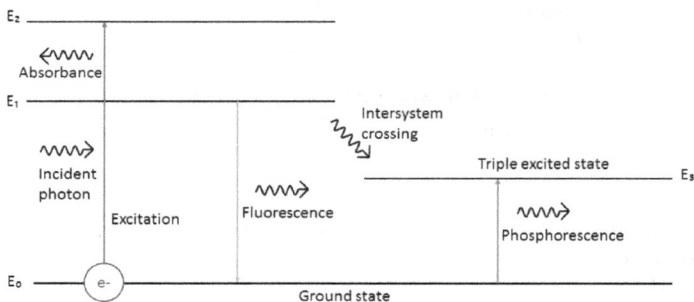

Figure 7: Variations of hardness (ZnO thin films) with deposition temperature, Copyright © 2013 Elsevier B.V. with permission from the reference [129]

4.3   Electrical properties

For the development of nanoelectronic applications of ZnO nanostructures, it is crucial to study their electrical properties. To study this property, we could examine the electrical transmission across ZnO nanowires and nanorods. A single crystalline ZnO nanowire, however, demonstrates high mobility of 80 cm^2 / V s [130].

In comparison to thin films, particles based on ZnO nanostructure can achieve faster performance speeds. ZnO's main deficiency is the difficulty of doping it with p-type electrons, making it unsuitable for widespread applications in electronics and photonics. A thin layer of ZnO p-type with low strength (0.5 cm) is formed by a Ga and N coding method [131]. When ZnO nanostructures are successfully doped with p-type metal, their potential applications in electronics and optoelectronics will be significantly enhanced. ZnO type p- and n-type nanowires can be used to make p-n junction diodes and light emitting diodes (LEDs). The field effect transistors (FETs) made from them can form complementary logic circuits. Besides creating an optical pore, the electrically driven nanowire laser also attempted to make a p-n junction within ZnO nanowires. To create this porous pattern, anodic aluminum membrane, was created with pores of an average size of approximately 40 nm. A two-stage vapor transfer growth was used, and boron was introduced as a p-type impurity. In addition to the electrical transport studies, an exhaustive analysis of the ZnO nanowires and nanorods electric field emission has also been conducted. There is a lot of evidence that nanomaterials that are pseudo-dimensional (Q1D) are ideal candidates for electron field emission. In fact, there have been numerous studies of the field emissions of vertically aligned ZnO nanowires and nanowires [27, 132-136].

4.4   Optical properties

Photonic devices rely heavily on ZnO nanostructures because of their inherent optical properties. The photoluminescence spectrum of ZnO nanostructures makes them ideal for photonic devices [137-140]. Excitonic emission from photoluminescence spectrum of ZnO nanorod has demonstrated how the quantum size constraint can enhance the exciton bond energy in significant ways. An emission peak appears at 380 nm due to the transfer of the band to the band and the green-yellow emission band associated with the oxygen-free space. However, these results are similar to those of bulky ZnO. It is interesting to note that with decreasing diameters of nanowires, the intensity of green emission increases. It is believed that the larger surface area of thinner nanowires in comparison to their volume results in an increased level of surface defects and recombination. An important characteristic of nanoscale systems is quantum confinement which leads to a change in water at a wavelength of near UV emission in nanofibers of ZnO [120]. Based on PL spectra, ZnO nanowire is a promising material for UV emission and UV laser applications. Moreover, zinc oxide nanowires and nanorods are an excellent candidate for optical waveguides because of their near-circular geometry and large refractive index [141, 142]. A threshold of 40 kW/cm^2 ~ 100 kW /cm^2 applies to these materials, although a higher crystal quality lowers the threshold. ZnO nanowire lasers have the advantage that the oxytonic recombination reduces the threshold, while quantum confinement creates a density of states at the edges of the strip and increases the radiation efficiency. One of the recent advances in these materials is the use of dielectric nanowires for light wave conduction. An investigation reported recently reported that ZnO nanowires can act as sub wavelength optical waveguides. In the experiment, optically pumped light emission was guided by ZnO nanowires and coupled to a SnO_2 nanoribbon. These findings suggest that ZnO nanostructures could serve as a building block for integrated electronic and optical circuits, according to these findings. Other than UV and laser light, ZnO nanowires can also be used for UV detection and optical switching. When the electric field component of incident light is parallel to the nanowire's long axis, the light current is maximized. It has been demonstrated that O_2 has a significant impact on the optical response, i.e., that the adsorption of O_2 on nanowires can speed up the relaxation speed of the optical current. The rest time of light in air is approximately 8 hours, but it is much longer in a vacuum. In this study, it was discovered that the O_2 desorption-absorption process affected the optical response of ZnO nanowires. By recombining electrons and holes on the adsorbed surface, the holes produced by light discharge the O_2 and significantly increase the conductivity of the adsorbed chemical surface. By turning off the light, the oxygen molecules are reabsorbed on the nanowire surface, which reduces its conductivity.

4.5   Chemical assay

Metals oxide when oxygen reaches their surface. The oxygen voids on the metal oxide surfaces are highly chemically and electrically active. They serve as n-type donors and greatly increase the conductivity of the metal oxide. The emission of electrons occurs by absorbing charge-accepting molecules in empty spots, such as in NO_2 and O_2. Conductivity of type n oxide is reduced by it after it is discharged from the conducting strip. In contrast, CO and H_2 molecules react with oxygen adsorbed on the surface and, as a result, increase conductivity due to the removal of oxygen. This is the principle behind metal oxide gas sensors. Among the main materials used for solid state gas sensors, ZnO thin and bulk layers can measure gasses such as CO, NH_3, alcohol, and H_2 at high temperatures, such as 400 °C.

According to this property, the chemical sensing performance of Q1D, ZnO nanowires and nanorods, is expected to make it superior to its thin film counterpart. Due to their small diameter, which is comparable to the length of Debye, chemicals that adsorb on their surfaces have an impact on the electronic structure. This makes Q1D ZnO more sensitive to thin films. Additionally, ZnO nanowires and nanorods can be used as two-terminal measurement devices or as field-effect transistors (FET), where the measurement property can be adjusted by a transverse electric field. Gate voltages can also be used to control the sensitivity range. Further, it is possible to restore nanowire conductivity by applying a negative gate potential greater than threshold voltage. The gas selectivity of NO_2 and NH_3 using the ZnO FET nanowires is also important as a gas detection function under gate refresh. As nanowires have a large surface-to-volume ratio, they not only improve gas sensor performance, but also facilitate the process of storing hydrogen [123, 124, 143-145].

4.6   Thermal properties

Grid parameters and a vast majority of semiconductor properties are affected by temperature changes. Thermal conductivity (K) is a kinetic property determined by vibration, rotation, and electronic transmission. Many semiconductors, including ZnO, have defects, and these affect the thermal conductivity significantly. Most measurements of thermal conductivity are based on studies of vapor phase materials that measure the polar shape of ZnO. Crystalline ZnO is a thermochromic, meaning that their color changes from white to yellow when heated and returns to white shading upon cooling. As a ceramic material, ZnO is advantageous due to its high thermal limit, thermal conductivity, low thermal expansion, and high liquefying temperature. Thermal conductivity of ZnO material can be influenced by the presence of point imperfections, such as vacancies in the material, impurities and isotope variations. Defects in ZnO materials such as voids, impurities, and isotope changes influence the material's thermal conductivity [146, 147].

Conclusion

A number of fascinating properties of ZnO make this material useful in new materials and applications, including the absorption of ultraviolet light, light stability, biocompatibility, and biodegradability. Consequently, the study of its various properties and methods of synthesis on a large industrial scale has become one of the most interested topics of the scientific community. Therefore, ZnO particles are synthesized in both a nano and micrometric size, and a variety of methods are employed in the process. In general, these methods can be classified into two categories: metallurgical and chemical methods. There are two processes by which ZnO is obtained in metallurgical processes: direct and indirect. Additionally, chemical methods were also used for the preparation of ZnO and can also be classified into two groups: dispersion methods and compaction methods. Through mechano-chemical processes, ZnO is produced by grinding suitable precursors; the resultant product may consist of very fine nanoparticles with an average size of 10-20 nanometers. A variety of compression methods are available, including sol-gel, hydro- and solvent-based methods, emulsions and micro-emulsions, and many others. Most of these methods involve the use of a homogeneous molecular solution that is exposed to the nucleation process and eventually grows as a result of the process. The materials that are obtained can be used to produce advanced and durable ceramics, transparent solar filters that block infrared and ultraviolet light, and catalysts. Additionally, these substances can be employed to diagnose, treat, and prevent a variety of diseases. This is illustrated by the notion of targeted drug delivery that delivers a medication directly into a patient's cell while at the same time causing the lowest side effects. In conclusion, it can be stated with confidence that ZnO can be classified as a multifunctional material, it can be expected that we will see new facilities developed for the inclusion of ZnO in the future.

Acknowledgements

This project has received funding from the European Union's Horizon 2020 research and innovation programme under the grant agreement No 857287.

References

[1] S.S. Kumar, P. Venkateswarlu, V.R. Rao, G.N. Rao, Synthesis, characterization and optical properties of zinc oxide nanoparticles, International Nano Letters, 3 (2013) 1-6. https://doi.org/10.1186/2228-5326-3-30

[2] K. Nakahara, H. Takasu, P. Fons, A. Yamada, K. Iwata, K. Matsubara, R. Hunger, S. Niki, Interactions between gallium and nitrogen dopants in ZnO films grown by radical-source molecular-beam epitaxy, Applied physics letters, 79 (2001) 4139-4141. https://doi.org/10.1063/1.1424066

[3] Y. Zhang, Y.-R. Leu, R.J. Aitken, M. Riediker, Inventory of engineered nanoparticle-containing consumer products available in the Singapore retail market and likelihood of release into the aquatic environment, International journal of environmental research and public health, 12 (2015) 8717-8743. https://doi.org/10.3390/ijerph120808717

[4] F. Piccinno, F. Gottschalk, S. Seeger, B. Nowack, Industrial production quantities and uses of ten engineered nanomaterials in Europe and the world, Journal of Nanoparticle Research, 14 (2012) 1-11. https://doi.org/10.1007/s11051-012-1109-9

[5] B. Das, S. Patra, Chapter 1 - Antimicrobials: Meeting the Challenges of Antibiotic Resistance Through Nanotechnology, in: A. Ficai, A.M. Grumezescu (Eds.) Nanostructures for Antimicrobial Therapy, Elsevier, 2017, pp. 1-22. https://doi.org/10.1016/B978-0-323-46152-8.00001-9

[6] C.Ş. Iosub, E. Olăreţ, A.M. Grumezescu, A.M. Holban, E. Andronescu, Toxicity of nanostructures-a general approach, in: Nanostructures for Novel Therapy, Elsevier, 2017, pp. 793-809. https://doi.org/10.1016/B978-0-323-46142-9.00029-3

[7] R. Kessler, Engineered nanoparticles in consumer products: understanding a new ingredient, in, National Institute of Environmental Health Sciences, 2011. https://doi.org/10.1289/ehp.119-a120

[8] M. Zare, K. Namratha, S. Ilyas, A. Sultana, A. Hezam, S. L, M.A. Surmeneva, R.A. Surmenev, M.B. Nayan, S. Ramakrishna, S. Mathur, K. Byrappa, Emerging Trends for ZnO Nanoparticles and Their Applications in Food Packaging, ACS Food Science & Technology, 2 (2022) 763-781. https://doi.org/10.1021/acsfoodscitech.2c00043

[9] G.R. Khan, Crystallographic, structural and compositional parameters of Cu-ZnO nanocrystallites, Applied Physics A, 126 (2020) 311. https://doi.org/10.1007/s00339-020-03480-y

22

[10] P.J. Borm, D. Robbins, S. Haubold, T. Kuhlbusch, H. Fissan, K. Donaldson, R. Schins, V. Stone, W. Kreyling, J. Lademann, The potential risks of nanomaterials: a review carried out for ECETOC, Particle and fibre toxicology, 3 (2006) 1-35. https://doi.org/10.1186/1743-8977-3-11

[11] C. Sioutas, R.J. Delfino, M. Singh, Exposure assessment for atmospheric ultrafine particles (UFPs) and implications in epidemiologic research, Environmental health perspectives, 113 (2005) 947-955. https://doi.org/10.1289/ehp.7939

[12] D. Dimova-Malinovska, Nanostructured ZnO thin films: properties and applications, in: Nanotechnological Basis for Advanced Sensors, Springer, 2011, pp. 157-166. https://doi.org/10.1007/978-94-007-0903-4_16

[13] Markets, Markets, Zinc Oxide Market by Process (French Process, Wet Process, American Process), Grade (Standard, Treated, USP, FCC), Application (Rubber, Ceramics, Chemicals, Agriculture, Cosmetics & Personal Care, Pharmaceuticals), Region-Global Forecast to 2024, in, MarketsandMarkets™ Research Private Ltd. Hadapsar, India, 2019.

[14] G. Heideman, R. Datta, J.W. Noordermeer, B.v. Baarle, Influence of zinc oxide during different stages of sulfur vulcanization. Elucidated by model compound studies, Journal of applied polymer science, 95 (2005) 1388-1404. https://doi.org/10.1002/app.21364

[15] D.P. Norton, Y. Heo, M. Ivill, K. Ip, S. Pearton, M.F. Chisholm, T. Steiner, ZnO: growth, doping & processing, Materials today, 7 (2004) 34-40. https://doi.org/10.1016/S1369-7021(04)00287-1

[16] K. Qi, B. Cheng, J. Yu, W. Ho, Review on the improvement of the photocatalytic and antibacterial activities of ZnO, Journal of Alloys and Compounds, 727 (2017) 792-820. https://doi.org/10.1016/j.jallcom.2017.08.142

[17] H. Yang, C. Liu, D. Yang, H. Zhang, Z. Xi, Comparative study of cytotoxicity, oxidative stress and genotoxicity induced by four typical nanomaterials: the role of particle size, shape and composition, Journal of applied Toxicology, 29 (2009) 69-78. https://doi.org/10.1002/jat.1385

[18] T.M. Allen, P.R. Cullis, Drug delivery systems: entering the mainstream, Science, 303 (2004) 1818-1822. https://doi.org/10.1126/science.1095833

[19] G.-C. Yi, C. Wang, W.I. Park, ZnO nanorods: synthesis, characterization and applications, Semiconductor science and technology, 20 (2005) S22. https://doi.org/10.1088/0268-1242/20/4/003

[20] Q. Zhao, M. Willander, R. Morjan, Q. Hu, E. Campbell, Optical recombination of ZnO nanowires grown on sapphire and Si substrates, Applied Physics Letters, 83 (2003) 165-167. https://doi.org/10.1063/1.1591069

[21] P. Gao, Y. Ding, Z. Wang, Crystallographic orientation-aligned ZnO nanorods grown by a tin catalyst, Nano Letters, 3 (2003) 1315-1320. https://doi.org/10.1021/nl034548q

[22] G. Shen, J.H. Cho, J.K. Yoo, G.-C. Yi, C.J. Lee, Synthesis and optical properties of S-doped ZnO nanostructures: nanonails and nanowires, The Journal of Physical Chemistry B, 109 (2005) 5491-5496. https://doi.org/10.1021/jp045237m

[23] E.A. Meulenkamp, Synthesis and growth of ZnO nanoparticles, The journal of physical chemistry B, 102 (1998) 5566-5572. https://doi.org/10.1021/jp980730h

[24] R. Wahab, S. Ansari, Y.-S. Kim, H.-K. Seo, H.-S. Shin, Room temperature synthesis of needle-shaped ZnO nanorods via sonochemical method, Applied Surface Science, 253 (2007) 7622-7626. https://doi.org/10.1016/j.apsusc.2007.03.060

[25] X.Y. Kong, Y. Ding, R. Yang, Z.L. Wang, Single-crystal nanorings formed by epitaxial self-coiling of polar nanobelts, Science, 303 (2004) 1348-1351. https://doi.org/10.1126/science.1092356

[26] Z.W. Pan, Z.R. Dai, Z.L. Wang, Nanobelts of semiconducting oxides, science, 291 (2001) 1947-1949. https://doi.org/10.1126/science.1058120

[27] Y. Xing, Z. Xi, Z. Xue, X. Zhang, J. Song, R. Wang, J. Xu, Y. Song, S.-L. Zhang, D. Yu, Optical properties of the ZnO nanotubes synthesized via vapor phase growth, Applied Physics Letters, 83 (2003) 1689-1691. https://doi.org/10.1063/1.1605808

[28] B. Zhang, N. Binh, K. Wakatsuki, Y. Segawa, Y. Yamada, N. Usami, M. Kawasaki, H. Koinuma, Formation of highly aligned ZnO tubes on sapphire (0001) substrates, Applied physics letters, 84 (2004) 4098-4100. https://doi.org/10.1063/1.1753061

[29] Y.-B. Hahn, Zinc oxide nanostructures and their applications, Korean Journal of Chemical Engineering, 28 (2011) 1797-1813. https://doi.org/10.1007/s11814-011-0213-3

[30] B. Nikoobakht, X. Wang, A. Herzing, J. Shi, Scalable synthesis and device integration of self-registered one-dimensional zinc oxide nanostructures and related materials, Chemical Society Reviews, 42 (2013) 342-365. https://doi.org/10.1039/C2CS35164A

[31] L. Tien, S. Pearton, D. Norton, F. Ren, Synthesis and microstructure of vertically aligned ZnO nanowires grown by high-pressure-assisted pulsed-laser deposition, Journal of Materials Science, 43 (2008) 6925-6932. https://doi.org/10.1007/s10853-008-2988-0

[32] J. Cui, Zinc oxide nanowires, Materials Characterization, 64 (2012) 43-52. https://doi.org/10.1016/j.matchar.2011.11.017

[33] T. Xu, P. Ji, M. He, J. Li, Growth and structure of pure ZnO micro/nanocombs, Journal of Nanomaterials, 2012 (2012). https://doi.org/10.1155/2012/797935

[34] S. Jeong, B. Park, S.-B. Lee, J.-H. Boo, Metal-doped ZnO thin films: synthesis and characterizations, Surface and Coatings Technology, 201 (2007) 5318-5322. https://doi.org/10.1016/j.surfcoat.2006.07.185

[35] T.M. Shang, J.H. Sun, Q.F. Zhou, M.Y. Guan, Controlled synthesis of various morphologies of nanostructured zinc oxide: flower, nanoplate, and urchin, Crystal Research and Technology: Journal of Experimental and Industrial Crystallography, 42 (2007) 1002-1006. https://doi.org/10.1002/crat.200710959

[36] V. Polshettiwar, B. Baruwati, R.S. Varma, Self-assembly of metal oxides into three-dimensional nanostructures: synthesis and application in catalysis, ACS nano, 3 (2009) 728-736. https://doi.org/10.1021/nn800903p

[37] Q. Xie, Z. Dai, J. Liang, L. Xu, W. Yu, Y. Qian, Synthesis of ZnO three-dimensional architectures and their optical properties, Solid state communications, 136 (2005) 304-307. https://doi.org/10.1016/j.ssc.2005.07.023

[38] J. Liu, X. Huang, Y. Li, K. Sulieman, F. Sun, X. He, Selective growth and properties of zinc oxide nanostructures, Scripta materialia, 55 (2006) 795-798. https://doi.org/10.1016/j.scriptamat.2006.07.010

[39] M. Bitenc, Z.C. Orel, Synthesis and characterization of crystalline hexagonal bipods of zinc oxide, Materials Research Bulletin, 44 (2009) 381-387. https://doi.org/10.1016/j.materresbull.2008.05.005

[40] Z.L. Wang, Nanostructures of zinc oxide, Materials today, 7 (2004) 26-33. https://doi.org/10.1016/S1369-7021(04)00286-X

[41] A. Umar, Y. Hahn, Aligned hexagonal coaxial-shaped ZnO nanocolumns on steel alloy by thermal evaporation, Applied physics letters, 88 (2006) 173120. https://doi.org/10.1063/1.2200472

[42] B.H. Juárez, P.D. García, D. Golmayo, A. Blanco, C. Lopez, ZnO inverse opals by chemical vapor deposition, Advanced Materials, 17 (2005) 2761-2765. https://doi.org/10.1002/adma.200500569

[43] L. Znaidi, Sol-gel-deposited ZnO thin films: A review, Materials Science and Engineering: B, 174 (2010) 18-30. https://doi.org/10.1016/j.mseb.2010.07.001

[44] Z. Yin, S. Wu, X. Zhou, X. Huang, Q. Zhang, F. Boey, H. Zhang, Electrochemical deposition of ZnO nanorods on transparent reduced graphene oxide electrodes for hybrid solar cells, small, 6 (2010) 307-312. https://doi.org/10.1002/smll.200901968

[45] S. Pearton, D. Norton, K. Ip, Y. Heo, T. Steiner, RETRACTED: Recent progress in processing and properties of ZnO, Progress in materials science, 50 (2005) 293-340. https://doi.org/10.1016/j.pmatsci.2004.04.001

[46] A. Mamalis, Recent advances in nanotechnology, Journal of Materials Processing Technology, 181 (2007) 52-58. https://doi.org/10.1016/j.jmatprotec.2006.03.052

[47] L. Shao, J. Chen, Synthesis and application of nanoparticles by a high gravity method, China Particuology, 3 (2005) 134-135. https://doi.org/10.1016/S1672-2515(07)60180-8

[48] A. Khaleel, P.N. Kapoor, K.J. Klabunde, Nanocrystalline metal oxides as new adsorbents for air purification, Nanostructured Materials, 11 (1999) 459-468. https://doi.org/10.1016/S0965-9773(99)00329-3

[49] M. Tokumoto, V. Briois, C.V. Santilli, S.H. Pulcinelli, Preparation of ZnO nanoparticles: structural study of the molecular precursor, Journal of Sol-Gel Science and Technology, 26 (2003) 547-551. https://doi.org/10.1023/A:1020711702332

[50] H. Oh, J. Krantz, I. Litzov, T. Stubhan, L. Pinna, C.J. Brabec, Comparison of various sol-gel derived metal oxide layers for inverted organic solar cells, Solar Energy Materials and Solar Cells, 95 (2011) 2194-2199. https://doi.org/10.1016/j.solmat.2011.03.023

[51] X. Zhao, B. Zheng, C. Li, H. Gu, Acetate-derived ZnO ultrafine particles synthesized by spray pyrolysis, Powder technology, 100 (1998) 20-23. https://doi.org/10.1016/S0032-5910(98)00047-3

[52] T. Tani, L. Mädler, S.E. Pratsinis, Homogeneous ZnO nanoparticles by flame spray pyrolysis, Journal of Nanoparticle Research, 4 (2002) 337-343. https://doi.org/10.1023/A:1021153419671

[53] X. Li, G. He, G. Xiao, H. Liu, M. Wang, Synthesis and morphology control of ZnO nanostructures in microemulsions, Journal of Colloid and Interface Science, 333 (2009) 465-473. https://doi.org/10.1016/j.jcis.2009.02.029

[54] Z.R. Dai, Z.W. Pan, Z. Wang, Novel nanostructures of functional oxides synthesized by thermal evaporation, Advanced Functional Materials, 13 (2003) 9-24. https://doi.org/10.1002/adfm.200390013

[55] I. Amarilio-Burshtein, S. Tamir, Y. Lifshitz, Growth modes of ZnO nanostructures from laser ablation, Applied Physics Letters, 96 (2010) 103104. https://doi.org/10.1063/1.3340948

[56] W.I. Park, C.H. Lee, J.H. Chae, D.H. Lee, G.C. Yi, Ultrafine ZnO nanowire electronic device arrays fabricated by selective metal-organic chemical vapor deposition, Small, 5 (2009) 181-184. https://doi.org/10.1002/smll.200800617

[57] L. Damonte, L.M. Zélis, B.M. Soucase, M.H. Fenollosa, Nanoparticles of ZnO obtained by mechanical milling, Powder Technology, 148 (2004) 15-19. https://doi.org/10.1016/j.powtec.2004.09.014

[58] S. Komarneni, M. Bruno, E. Mariani, Synthesis of ZnO with and without microwaves, Materials research bulletin, 35 (2000) 1843-1847. https://doi.org/10.1016/S0025-5408(00)00385-8

[59] M. Bitenc, P. Podbršček, Z. Crnjak Orel, M.A. Cleveland, J.A. Paramo, R.M. Peters, Y.M. Strzhemechny, Correlation between morphology and defect luminescence in precipitated ZnO nanorod powders, Crystal Growth and Design, 9 (2009) 997-1001. https://doi.org/10.1021/cg8008078

[60] P. Banerjee, S. Chakrabarti, S. Maitra, B.K. Dutta, Zinc oxide nano-particles-sonochemical synthesis, characterization and application for photo-remediation of heavy metal, Ultrasonics sonochemistry, 19 (2012) 85-93. https://doi.org/10.1016/j.ultsonch.2011.05.007

[61] L. Pyskło, W. Parasiewicz, P. Pawłowski, K. Niciński, Zinc Oxide in Rubber Compounds, Instytut Przemyslu Gumowego: Piastow, Poland, (2007).

[62] S. Mahmud, M. Johar Abdullah, G.A. Putrus, J. Chong, A. Karim Mohamad, Nanostructure of ZnO fabricated via French process and its correlation to electrical properties of semiconducting varistors, Synthesis and Reactivity in Inorganic and Metal-Organic and Nano-Metal Chemistry, 36 (2006) 155-159. https://doi.org/10.1080/15533170500524462

[63] T. Tsuzuki, P.G. McCormick, Mechanochemical synthesis of nanoparticles, Journal of materials science, 39 (2004) 5143-5146. https://doi.org/10.1023/B:JMSC.0000039199.56155.f9

[64] Z. Zhou, J. Wang, C.G. Jhun, ZnO Nanospheres Fabricated by Mechanochemical Method with Photocatalytic Properties, Catalysts, 11 (2021) 572. https://doi.org/10.3390/catal11050572

[65] F. Elmi, H. Alinezhad, Z. Moulana, F. Salehian, S. Mohseni Tavakkoli, F. Asgharpour, H. Fallah, M.M. Elmi, The use of antibacterial activity of ZnO nanoparticles in the treatment of municipal wastewater, Water Science and Technology, 70 (2014) 763-770. https://doi.org/10.2166/wst.2014.232

[66] H. Çolak, E. Karaköse, Y. Derin, Properties of ZnO nanostructures produced by mechanochemical-solid state combustion method using different precursors, Materials Chemistry and Physics, 193 (2017) 427-437. https://doi.org/10.1016/j.matchemphys.2017.03.009

[67] A. Kołodziejczak-Radzimska, T. Jesionowski, A. Krysztafkiewicz, Obtaining zinc oxide from aqueous solutions of KOH and Zn (CH3COO) 2, Physicochemical Problems of Mineral Processing, 44 (2010) 93-102.

[68] R. Hong, T. Pan, J. Qian, H. Li, Synthesis and surface modification of ZnO nanoparticles, Chemical Engineering Journal, 119 (2006) 71-81. https://doi.org/10.1016/j.cej.2006.03.003

[69] J. Xu, Q. Pan, Z. Tian, Grain size control and gas sensing properties of ZnO gas sensor, Sensors and Actuators B: Chemical, 66 (2000) 277-279. https://doi.org/10.1016/S0925-4005(00)00381-6

[70] A.S. Lanje, S.J. Sharma, R.S. Ningthoujam, J.-S. Ahn, R.B. Pode, Low temperature dielectric studies of zinc oxide (ZnO) nanoparticles prepared by precipitation method, Advanced Powder Technology, 24 (2013) 331-335. https://doi.org/10.1016/j.apt.2012.08.005

[71] Y. Wang, C. Zhang, S. Bi, G. Luo, Preparation of ZnO nanoparticles using the direct precipitation method in a membrane dispersion micro-structured reactor, Powder Technology, 202 (2010) 130-136. https://doi.org/10.1016/j.powtec.2010.04.027

[72] W. Jia, S. Dang, H. Liu, Z. Zhang, C. Yu, X. Liu, B. Xu, Evidence of the formation mechanism of ZnO in aqueous solution, Materials Letters, 82 (2012) 99-101. https://doi.org/10.1016/j.matlet.2012.05.013

[73] Z. Cao, Z. Zhang, F. Wang, G. Wang, Synthesis and UV shielding properties of zinc oxide ultrafine particles modified with silica and trimethyl siloxane, Colloids and surfaces A: physicochemical and engineering aspects, 340 (2009) 161-167. https://doi.org/10.1016/j.colsurfa.2009.03.024

[74] S. Bu, Z. Jin, X. Liu, T. Yin, Z. Cheng, Preparation of nanocrystalline TiO2 porous films from terpineol-ethanol-PEG system, Journal of materials science, 41 (2006) 2067-2073. https://doi.org/10.1007/s10853-006-8000-y

[75] R. Campostrini, M. Ischia, L. Palmisano, Pyrolysis study of sol-gel derived TiO 2 powders: part I. TiO 2-anatase prepared by reacting titanium (IV) isopropoxide with formic acid, Journal of thermal analysis and calorimetry, 71 (2003) 997-1010.

[76] R. Campostrini, M. Ischia, L. Palmisano, Pyrolysis study of Sol-gel derived TiO 2 Powders: Part II. TiO 2-anatase prepared by reacting titanium (IV) isopropoxide with oxalic acid, Journal of thermal analysis and calorimetry, 71 (2003) 1011-1022.

[77] A. Peng, E. Xie, C. Jia, R. Jiang, H. Lin, Photoluminescence properties of TiO2: Eu3+ thin films deposited on different substrates, Materials letters, 59 (2005) 3866-3869. https://doi.org/10.1016/j.matlet.2005.07.028

[78] S. Sivakumar, P.K. Pillai, P. Mukundan, K. Warrier, Sol-gel synthesis of nanosized anatase from titanyl sulfate, Materials letters, 57 (2002) 330-335. https://doi.org/10.1016/S0167-577X(02)00786-3

[79] S.S. Watson, D. Beydoun, J.A. Scott, R. Amal, The effect of preparation method on the photoactivity of crystalline titanium dioxide particles, Chemical Engineering Journal, 95 (2003) 213-220. https://doi.org/10.1016/S1385-8947(03)00107-4

[80] W. Xu, W. Hu, M. Li, Sol-gel derived hydroxyapatite/titania biocoatings on titanium substrate, Materials Letters, 60 (2006) 1575-1578. https://doi.org/10.1016/j.matlet.2005.11.072

[81] U. Schubert, Chemistry and fundamentals of the sol-gel process, The Sol-Gel Handbook, (2015) 1-28. https://doi.org/10.1002/9783527670819.ch01

[82] J.D. Mackenzie, E.P. Bescher, Chemical routes in the synthesis of nanomaterials using the sol-gel process, Accounts of chemical research, 40 (2007) 810-818. https://doi.org/10.1021/ar7000149

[83] T. Mahato, G. Prasad, B. Singh, J. Acharya, A. Srivastava, R. Vijayaraghavan, Nanocrystalline zinc oxide for the decontamination of sarin, Journal of hazardous materials, 165 (2009) 928-932. https://doi.org/10.1016/j.jhazmat.2008.10.126

[84] H. Benhebal, M. Chaib, T. Salmon, J. Geens, A. Leonard, S.D. Lambert, M. Crine, B. Heinrichs, Photocatalytic degradation of phenol and benzoic acid using zinc oxide powders prepared by the sol-gel process, Alexandria Engineering Journal, 52 (2013) 517-523. https://doi.org/10.1016/j.aej.2013.04.005

[85] M. Ristić, S. Musić, M. Ivanda, S. Popović, Sol-gel synthesis and characterization of nanocrystalline ZnO powders, Journal of Alloys and Compounds, 397 (2005) L1-L4. https://doi.org/10.1016/j.jallcom.2005.01.045

[86] S. Yue, Z. Yan, Y. Shi, G. Ran, Synthesis of zinc oxide nanotubes within ultrathin anodic aluminum oxide membrane by sol-gel method, Materials Letters, 98 (2013) 246-249. https://doi.org/10.1016/j.matlet.2013.02.037

[87] P. Dhiman, J. Chand, A. Kumar, R.K. Kotnala, K.M. Batoo, M. Singh, Synthesis and characterization of novel Fe@ZnO nanosystem, Journal of Alloys and Compounds, 578 (2013) 235-241. https://doi.org/10.1016/j.jallcom.2013.05.015

[88] T. Sharma, M. Garg, Optical and morphological characterization of ZnO nano-sized powder synthesized using single step sol-gel technique, Optical Materials, 132 (2022) 112794. https://doi.org/10.1016/j.optmat.2022.112794

[89] S.A. Ayon, S. Hasan, M.M. Billah, S.S. Nishat, A. Kabir, Improved luminescence and photocatalytic properties of Sm3+-doped ZnO nanoparticles via modified sol-gel route: A unified experimental and DFT+U approach, Journal of Rare Earths, (2022). https://doi.org/10.21203/rs.3.rs-1079490/v1

[90] S. Arya, P. Mahajan, S. Mahajan, A. Khosla, R. Datt, V. Gupta, S.-J. Young, S.K. Oruganti, Review-Influence of Processing Parameters to Control Morphology and Optical Properties of Sol-Gel Synthesized ZnO Nanoparticles, ECS Journal of Solid State Science and Technology, 10 (2021) 023002. https://doi.org/10.1149/2162-8777/abe095

[91] K. Omri, I. Najeh, R. Dhahri, J. El Ghoul, L. El Mir, Effects of temperature on the optical and electrical properties of ZnO nanoparticles synthesized by sol-gel method, Microelectronic Engineering, 128 (2014) 53-58. https://doi.org/10.1016/j.mee.2014.05.029

[92] Y. Li, L. Xu, X. Li, X. Shen, A. Wang, Effect of aging time of ZnO sol on the structural and optical properties of ZnO thin films prepared by sol-gel method, Applied Surface Science, 256 (2010) 4543-4547. https://doi.org/10.1016/j.apsusc.2010.02.044

[93] A. B Djurisic, X. Y Chen, Y. H Leung, Recent progress in hydrothermal synthesis of zinc oxide nanomaterials, Recent patents on nanotechnology, 6 (2012) 124-134. https://doi.org/10.2174/187221012800270180

[94] B. Innes, T. Tsuzuki, H. Dawkins, J. Dunlop, G. Trotter, M. Nearn, P. McCornick, Nanotechnology and Cosmetic Chemistry, Cosmetics, Aerosols & Toiletries in Australia, 15 (5), 10, 24 (2002).

[95] R. Jing, A. Ibni Khursheed, J.a. Song, L. Sun, Z. Yu, Z. Nie, E. Cao, A comparative study on the acetone sensing properties of ZnO disk pairs, flowers, and walnuts prepared by hydrothermal method, Applied Surface Science, 591 (2022) 153218. https://doi.org/10.1016/j.apsusc.2022.153218

[96] A. Kołodziejczak-Radzimska, E. Markiewicz, T. Jesionowski, Structural characterisation of ZnO particles obtained by the emulsion precipitation method, Journal of Nanomaterials, 2012 (2012). https://doi.org/10.1155/2012/656353

[97] R. Hong, T. Pan, J. Qian, H. Li, Synthesis and surface modification of ZnO nanoparticles, Chemical Engineering Journal, 119 (2006) 71-81. https://doi.org/10.1016/j.cej.2006.03.003

[98] Y. Wang, C. Zhang, S. Bi, G. Luo, Preparation of ZnO nanoparticles using the direct precipitation method in a membrane dispersion micro-structured reactor, Powder Technology, 202 (2010) 130-136. https://doi.org/10.1016/j.powtec.2010.04.027

[99] Z. Cao, Z. Zhang, F. Wang, G. Wang, Synthesis and UV shielding properties of zinc oxide ultrafine particles modified with silica and trimethyl siloxane, Colloids and Surfaces A: Physicochemical and Engineering Aspects, 340 (2009) 161-167. https://doi.org/10.1016/j.colsurfa.2009.03.024

[100] Z.M. Khoshhesab, M. Sarfaraz, Z. Houshyar, Influences of Urea on Preparation of Zinc Oxide Nanostructures Through Chemical Precipitation in Ammonium Hydrogencarbonate Solution, Synthesis and Reactivity in Inorganic, Metal-Organic, and Nano-Metal Chemistry, 42 (2012) 1363-1368. https://doi.org/10.1080/15533174.2012.680119

[101] K. Mohan Kumar, B.K. Mandal, E. Appala Naidu, M. Sinha, K. Siva Kumar, P. Sreedhara Reddy, Synthesis and characterisation of flower shaped Zinc Oxide nanostructures and its antimicrobial activity, Spectrochimica Acta Part A: Molecular and Biomolecular Spectroscopy, 104 (2013) 171-174. https://doi.org/10.1016/j.saa.2012.11.025

[102] W. Ao, J. Li, H. Yang, X. Zeng, X. Ma, Mechanochemical synthesis of zinc oxide nanocrystalline, Powder Technology, 168 (2006) 148-151. https://doi.org/10.1016/j.powtec.2006.07.014

[103] A. Stanković, L. Veselinović, S.D. Škapin, S. Marković, D. Uskoković, Controlled mechanochemically assisted synthesis of ZnO nanopowders in the presence of oxalic

acid, Journal of Materials Science, 46 (2011) 3716-3724.
https://doi.org/10.1007/s10853-011-5273-6

[104] T. Tsuzuki, P.G. McCormick, ZnO nanoparticles synthesised by mechanochemical processing, Scripta Materialia, 44 (2001) 1731-1734. https://doi.org/10.1016/S1359-6462(01)00793-X

[105] M. Ristić, S. Musić, M. Ivanda, S. Popović, Sol-gel synthesis and characterization of nanocrystalline ZnO powders, Journal of Alloys and Compounds, 397 (2005) L1-L4. https://doi.org/10.1016/j.jallcom.2005.01.045

[106] T.H. Mahato, G.K. Prasad, B. Singh, J. Acharya, A.R. Srivastava, R. Vijayaraghavan, Nanocrystalline zinc oxide for the decontamination of sarin, Journal of Hazardous Materials, 165 (2009) 928-932. https://doi.org/10.1016/j.jhazmat.2008.10.126

[107] A.A. Ismail, A. El-Midany, E.A. Abdel-Aal, H. El-Shall, Application of statistical design to optimize the preparation of ZnO nanoparticles via hydrothermal technique, Materials Letters, 59 (2005) 1924-1928. https://doi.org/10.1016/j.matlet.2005.02.027

[108] S. Musić, Đ. Dragčević, S. Popović, M. Ivanda, Precipitation of ZnO particles and their properties, Materials Letters, 59 (2005) 2388-2393. https://doi.org/10.1016/j.matlet.2005.02.084

[109] S.-J. Chen, L.-H. Li, X.-T. Chen, Z. Xue, J.-M. Hong, X.-Z. You, Preparation and characterization of nanocrystalline zinc oxide by a novel solvothermal oxidation route, Journal of Crystal Growth, 252 (2003) 184-189. https://doi.org/10.1016/S0022-0248(02)02495-8

[110] J.J. Schneider, R.C. Hoffmann, J. Engstler, A. Klyszcz, E. Erdem, P. Jakes, R.-A. Eichel, L. Pitta-Bauermann, J. Bill, Synthesis, Characterization, Defect Chemistry, and FET Properties of Microwave-Derived Nanoscaled Zinc Oxide, Chemistry of Materials, 22 (2010) 2203-2212. https://doi.org/10.1021/cm902300q

[111] S.A. Vorobyova, A.I. Lesnikovich, V.V. Mushinskii, Interphase synthesis and characterization of zinc oxide, Materials Letters, 58 (2004) 863-866. https://doi.org/10.1016/j.matlet.2003.08.008

[112] M. Singhai, V. Chhabra, P. Kang, D.O. Shah, Synthesis of ZnO nanoparticles for varistor application using Zn-substituted aerosol ot microemulsion, Materials Research Bulletin, 32 (1997) 239-247. https://doi.org/10.1016/S0025-5408(96)00175-4

[113] Ö.A. Yıldırım, C. Durucan, Synthesis of zinc oxide nanoparticles elaborated by microemulsion method, Journal of Alloys and Compounds, 506 (2010) 944-949. https://doi.org/10.1016/j.jallcom.2010.07.125

[114] R. Moleski, E. Leontidis, F. Krumeich, Controlled production of ZnO nanoparticles from zinc glycerolate in a sol-gel silica matrix, Journal of Colloid and Interface Science, 302 (2006) 246-253. https://doi.org/10.1016/j.jcis.2006.07.030

[115] F.-Q. He, Y.-P. Zhao, Growth of ZnO nanotetrapods with hexagonal crown, Applied physics letters, 88 (2006) 193113. https://doi.org/10.1063/1.2202003

[116] Y. Dai, Y. Zhang, Q. Li, C. Nan, Synthesis and optical properties of tetrapod-like zinc oxide nanorods, Chemical Physics Letters, 358 (2002) 83-86. https://doi.org/10.1016/S0009-2614(02)00582-1

[117] Z.L. Wang, X. Kong, J. Zuo, Induced growth of asymmetric nanocantilever arrays on polar surfaces, Physical review letters, 91 (2003) 185502. https://doi.org/10.1103/PhysRevLett.91.185502

[118] H. Yan, R. He, J. Johnson, M. Law, R.J. Saykally, P. Yang, Dendritic nanowire ultraviolet laser array, Journal of the American Chemical Society, 125 (2003) 4728-4729. https://doi.org/10.1021/ja034327m

[119] P.X. Gao, Z.L. Wang, Nanoarchitectures of semiconducting and piezoelectric zinc oxide, Journal of Applied Physics, 97 (2005) 044304. https://doi.org/10.1063/1.1847701

[120] X. Wang, Y. Ding, C.J. Summers, Z.L. Wang, Large-scale synthesis of six-nanometer-wide ZnO nanobelts, The Journal of Physical Chemistry B, 108 (2004) 8773-8777. https://doi.org/10.1021/jp048482e

[121] L.M. Kukreja, S. Barik, P. Misra, Variable band gap ZnO nanostructures grown by pulsed laser deposition, Journal of crystal growth, 268 (2004) 531-535. https://doi.org/10.1016/j.jcrysgro.2004.04.086

[122] J. Chiou, K.K. Kumar, J. Jan, H. Tsai, C. Bao, W.-F. Pong, F. Chien, M.-H. Tsai, I.-H. Hong, R. Klauser, Diameter dependence of the electronic structure of ZnO nanorods determined by x-ray absorption spectroscopy and scanning photoelectron microscopy, Applied physics letters, 85 (2004) 3220-3222. https://doi.org/10.1063/1.1802373

[123] Z. Fan, J.G. Lu, Gate-refreshable nanowire chemical sensors, Applied Physics Letters, 86 (2005) 123510. https://doi.org/10.1063/1.1883715

[124] A. Kolmakov, M. Moskovits, Chemical sensing and catalysis by one-dimensional metal-oxide nanostructures, Annu. Rev. Mater. Res., 34 (2004) 151-180. https://doi.org/10.1146/annurev.matsci.34.040203.112141

[125] Y. Zhang, A. Kolmakov, S. Chretien, H. Metiu, M. Moskovits, Control of catalytic reactions at the surface of a metal oxide nanowire by manipulating electron density inside it, Nano Letters, 4 (2004) 403-407. https://doi.org/10.1021/nl034968f

[126] A. Asthana, K. Momeni, A. Prasad, Y.K. Yap, R.S. Yassar, In situ observation of size-scale effects on the mechanical properties of ZnO nanowires, Nanotechnology, 22 (2011) 265712. https://doi.org/10.1088/0957-4484/22/26/265712

[127] T.-H. Fang, W.-J. Chang, C.-M. Lin, Nanoindentation characterization of ZnO thin films, Materials Science and Engineering: A, 452-453 (2007) 715-720. https://doi.org/10.1016/j.msea.2006.11.008

[128] X. Bai, P. Gao, Z.L. Wang, E. Wang, Dual-mode mechanical resonance of individual ZnO nanobelts, Applied Physics Letters, 82 (2003) 4806-4808. https://doi.org/10.1063/1.1587878

[129] T.-T. Cao, Y.-J. Wang, Y.-Q. Zhang, Effect of Strongly Alkaline Electrolyzed Water on Silk Degumming and the Physical Properties of the Fibroin Fiber, PLOS ONE, 8 (2013) e65654. https://doi.org/10.1371/journal.pone.0065654

[130] P.-C. Chang, Z. Fan, C.-J. Chien, D. Stichtenoth, C. Ronning, J.G. Lu, High-performance ZnO nanowire field effect transistors, Applied physics letters, 89 (2006) 133113. https://doi.org/10.1063/1.2357013

[131] K.-K. Kim, H.-S. Kim, D.-K. Hwang, J.-H. Lim, S.-J. Park, Realization of p-type ZnO thin films via phosphorus doping and thermal activation of the dopant, Applied Physics Letters, 83 (2003) 63-65. https://doi.org/10.1063/1.1591064

[132] D. Banerjee, S.H. Jo, Z.F. Ren, Enhanced field emission of ZnO nanowires, Advanced Materials, 16 (2004) 2028-2032. https://doi.org/10.1002/adma.200400629

[133] Y.K. Tseng, C.J. Huang, H.M. Cheng, I.N. Lin, K.S. Liu, I.C. Chen, Characterization and field-emission properties of needle-like zinc oxide nanowires grown vertically on conductive zinc oxide films, Advanced functional materials, 13 (2003) 811-814. https://doi.org/10.1002/adfm.200304434

[134] H. Zhang, R. Wang, Y. Zhu, Effect of adsorbates on field-electron emission from ZnO nanoneedle arrays, Journal of applied physics, 96 (2004) 624-628. https://doi.org/10.1063/1.1757653

[135] Y. Li, Y. Bando, D. Golberg, ZnO nanoneedles with tip surface perturbations: Excellent field emitters, Applied Physics Letters, 84 (2004) 3603-3605. https://doi.org/10.1063/1.1738174

[136] T. Kim, J. Kim, S. Lee, H. Shim, E. Suh, K. Nahm, Characterization of ZnO needle-shaped nanostructures grown on NiO catalyst-coated Si substrates, Synthetic metals, 144 (2004) 61-68. https://doi.org/10.1016/j.synthmet.2004.01.010

[137] P. Yang, H. Yan, S. Mao, R. Russo, J. Johnson, R. Saykally, Nathan, Morris, J. Pham, R. He, and HJ Choi, Adv. Funct. Mater, 12 (2002) 323. https://doi.org/10.1002/1616-3028(20020517)12:5<323::AID-ADFM323>3.0.CO;2-G

[138] Z.L. Wang, X.Y. Kong, Y. Ding, P. Gao, W.L. Hughes, R. Yang, Y. Zhang, Semiconducting and piezoelectric oxide nanostructures induced by polar surfaces, Advanced Functional Materials, 14 (2004) 943-956. https://doi.org/10.1002/adfm.200400180

[139] S. Han, X. Feng, Z. Lu, D. Johnson, R. Wood, Erratum:"Transparent-cathode for top-emission organic light-emitting diodes"[Appl. Phys. Lett. 82, 2715 (2003)], Applied Physics Letters, 83 (2003) 2719-2719. https://doi.org/10.1063/1.1614436

[140] W. Lee, M.-C. Jeong, J.-M. Myoung, Evolution of the morphology and optical properties of ZnO nanowires during catalyst-free growth by thermal evaporation, Nanotechnology, 15 (2004) 1441. https://doi.org/10.1088/0957-4484/15/11/010

[141] M.H. Huang, S. Mao, H. Feick, H. Yan, Y. Wu, H. Kind, E. Weber, R. Russo, P. Yang, Room-temperature ultraviolet nanowire nanolasers, science, 292 (2001) 1897-1899. https://doi.org/10.1126/science.1060367

[142] C. Liu, J.A. Zapien, Y. Yao, X. Meng, C.S. Lee, S. Fan, Y. Lifshitz, S.T. Lee, High-Density, ordered ultraviolet light-emitting ZnO nanowire arrays, Advanced materials, 15 (2003) 838-841. https://doi.org/10.1002/adma.200304430

[143] L. Liu, X. Liu, Roles of drug transporters in blood-retinal barrier, Drug Transporters in Drug Disposition, Effects and Toxicity, (2019) 467-504. https://doi.org/10.1007/978-981-13-7647-4_10

[144] Z. Fan, D. Wang, P.-C. Chang, W.-Y. Tseng, J.G. Lu, ZnO nanowire field-effect transistor and oxygen sensing property, Applied Physics Letters, 85 (2004) 5923-5925. https://doi.org/10.1063/1.1836870

[145] Q. Li, T. Gao, Y. Wang, T. Wang, Adsorption and desorption of oxygen probed from ZnO nanowire films by photocurrent measurements, Applied Physics Letters, 86 (2005) 123117. https://doi.org/10.1063/1.1883711

[146] D.I. Florescu, L. Mourokh, F.H. Pollak, D.C. Look, G. Cantwell, X. Li, High spatial resolution thermal conductivity of bulk ZnO (0001), Journal of applied physics, 91 (2002) 890-892. https://doi.org/10.1063/1.1426234

[147] V. Coleman, Zinc Oxide Bulk, Thin Films and Nanostructures-Processing, Properties and Applications edited by C. Jagadish, S. Pearton, in, Elsevier, Amsterdam, 2007. https://doi.org/10.1016/B978-008044722-3/50001-4

ZnO and Their Hybrid Nano-Structures

Materials Research Foundations 146 (2023) 35-66

Materials Research Forum LLC

https://doi.org/10.21741/9781644902394-2

Chapter 2

Intrinsic Point Defects in ZnO

Vikas Dhiman [1,2], Prashant Choudhary [1], Neha Kondal [1,*]

[1] Department of Physics, Chandigarh University, Gharuan, Mohali, Punjab, India

[2] Govt. College Dhaliara, Distt. Kangra, Himachal Pradesh, India

* nehakondal91@gmail.com

Abstract

ZnO has attracted extraordinary interest as a premier host for dilute magnetic semiconductors due to its potential usage in optoelectronics, spin light emitting diodes, photovoltaics, photodetectors, and solar cells. It is a prospective material for spin light emitting diodes due to its broad band gap of 3.37 eV and high exciton binding energy. The extraordinary properties of ZnO results from the interactions of intrinsic defects in ZnO. The redistribution of these defects within the grain using sintering, annealing or extrinsic doping helps in tailoring the optical, electrical and magnetic properties of ZnO. Therefore, in this chapter, we will discuss about various intrinsic point defects present in ZnO, defect creation techniques, defect characterization, the factor affecting defect formation. The Zinc interstitial, Oxygen vacancies and their complexes are responsible for various band emissions and hence luminescent properties of various defects in ZnO are discussed.

Keywords

ZnO, Native Defects, Photoluminescence

Contents

1.   Introduction

Zinc oxide (ZnO) is a fundamentally important substance that has piqued the interest of both academic and industrial groups. The unique properties for which ZnO has drawn intense attention are large band gap (E_{gap} =3.37 eV), strong exciton binding energy (E_{EB} = 60 meV), great UV & Vis light luminescence at room temperature (RT), piezoelectric behavior and better mobility of charge carrier (μ_e & μ_h), transparent behavior & dielectric constant [1, 2]. Larger E_{EB} of ZnO compared to 25 meV (thermal energy equivalent to RT) suggests that ZnO has effective excitonic emission even at or beyond RT, making it a good choice for UV lasers [3]. ZnO's strong visible luminosity makes it an ideal foundation material for light-emitting-diodes (LEDs) [4]. Moreover, the ZnO has been found to have photoconductive functionalities and ultrafast nonlinearities [5]. High transparency allows ZnO thin film for use as transparent electrode or conducting layer in LEDs, touch panels, electronic paper displays, plasma display panels, liquid crystal displays, etc. [6]. ZnO also has notable piezoelectricity, allowing for research into ZnO-based electromechanical coupling devices [7]. Acoustic wave devices are prospective application field for ZnO because of its significant electromechanical coupling [8]. The high radiation resistivity of ZnO makes it suitable semiconducting material for devices operating in space, nuclear researches, extraction uranium ores and their treatment where harsh radiation conditions involving high energy particles are generally present [9]. Devices combining optical and magnetic effects, such as spin LEDs, spin-polarized solar cells, and magneto-optical switch, are also possible using ZnO [10]. The Food and Drug Administration (FDA) has classified ZnO as a GRAS (Generally Recognized As Safe) ingredient, and it is thus utilized as a food additive [11]. In fact, ZnO is widely accepted in industries due to its non-toxic nature, good chemical and thermal stability, low cost large scale production, availability of large single crystal [12]. Apart from semiconductor applications, ZnO is also used in photocatalysis, cigarette filters, rubber, paint, agricultural, pharmaceutical, cosmetics, and textile industries [13,14].

The capacity of ZnO to host strong fluorescent defects that light over the visible spectrum is one of its most fascinating features [15]. In fact, ZnO is useful for phosphor applications due of its intense fluorescence in the visible spectrum. Despite the fact that a lot of theoretical and experimental work has been done in finding and characterizing the major defect species in ZnO and their characteristics, the final picture, mainly about the intrinsic defects of ZnO, is not so clear [15-18].

For optimal semiconductor application, knowledge about the behavior of native point defects is essential. These flaws frequently affect the luminescence efficiency, minority carrier lifespan, and may even compensate for doping, either directly or indirectly. Native point defects, such as self-interstitials, vacancies, and antisites, can have a big influence on doping, which is the backbone of semiconductor technology. For example, we want to dope a semiconductor with a p-type dopant. In such circumstances, some native defects that act as donors may naturally occur and compensate for the acceptor added on purpose. This self-compensatory role of native defects in ZnO has already been reported [19]. Therefore, the incorporation of such features into ZnO requires careful defect engineering. Incorporating these properties necessitates meticulous defect engineering in ZnO. The features resulted due to the defects in ZnO are either used directly or have a significant impact on the various uses of ZnO in different field such as catalysis and electronics. As a result, rather than perfect ZnO, it is vital to discuss the different physical properties of defect rich ZnO from a practical standpoint because ZnO with a high defect content has a wide range of applications. For example, the optical signature, or luminescence, varies depending on the type of imperfection [15-17]. From a magnetic aspect, ZnO is diamagnetic; yet, in the presence of defects, ZnO shows ferromagnetism [20-22]. The existence of various sorts of defects in ZnO allows for high conductivity [23]. As a result, comprehending the function of defects in ZnO is crucial, and it is a current emphasis since it may broaden the domain of ZnO in terms of real-world applications. A brief idea about it would help in this regard. Therefore, in this chapter, we will discuss about various intrinsic point defects present in ZnO, defect creation techniques, defect characterization, the factor affecting defect formation and luminescent properties of various defects in ZnO.

2.   Basic properties of ZnO

ZnO is a white, water-insoluble compound with the semiconducting properties of a wide direct bandgap (E_{gap}). It has ionicity located between ionic and covalent semiconductors [1]. In terms of structure, ZnO has three polymorphs known as Wurtzite, Rock salt and Zinc blende (Fig.1). However, due to its great stability at ambient temperatures, wurtzite is the most common structure of ZnO [24].

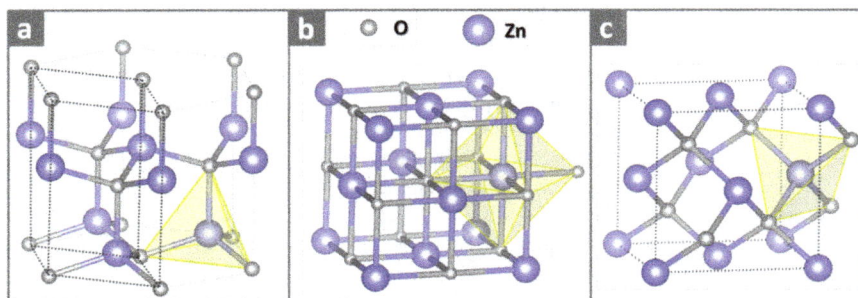

Figure 1: (a) Wurtzite, (b) Rock salt, and (c) Zinc blende structures of ZnO drawn by VESTA [25].

Wurtzite structure (Fig.1(a)) possesses a hexagonal close-packed (HCP) array of anions in which cations cover half of the tetrahedral interstices. It can be considered to be made up of two HCP lattices that are interpenetrating, with four atoms of two different sorts in each unit cell, resulting in two molecules per cell [1]. When Zn^{2+} and O^{2-} ions alternately stack along the c-axis, with Zn^{2+} ions on top and O^{2-} ions on the bottom, the Wurtzite structure is created. Two lattice constants, a (basal plane lattice parameter) and c (axial lattice parameter), make up the hexagonal unit cell. The lattice constants for wurtzite ZnO are $a = 3.249$ Å and $c = 5.206$ Å with 1.633 as the ratio of c/a. However, owing to defects, doping, and, lattice strain these lattice properties fluctuate [26-27].

The rock salt structure (Fig. 1(b)) is an ionic structure made up of anions arranged in a face centered cubic (FCC) arrangement with cations in the octahedral voids. It has six coordination number and $Fm\bar{3}m$ space group symmetry. Under ambient circumstances, the rock salt structure of ZnO is unstable, and it is only observed to exist at high pressures ($\geq$ 9 GPa) [28]. Studies show that the standard formation enthalpy (ΔH_F) of rocksalt ZnO from its polymorph wurtzite has the value -326.69 kJ/mol [31].

The zinc blende structure (Fig. 1(c)) has an FCC close packed array of anions, with cations occupying half of the tetrahedral interstices. In actuality, the unit cell is made up of two FCC sublattices, one of which is displaced along the body diagonal (1/4 of its length) of the other. It has four coordination number and $F\bar{4}3m$ space group symmetry. Similar to the rock salt structure, the zinc blende polymorph of ZnO is also unstable, and can only be seen when ZnO is formed on a cubic lattice substrate [24]. The typical characteristics of three polymorphs of ZnO are listed in Table 1.

Table 1: The typical characteristics of different ZnO polymorphs [1, 29-31].

Property	Value for ZnO		
	Hexagonal wurtzite	**Rocksalt**	**Zinc blende**
Lattice parameters (in Å)	$a = 3.2495$ & $c = 5.2069$	$a = 4.271; 4.283$	$a = 4.580; 4.400$
Space group symmetry	$P6_3mc$ (or C^4_{6v})	$Fm\bar{3}m$ (or O^5_h)	$F\bar{4}3m$ (or T^2_d)
Point group symmetry	6mm (or C_{6v})	$m\bar{3}m$ (or O_h)	$\bar{4}3m$ (or T_d)
Volume (in Å³)	23.829; 23.785	19.60; 19.40	----------
ZnO bond energy (DFT calculations)	-7.692 eV	-7.679 eV	-7.455 eV

3.   Defects in materials

Solid materials that may be found in nature are crystalline and amorphous solid materials [32]. The atoms, molecules, and ions that make up crystalline solids are arranged in a highly organized microscopic structure with long-range order in all directions. However, in amorphous materials the long-range ordering is absent, and only short-range ordering is present. In addition to this, the regular single crystalline patterns with randomly aligned grains make up polycrystalline materials (not fully random, but some preferred directions are favored, which is a property of the material).

Although a faultless crystal structure is the most energetically stable configuration of every material, but in fact, the materials contains faults in its lattice, which are known as defects. The number of such faults ($N_{Defects}$) under thermal equilibrium condition may be calculated using Boltzmann statistics [33].

$$N_{Defects} = \beta N_{Sites} e^{-\frac{\alpha E}{k_B T}}$$ (1)

Here N_{Sites} = total number of sites, E = energy needed for defect creation, T = temperature, k_B = Boltzmann's constant, and α & β = dimensionless constant having order 1. As per eqn.1, the room-temperature equilibrium percentage of such faults can approach 10^6 if their energy of formation is 0.35 eV, for example, which is substantial enough to have an effect.

In Fig.2, the variation in Gibbs free energy ($\Delta G = \Delta H - T\Delta S$; where ΔH represents enthalpy change, T represents temperature, and ΔS represents entropy change) has been plotted versus defect concentration (Dc). As seen in Fig.2(a), the existence of defects of a certain concentration decreases the lattice's free energy [34]. Further, the concentration at

Materials Research Forum LLC
https://doi.org/10.21741/9781644902394-2

which G is minimal, is the equilibrium defect concentration (Dc^{eq}) of a crystal at any temperature, and its value fluctuates with temperature as shown in Fig.2(b).

In complex systems, however, there may be more than one local minimum, indicating the presence of a variety of defects [35]. Following that, the system will settle into one of the local minima. However, the system can equilibrate to a new local minimum after a little perturbation suggesting a system with a distinct defect configuration.

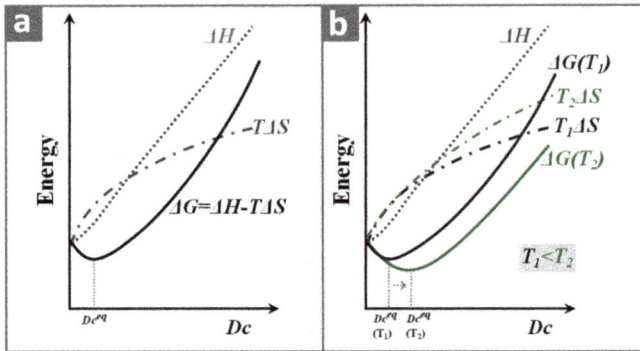

Figure 2: (a) General variation of ΔG with Dc, and (b) effect of temperature on Dc^{eq} [34].

Defects of many kinds can be found in crystalline solids. As shown in Fig.3, they are categorized as zero-dimensional (0D; Pont defects), one-dimensional (1D; Line defects), two-dimensional (2D; Surface defects) and three-dimensional (3D; Volume defects).

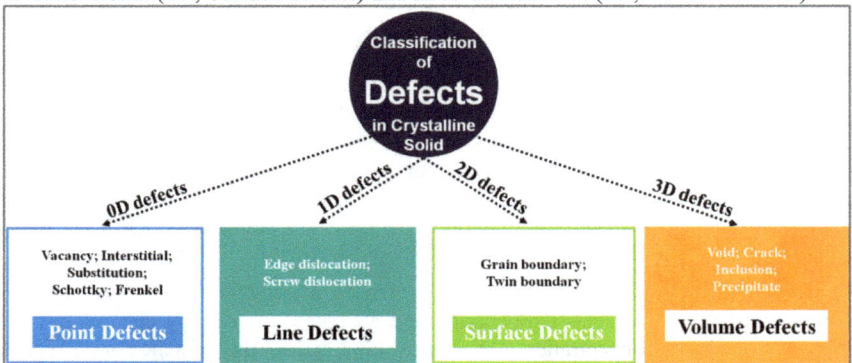

Figure 3: Classification of defects in crystalline solid.

Point defects are the 0D defects always present in any material system. The absence or erroneous atom placement in the crystal lattice is referred to as a point defect. Voids, interstitials, antisites, and impurities are examples of point defects (interstitial and substitutional).

Point defects include: Vacancy (void), interstitial (self or impurity atom), antisite, and substitution by impurity atom [36]. Fig.4 represents different types of point defects in mono-atomic lattice and intrinsic point defects in a diatomic (compound AB) crystal lattice.

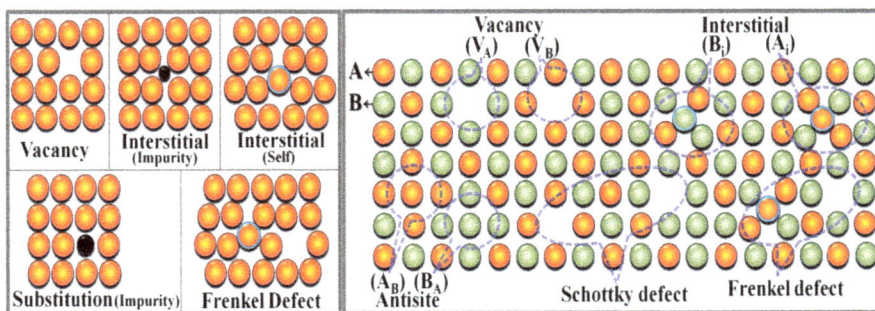

Figure 4: Possible point defects in mono-atomic lattice (Left side) and intrinsic point defects in a compound AB (Right side).

Since zinc oxide is a diatomic molecule, the potential internal point defects in ZnO will be similar to those of the AB compound as shown in Fig. 4. This means that ZnO can have three types of simple intrinsic point defects (vacancy, interstitial and antisite), as well as complex intrinsic point defects such as Schottky defects, Frenkel defects, etc. Point defect vacancy refers to the disappearance of a zinc or oxygen atom from their original lattice site, so the two types of vacancies in ZnO are zinc vacancy (V_{Zn}) and oxygen vacancy (V_O). Further, the point defect interstitial refers to the existence of an extra atom at a place in the lattice that is normally unoccupied by atoms, resulting in oxygen interstitial (O_i) and zinc interstitial (Z_i) defects. Furthermore, the antisite point defect refers to an O atom occupying the Zn lattice or vice versa, resulting in antisite Oxygen (O_{Zn}) or antisite Zinc (Zn_O) defects. In addition to single point defects, complex intrinsic point defects like Schottky defects (V_{Zn}-V_O, $2V_{Zn}$-V_O) and Frenkel defects (Zn_i-V_O, O_i-V_{Zn}) can also be present in crystal lattice. Apart from these internal point defects, external point defects such as impurity substitution (X_O and X_{Zn} where X indicates impurity) and interstitial impurity (X_i) are also point defects related to ZnO.

4.   Defect creation techniques

Various approaches are used to incorporate defects into solid materials [37-45]. Each approach has its own set of benefits as well as drawbacks. The type of defect production approach employed is determined by the nature of the defects we want to incorporate in the material. Annealing, mechanical machining, substitution, and ion beam irradiation are the common methods for creating defects. The recovery, change of previously existing defects, and defect rearrangement also occur during defect inclusion using any of the approaches. The commonly used defects incorporation techniques are discussed in this section.

Annealing: One of the most often utilized defect integration techniques in solids is annealing [37-38]. Many researchers, on the other hand, have also employed annealing to recover undesired flaws [39]. Appropriate annealing temperature, rate of heating, annealing environment, and annealing duration have all been shown to help controllably improve material characteristics. All locations of the crystal lattice are altered during annealing; however, annealing at various temperature ranges can have a variety of effects on a given site [37-38].

Mechanical milling: Mechanical milling can be used as a top-to-bottom strategy for manufacturing nanomaterials to reduce particle size as well as grain size [40]. The grain surface to volume ratio of such nanomaterials is quite high. Following milling, a substantial number of defect species are formed on the grain surfaces because the grain surfaces are extremely defect-rich. By altering the milling environment, rotating speed, and ball-to-mass ratio, we can control the type of defects and their concentration [41-42].

Substitution approach: In this approach, some of the parent atoms are replaced by foreign atoms of similar sort using the substitution approach [43]. The band structure is altered owing to the presence of the foreign atom, which is reflected in material characteristics. Ionic radius, valence state, and solubility are essential characteristics in Substitution [44]. One of the most appealing features of this fault generation approach is that it may be used to dope a specific spot.

Ion beam irradiation approach: Irradiating a material with a high-energy ion beam can also be used to create/modify defects (Fig.5). The approaches outlined above are used to construct an equilibrium defect configuration that is stable based on free energy minimization; however, ion beam irradiation can be used to efficiently generate non-equilibrium defect states in a material [45,46]. Thermodynamically, the defect states created in this approach are frequently discovered to be inaccessible. Irradiation causes changes in surface as well as bulk defect states which depend upon the type of ions, their flux as well as energy. The major advantages of this approach are the ability to produce defects in controlled way and doping at specific area, both of which are beneficial to device technology. However, the inability to include site-specific defects using ion beam irradiation is the fundamental downside of this method [45].

5.   Native defects in ZnO

Because no semiconductor material can achieve perfect order, certain intrinsic flaws will always remain in its crystal lattice. These innate defects have a substantial impact on the host material's electrical and optical properties. Intrinsic defects in ZnO originate from Zn or O atoms migrating, diffusing, or being displaced from their original positions in the crystal lattice, and results in interstitials, antisites, or vacancies formation, as discussed in Section-3. It is crucial to keep knowledge of how point defects behave if you want to use a semiconductor successfully as these faults often govern doping, minority transporter longevity, and luminescence efficacy. Furthermore, when the kind and concentration of defects in ZnO vary, the characteristics of the material change, which has an impact on ZnO applications [15-19].

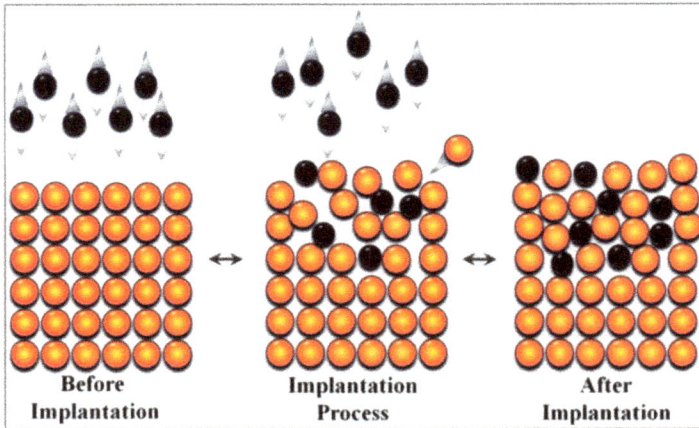

Figure 5: Defects incorporation by ion implantation techniques.

In addition to direct effects the innate flaws also causes a slew of problems during the exogenous doping of ZnO [46]. For example, the ZnO generally has n-type conductivity and the origins of this n-type conductivity are assumed to be V_O and Zn_i, which are having significantly lower formation energies [47]. As a result, the n-type inherent defects compensate for the externally introduced acceptor type dopants, making p-type ZnO production difficult. Despite several investigations using density functional theory (DFT), local density approximation (LDA), generalized gradient approximation (GGA), and the Hubbard parameter to probe the electronic level of native defects, no reasonable coherence on their origin and placement within the bandgap has been revealed [19].

All inherent faults are determined by the material's growth kinetics and processing procedures. However, the presence of various defects is drastically affected by their formation energy (E_F). The concentration of point defects (Dc) under thermodynamic equilibrium considering the low defect concentrations (dilute regime), i.e. neglecting interaction among different defects, is expressed as [19]:

$$Dc = N_{\text{sites}} \exp\left(-\frac{E_F}{k_B T}\right) \tag{2}$$

Here N_{Sites} = total number of sites, T = temperature, and k_B = Boltzmann's constant. Equation 2 clearly indicates that the defects having higher value of E_F will have low concentration. It should be emphasized that the contribution in formation energy by formation volume resulted by the change in volume after adding defect and the formation entropy can be ignored when computing Dc under dilute regimes and low temperatures. However, the formation volume under the high pressure regime should be considered and the formation entropy should be given considerable attention in the case of high temperature studies.

Oba et al. [48] used first principles calculations with plane-wave pseudopotential under GGA to calculate E_F and examine electronic structure of native defects in ZnO. In case of a defect having charge q the formation energy was expressed as:

$$E_F = E_T(q) - n_{Zn}\mu_{Zn} - n_O\mu_O + qE_f \tag{3}$$

where $E_T(q)$ = supercell energy with a defect having charge state q, μ_O & μ_{Zn} = atomic chemical potentials of O & Zn, n_O & n_{Zn} = number of O & Zn atom in the supercell, and E_f = Fermi energy. They observed that the amount of E_F necessary to create donor-type defects in the p-type condition is relatively low. As a result, the donor-type inherent defects self-compensate for the externally introduced p-type dopants, making p-type ZnO production difficult. Furthermore, they pointed out that the electronic structure of the native donor implies that in the case of undoped ZnO the Zn_i or Zn_O can explain the n-type electrical conductivity, but in the n-type conditions, the V_O with the lowest E_F is dominant. Jannoti et al. [19] used two distinct techniques to calculate the E_F for different defect states in ZnO under Zn-rich environment: LDA and LDA+U. The theoretically calculated values of E_F for various defects in ZnO are listed in Table 2. For the n-type samples, they found that V_O is the deep donor with high E_F and can reimburse the p-type doping; whereas Zn_i is a shallow donor with high E_F and quick diffusion, therefore isolated Zn_i barely persist as point defects. Under the same circumstances, Zn_O acted as a shallow donor, causing significant off-site displacement and produced a significant local lattice relaxation. However under p-type conditions, the neutral splitting interstitial was found to be more congruent. O_{Zn} was determined to have the greatest E_F among different acceptor-type intrinsic point defects. In fact, Zn_O acted as a deep acceptor with significant off-site displacement.

Table 2: Calculated E_F values for various defects in ZnO under Zn-rich environment

Defects	V_O			V_{Zn}		
Charge state	0	+1	+2	0	-1	-2
E^{LDA} (in eV)	0.69	0.64	-0.37	5.94	6.02	6.31
E^{LDA+U} (in eV)	1.34	0.81	-0.60	6.39	6.49	6.94

Defects	Zn_i			O_i		
Charge state	0	+1	+2	0	-1	-2
E^{LDA} (in eV)	2.76	1.32	-0.10	6.36	6.63	7.49
E^{LDA+U} (in eV)	3.62	1.56	-0.45	6.83	6.83	8.28

Defects	Zn_O					O_{Zn}		
Charge state	0	+1	+2	+3	+4	0	-1	-2
E^{LDA} (in eV)	3.43	1.81	0.22	0.48	0.14	9.94	10.53	11.08
E^{LDA+U} (in eV)	4.98	2.74	0.53	0.44	-0.13	10.04	10.88	11.76

Guan et al. [49] used first-principle calculations to study the influence of inherent point defects on the absorption spectrum and electronic structure of ZnO. They reported that under Zn-rich conditions, V_O and Zn_i had the lowest E_F and the most stable structure, whereas under O-rich conditions, V_{Zn} and O_i have the lowest E_F and the most stable structure. Sokol et al. [50] investigated intrinsic point defects in ZnO using an embedded cluster hybrid quantum mechanical/molecular mechanical method. They estimated the defect energy for each of the primary oxidation states of the faults. Table 3 gives a summary of the defect energetics determined by their computations. They discovered that the O_i has the lowest value of E_F among all intrinsic defects and is the dominant species under oxidizing circumstances. However, under reducing circumstances, the defects Zn_i and V_O were discovered to have low but almost identical E_F values, indicating that these two defects must have dominance. Further, they proposed that Zn_i with 0 and +1 charge gives rise to the donor bands observed at 0.12 eV and 0.29 eV, respectively. V_{Zn} was shown to be stable as a majority acceptor in different charge states (-2, -1, 0, +1, +2) with source for major PL bands: UV emission at 3.2 eV, green emission at 2.5 eVand red emission at ~2 eV. They found that green and blue emission can occur as a result of excitonic and donor–acceptor pair recombination from Zn_i to O_i (in the split-interstitial peroxy state with charge 0). They also said that while V_O cannot contribute to green emission, it may contribute to near E_{Gap} and orange-red emission by exciton recombination process.

Table 3: Possible defect formation reaction and corresponding energies [50]

Defect	V_O		O_i
Possible Formation Reaction {Energy in eV}	$O_O^x \rightarrow V_O^x + 1/2\, O_2(g)$ {5.296} $O_O^x \rightarrow V_O^{\bullet} + e' + 1/2\, O_2(g)$ {5.275} $O_O^x \rightarrow V_O^{\bullet\bullet} + 2e' + 1/2\, O_2(g)$ {5.403}		$V_i^x + 1/2\, O_2(g) \rightarrow O_i^x$ {1.652} $V_i^x + 1/2\, O_2(g) \rightarrow O_i' + h^{\bullet}$ {6.129} $V_i^x + 1/2\, O_2(g) \rightarrow O_i'' + 2h^{\bullet}$ {10.345}

Defect	V_{Zn}		Zn_i
Possible Formation Reaction {Energy in eV}	$Zn_{Zn}^x \rightarrow V_{Zn}^{\bullet\bullet} + 2e' + Zn(s)$ {10.895} $Zn_{Zn}^x \rightarrow V_{Zn}^{\bullet} + e' + Zn(s)$ {8.776} $Zn_{Zn}^x \rightarrow V_{Zn}^x + Zn(s)$ {7.280} $Zn_{Zn}^x \rightarrow V_{Zn}' + h^{\bullet} + Zn(s)$ {9.514} $Zn_{Zn}^x \rightarrow V_{Zn}'' + 2h^{\bullet} + Zn(s)$ {12.077}		$V_i^x + Zn(s) \rightarrow Zn_i^x$ {4.722} $V_i^x + Zn(s) \rightarrow Zn_i^{\bullet} + e'$ {3.563} $V_i^x + Zn(s) \rightarrow Zn_i^{\bullet\bullet} + 2e'$ {2.183}

Defect	O sublattice Frenkel defect	Zn sublattice Frenkel defect
Possible Formation Reaction {Energy in eV}	$O_O^x + V_i^x \rightarrow V_O^x + O_i^x$ {6.948} $O_O^x + V_i^x \rightarrow V_O^{\bullet} + O_i'$ {7.964} $O_O^x + V_i^x \rightarrow V_O^{\bullet\bullet} + O_i''$ {8.868}	$Zn_{Zn}^x + V_i^x \rightarrow V_{Zn}^x + Zn_i^x$ {12.002} $Zn_{Zn}^x + V_i^x \rightarrow V_{Zn}' + Zn_i^{\bullet}$ {9.637} $Zn_{Zn}^x + V_i^x \rightarrow V_{Zn}'' + Zn_i^{\bullet\bullet}$ {7.380} $Zn_{Zn}^x + V_i^x \rightarrow V_{Zn}^x + Zn_i^{\bullet\bullet} + 2e'$ {9.463}

Defect	Schottky defect	
Possible Formation Reaction {Energy in eV}	$Zn_{Zn}^x + O_O^x \rightarrow V_{Zn}^x + V_O^x + ZnO(s)$ {8.872} $Zn_{Zn}^x + O_O^x \rightarrow V_{Zn}' + V_O^{\bullet} + ZnO(s)$ {7.645} $Zn_{Zn}^x + O_O^x \rightarrow V_{Zn}'' + V_O^{\bullet\bullet} + ZnO(s)$ {6.896}	

Zhang et al. [51] studied the physics of both intrinsic (Zn_i, V_O, Zn_O, V_{Zn} & O_i) as well as impurity (F & Al) generated point defect in ZnO. They discovered that ZnO is n-type in Zn-rich sample because Zn_i (which act as shallow donor) is abundant due to its low enthalpy of formation under both Zn as well as O-rich conditions. In addition to this, the defects V_{Zn} or O_i that can compensate for Zn_i in Zn-rich ZnO have a high enthalpy of formation. Furthermore, they observed that the acceptor-type native defects O_i and V_{Zn} cannot result in p-type ZnO because the donor-type defects (Zn_i, V_O, and Zn_O), which have a low enthalpy of formation, compensate for their influence under both Zn and O-rich circumstances. Furthermore, the green emission was discovered to be caused by radiative recombination of electron-hole (e^-&h^+) at V_O.

The further detailed discussion on individual point defect observed in ZnO is given below in subsequent subsections.

5.1 Zinc interstitial

Zn_i is recognized as a zinc atom in an interstitial location; both hypothetical and experimental results reveal that Zn_i are shallow donors [52]. The Zn interstitial sites are found to be stable at octahedral sites of the wurtzite structure [46]. Another probable site, namely the tetrahedral site has been found to be inherently less favorable or energetically unstable [19, 53]. The Zn_i are rarely movable at ambient temperature due to their high formation energy of the order of 3.62 eV and 1.56 eV for Zn_i and Zn_i^+ respectively and comparatively low migration barrier of 0.57 eV even under zinc-rich conditions. Because of the low migratory barriers, it is anticipated that they will have a strong potential to disperse or bind with various dopants [19]. The experimental value of migration barrier reported by Thomas is found to be 0.55 eV [54]. Especially in wurtzite structure, Zn_i consist two types of interstitial sites, namely octahedral and tetrahedral sites. Oba et al. [55] reported that Zn_i occupy Octahedral sites due to high stability whereas tetrahedral sites are unfavorable and unstable reported by Sokol et al. [50]. The octahedral site consist three Zn and O atoms separated by about 1.07 d_0 where d_0 is the Zn-O bond length along c-axis whereas in case of tetrahedral site the bond length is around 0.833 d_0 [19]. As a result of the less geometric limitations, octahedral locations are projected to be more desirable. The formation energy of Zn_i is regarded to be high in n-type, although its concentration is thought to below. Zn_i can be stabilized in the presence of a large proportion of oxygen vacancies, according to Kim et al. [56]. But due to low formation energy of p-type ZnO holes can be compensated by Zn_i.

5.2 Zinc vacancy

V_{Zn} is produced when zinc is depleted or transferred from its native place or site to an interstitial site. Theoretically, the V_{Zn} has been proposed as a dominating acceptor-type defect in zinc oxide [46]. Tuomisto et al. [57] reported same results with positron annihilation. Zinc vacancy has been reported as the dominant defect in ZnO by Erhart et al. [58]. Zinc vacancies are coordinated with four oxygen atoms by dangling bonds and initiate partially filled states in E_{gap}. The V_{Zn} exists in three states, namely V_{Zn}, V_{Zn}^{-1}, and V_{Zn}^{-2}, since these dangling oxygen bonds interact to generate a dualy charged state near to the valence band, resulting in three partially filled states. The formation energy of zinc vacancies is very high i.e. in the order of 6.39 eV even under zinc rich environment; however in oxygen-rich environment, it is 3.7 eV. As zinc vacancy is connected to zinc insufficiency, the formation energy for zinc vacancy is lowest under O-rich environments. According to Kohan et al. [53], green colour in photoluminescence spectra comes from zinc vacancy levels due to acceptor defects and n-type conductivity.

5.3 Zinc antisite

A zinc antisite is a Zn atom that has been positioned in the incorrect lattice or in the location of an O atom. Zinc antisites are shallow donors with a high formation energy of 4.98 eV. Although Kohan et al. [53] determined that the energy in zinc antisite defects is larger than in other defects and remain unstable even under zinc rich environment because of their high formation energy. The Zn antisite is thought to be a complex comprising V_o and Zn_i with extremely deep and shallow levels. In the relaxed geometry of ZnO, zinc is displaced more than one Å away from the oxygen site. The formation energy of these complexes in n-type ZnO is greater than Zn_i.

5.4   Oxygen interstitial and antisite

Oxygen interstitial (O_i) is the name given to an additional oxygen atom in a lattice that is positioned in the interstitial location. O_i has a high formation energy of about 6.83 eV and a migration barrier up to 0.9 eV. Few configurations in oxygen interstitial were reported by Erhat et al. such as octahedral, dumbbell or split, and molecule of O_2 like configuration [58]. The formation energy of oxygen interstitial in octahedral configuration was shown to be greater than that of Zn vacancy [19, 59]. The formation energy of split configuration is higher than that of octahedral configuration in n-type ZnO. Electrically, across the fermi level range they play no function in neutral charge states.

Oxygen antisite is defined as an oxygen atom placed at a zinc site. When compared to oxygen antisite and other defects; Oxygen antisite has a greater formation energy of 10.04 eV. As a result, it is possible to deduce that O_i and O antisite have very high formation energies and are electrically neutral or inactive.

5.5   Oxygen vacancy

A missing oxygen atom from its lattice site is referred as oxygen vacancy (V_o). The formation energy of V_o^+ is very low i.e 0.81 eV as compare to other defects even in Zn rich conditions, implying that creation of V_o is simpler and concentration will be higher under equilibrium conditions. V_o contains four bonds and two electrons and occurs in three states: V_o, V_o^{+1}, V_o^{+2}, with V_o^+ being the least stable of the three. These oxygen vacancies, on the other hand, are deep donors.

Based on few experimental results of electron paramagnetic resonance (EPR) by Vanheusden et al. [60], there have been arguments over the source of green luminescence. Leiter et al. [61] ascribe V_o to green luminescence. Vo has also been blamed for green emissions, according to Hofmann et al. Janotti et al. calculated the formation energy of neutral V_o, which suggests that V_o has a low formation energy and that the migration barrier in n-type is 2.04 eV. V_o has low diffusion at normal temperature, however higher temperatures can cause significant movements of V_o, which is much lower in p-type ZnO. As a result, oxygen vacancies serve as a possible source of compensation for p-type ZnO.

6.   Defect characterization techniques

For a better understanding of defect dynamics inside any material, proper characterization of various defects present it is very essential. To detect defects in materials, researchers use a range of characterization techniques. All characterization approaches, however, do not identify all sorts of defects. As a result, a variety of strategies for appropriately assigning faults should be used. Depending on the nature of the material and the researcher's interest, some techniques are given a higher priority in each situation. Here in this section, the benefits and downsides of various characterization schemes will be highlighted.

Photoluminescence (PL) of ZnO has been widely investigated to verify its luminescence qualities and, more significantly, to study the defects [62]. Low temperature PL (LT-PL) spectroscopy is useful for detecting defects involving shallow donor and acceptor levels [63]. The technique of cathodoluminescence (CL) spectroscopy is quite similar, but it has the benefit of allowing for depth resolved characterization [64, 65]. However, during CL measurements, electron excitation can harm the sample surface, whereas PL is a non-destructive instrument. Another technique for studying electrically active defects is deep level transient spectroscopy [66]. Defects containing unpaired electrons are susceptible to electron paramagnetic spectroscopy (EPR) [67]. The kind, concentration, and movement of free carriers may all be determined using hall measurements [68]. The Rutherford back scattering based measurements are extremely sensitive to interstitial defects, whereas the positron annihilation spectroscopy related experiment is employed to identify vacancy type defects [69, 70]. UV-visible absorption spectroscopy, on the other hand, may be used to examine the system's overall flaws. Raman spectroscopy and X-ray diffraction (XRD) can be used to determine structure defect information [71]. This sort of information may also be used to identify flaws in material. A range of electron microscopic methods, in addition to these, may be employed to directly analyses the crystal disorder such as grain boundaries and dislocation loops [72].

6.1   PL for identification of defects in ZnO

PL spectroscopy is a highly sensitive and nondestructive method to identify extrinsic as well as intrinsic native point defects present in a material. As we know that the electrical band structure of any semiconductor contains a valence band (VB) refers to the highest occupied orbital and conduction band (CB) refers to the lowest unoccupied orbital. The region lying between the top of VB and the bottom of CB is known as band gap. In case of a perfect crystal of material without any defect this band does not contain any energy state and is therefore also known as forbidden energy gap. In the case of a perfect crystal of a material with no defects, this band has no energy state and is therefore also known as the forbidden energy gap. But the presence of defects in crystal structure of material results in the appearance of some defect states (DS) in the region of forbidden energy gap as shown in Fig.6. By photo-exciting electrons from VB to CB with varied energies to different excited states, we may produce e⁻&h⁺ pairs. But these excited electrons in are exceedingly unstable and try to return to the VB through a specific path to recombine with the holes. These photoinduced e⁻&h⁺ can have recombination through three different processes [73].

Process-1 (Non-radiative recombination): In this case, the electrons lying at the base of CB can come back to VB without the release of radiations. Such non-radiative transitions results the recombination of e⁻&h⁺ through phonon rather than a photon. In phosphors and optoelectronics, the non-radiative recombination is an undesirable phenomenon that increase the heat loss in system and decrease the efficiency to generate light.

Process-2 (Near-band-edge PL emission): In this case, the electrons lying at the base of CB can directly come back to VB with the release of radiations i.e. photon. The energy of the released photon is generally equal to the E_{gap}; however, in some cases photons with energy more than E_{gap} can also be released due to the transition of electrons from a level lying above the bottom of CB.

Process-3 (Defect related emission): In this case the photo-excited electron undergoes a transition from base of CB to one of DS without release of radiation (i.e. non-radiative transfer) and then electron from DS undergoes a radiative transition to recombine with the holes lying in VB. As a result, the photon released in this situation has less energy than the E_{gap}. This emitted radiation is referred to as the excitonic PL signal, and it indicates the presence of numerous native defects in the material.

Process-2 and Process-3 can both induce PL phenomena, with the former being connected with band-to-band PL and the latter with excitonic PL. Surface imperfections may be detected using the excitonic PL.

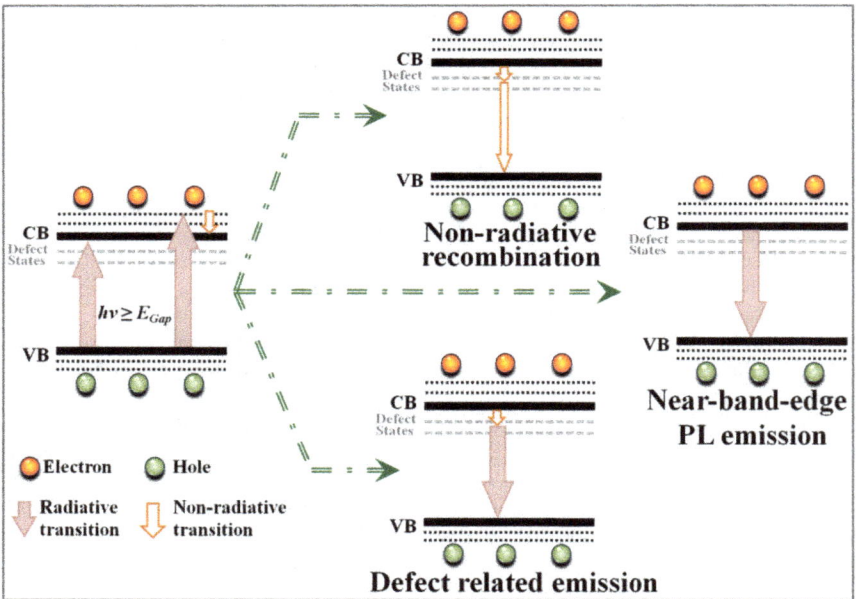

Figure 6: The major processes involved in PL-based defect analysis of materials.

PL spectroscopy may be used to investigate the existence of different flaws in ZnO. The PL spectrum of ZnO may be separated into two halves, as illustrated in Fig.7: (1) Near-band-edge PL emission in the UV is typically caused by the grain's interior. This emission is either a direct transition from CB to VB, or a transition to/from donor and acceptor levels near VB and CB, respectively. (2) Defect related emission caused by surface imperfections that is observable in the visible zone. The participation of deep levels in the band gap of ZnO causes these emissions.

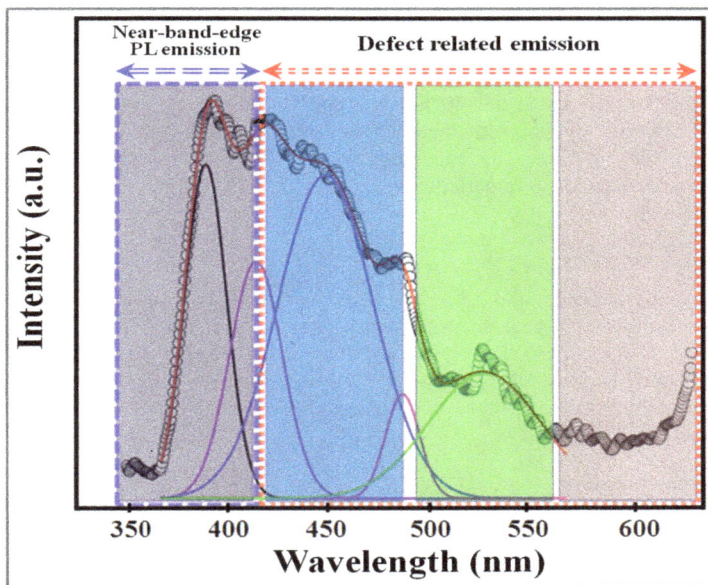

Figure 7: Near-band and defect-related emission in a typical PL of ZnO [74].

It is critical to examine the origin of luminous centres, which are caused by intrinsic or extrinsic defects in the semiconductor, for any possible material for optoelectronics. Modifying the distribution of these flaws inside the grains may be used to tune and regulate the optical characteristics. The genesis of luminescence in ZnO and the flaws that cause it are currently being debated. In the section below, the defects that cause emissions in the visible range are described.

6.1.1 Blue emission

The blue luminescence of ZnO makes it a potential choice for applications such as optoelectronics and biological fluorescent labeling due to its excellent electrical characteristics and biocompatible nature. According to Wu et al. [77,78], blue emission of ZnO is unusual [75,76]. However, non-equilibrium routes, for instance laser ablation, have been reported to result a high occurrence of blue luminescence in ZnO. This emission is thought to be generated by energies close to E_{gap}, although they can also be activated by energy lesser to E_{gap}.

Several intrinsic defects are found inside the ZnO forbidden band gap, but two defects, Zn_i & V_O, are prevalent, according to some researchers [52,79]. According to Ischencko et al. [80], non-equilibrium routes may result in large concentrations of specific faults, which aids in the creation of a better knowledge of visible light emission. The influence of annealing on blue luminescence by ZnO was investigated by Zeng et al. [77]. High blue emission with peak located at 440 nm was found, which red shifted to 455 nm after annealing ZnO in air but remained steady after annealing in nitrogen [77]. As a consequence, they arrived to the tentative hypothesis that the Zn_i defects were the cause of PL emission. He went on to remark that non-equilibrium faults have a significant influence on ZnO luminescence.

6.1.2 Yellow emission

The emission corresponding to 2.1 eV is caused by the yellow luminescence band in ZnO. In case of hollow spherical shaped nanostructures ZnO, Zang et.al [81] discovered significant yellow luminescence which they attributed to the singly-ionized V_O. Peng et al. [82], on the other hand, ascribe yellow luminescence found at 580 nm to O_i. According to Zwinge, recombining donor and acceptor pair, as well as Li acceptor, also results in yellow luminescence [83].

6.1.3 Green emission

One of the most disputed emissions in ZnO has been its luminescence in the green region of the visible spectrum. A lot of authors have suggested many hypotheses on the subject, but no clear consensus has emerged. Many publications ascribe the green glow to oxygen vacancy (V_O) [60,84]. Zhao et al. [85] reported that Zn vacancy (V_{Zn}) induces green luminescence, which was also observed by Kohan et al. [53]. Green luminescence of ZnO, on the other hand, was ascribed by Liu et al. [86] to interstitial native faults (Zn_i and O_i). As a result, if we see the literature several centers, such as V_O, V_{Zn}, Zn_i and O_i, may be believed to be responsible for the green luminescence of ZnO. The condition might emerge as a result of material production technique, flaws in ZnO crystals, or variations in these imperfections. However, Zn_i and V_O defects are mentioned as viable candidates resulting in green emission of ZnO [87].

Several research groups looked at the conditions for making ZnO and saw green emission, which they attributed to V_O with unit charge [60]. Some investigations have shown that Cu as impurity can also be the source of green luminescence of doped ZnO [88]. But, further studies revealed no link between the copper ion concentration and green emission intensity, indicating that the green emission was caused by V_O [61]. Shan et al. [84] conducted extensive investigation by depositing ZnO thin film on sapphire and came to the conclusion that oxygen vacancies were responsible for intraband emission at $\lambda = 510$ nm.

6.1.4 Red emission

In an LT-PL examination of pure ZnO annealed in air at 700 °C, Özgür et al. [1] found a red emission. A Gaussian shaped red emission band of 0.5 eV FWHM was seen at 15 K; however, this red emission band was suppressed at higher temperatures (80 to 200 K), accompanied by the appearance of a green emission band. This change was ascribed to the struggle for holes among the acceptors responsible for red and green emission. Further, the green emission was quickly quenched when the temperature rose over 200 K, and only red emission was visible at RT.

6.1.5 Factors influencing PL of ZnO

In fact, the PL of ZnO is quite complex, and many factors such as excitation source (wavelength, continues or pulsed nature, intensity), size of ZnO particles, external doping in ZnO, form of ZnO (powder or film) and post synthesis annealing of ZnO. Some of the major factors affecting the PL of ZnO are discussed in this section.

Effect of excitation source: In general, the PL of a material is determined not only by its surface quality, but also by the excitation source used in PL studies. However, according to Timothy H. Gfroerer, the quality of the surface has a significant impact on PL's reliance on the excitation source. They found that the PL spectrum varies with the energy of the excitation source only if the surface is of low quality, but that PL is independent of the energy of the excitation source for surfaces of excellent quality [89].

Jing et al. [90] used 300 and 350 nm light to investigate the influence of excitation wavelength on the PL of ZnO nanoparticles (Fig.8). Despite the fact that the energy of incident photons is more than E_{gap} of ZnO in both cases, the PL spectra are not fully identical, as seen in Fig.8. This variation in PL was discovered because incoming photons of different energy can stimulate electrons from VB to various levels of CB, and these excited electrons can return to VB through various paths, revealing the PL process' complexity. However, it was observed that the positions of the peaks in the PL spectrum were nearly fixed, indicating that some flaws have very high stability and are constantly engaged in PL emissions.

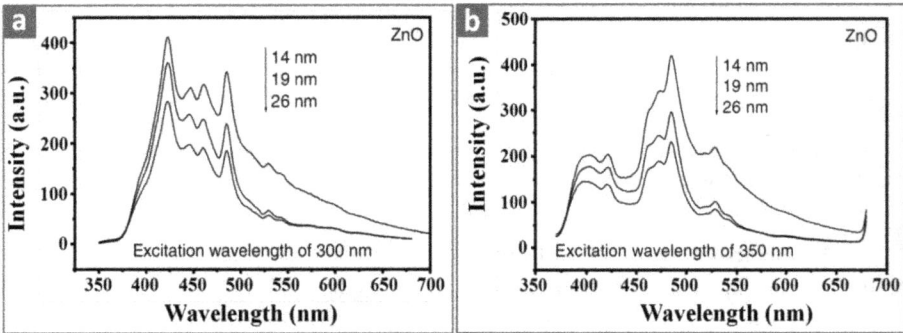

Figure 8: Effect of wavelength of incident light on PL of ZnO [90].

In addition to wavelength, also note that continuous-wave (CW) excitation can produce different experimental results from pulsed excitation. Lee et al. [91] compared the PL spectrum of ZnO nanorods using a CW laser (HeCd laser) and a pulsed laser (35 ps pulse width) as the excitation source. They observed that in both cases the PL spectrum showed a band–band transition in the UV region and a band due to defect-related emission in the visible region; however, the peak observed using a CW laser in PL with a 160 meV FWHM at 3.25 eV in the UV region was observed with a 35 meV FWHM at 3.20 eV when pulsed excitation source was used. Furthermore, the excitation power density and pulse duration/repetition rate can also be experimental variables.

Many research groups have looked into the influence of excitation source intensity (EsI) on ZnO PL spectra, and they have determined that EsI has a considerable impact. P.R. Talakonda investigated the effect of EsI on ZnO PL at room temperature and liquid nitrogen temperatures [95]. They detected a blue shift in the UV emission band in some samples when the EsI was reduced at both temperatures, indicating that the PL of ZnO is highly influenced by the EsI. Using Raman spectroscopy, they also found that the heating effect resulted by change in EsI alone cannot explain the detected excitation intensity dependence of PL.

Das et al. [96] investigated the effect of EsI on PL of different morphologies of ZnO, i.e. nanoparticles (NPs) and nanorods (NRs). Room-temperature spectrum of ZnO NPs (Fig.9(a)) and ZnO NRs (Fig.9(b)) at excitation intensities 0.007, 0.01, 0.02, 0.03, 0.08, 0.3, 0.6 and 1.3 W/cm^2 showed that the PL intensity has a direct dependence on EsI. However, the magnitude of this direct dependence was found to depend on the morphology of ZnO. The intensity of PL in ZnO NPs became saturated at EsI of 0.3 $W/cm2$ and remained nearly constant up to EsI of 1.3 W/cm^2, but in ZnO NRs, the intensity of PL increased continuously with increasing EsI.

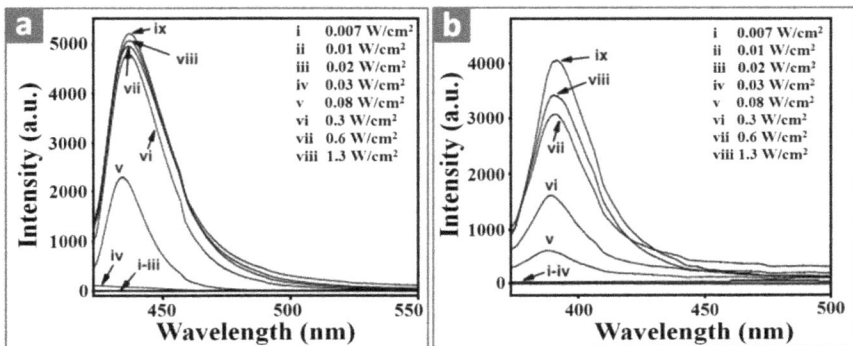

Figure 9: Dependence of PL intensity on EsI for **(a)** *ZnO NPs, and* **(b)** *ZnO NRs [96].*

Further, the variation PL intensity at RT with EsI also depends upon the annealing of ZnO nanoparticles. Fig.10 shows that the photoluminescence intensity is quite high for ZnO nanoparticles annealed at 600 °C for ten hours in atmosphere compared to ZnO nanoparticles without annealing. It demonstrates that annealing has an influence on defect concentration.

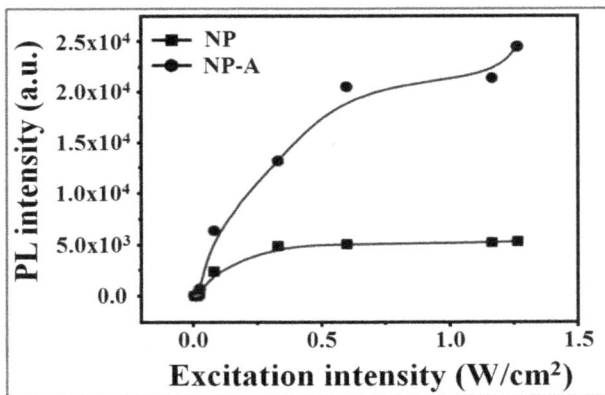

Figure 10: Variation of room temperature PL intensity on EsI for ZnO-NPs with (NP-A) and without annealing (NP) [96].

Effect of size of ZnO particles: Jing et al. [90] obtained the PL spectra of differently sized ZnO NPs using excitation photons with energy greater than the E_{gap} of ZnO. They reported that the excitonic PL intensity of ZnO NPs reduces as particle size grows (Fig.8). This was due to a reduction in the concentration of surface imperfections as particle size increased. This demonstrates that a bulk ZnO material has less surface flaws than tiny nanostructured ZnO.

Effect of external doping in ZnO: Metal ion doping is a typical method of modifying ZnO to improve its photocatalytic and photoelectric characteristics. However, various dopants have distinct influencing processes on photo-induced carriers, resulting in diverse PL spectrum effects. The PL characteristics of Co-doped ZnO hollow microspheres were investigated by Choi et al. [92]. PL spectra of ZnO microspheres with 0, 3, 6, 9 and 12 % doped Co is shown in Fig. 9. Fig.9 shows that when the concentration of Co rose, the intensity of UV emission decreased while the intensity of visible emission increased. The decrease in UV emission intensity was due to an increase in impurities and defects.

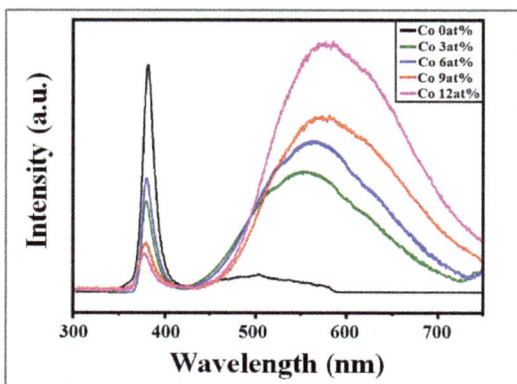

Figure 11: PL spectrum of Co doped ZnO nanostructures [92].

Fig.12 shows PL of ZnO doped with different concentrations of magnesium (Mg) recorded using excitation wavelength 325 nm [93]. It can be clearly seen from the spectrum that the defect related emission increased initially with the addition of Mg in ZnO i.e. up to 5% Mg in ZnO), although thereafter there was a drastic decrease in the PL intensity corresponding to the defect related emission. This means, the concentration of dopant also plays an important role for deciding the defect related PL.

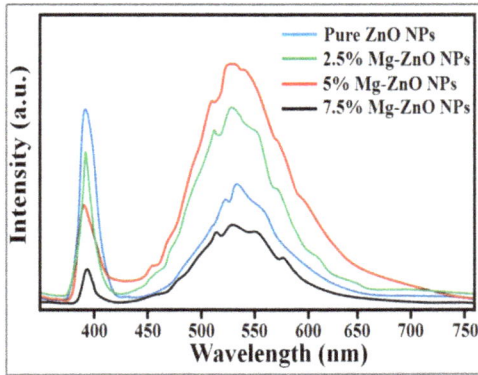

Figure 12: PL spectrum of Mg doped ZnO nanostructures [93].

Alam et al. [94] performed a comparative study on the PL of ZnO doped with different rare earth metals, viz. P, Dy, Sm, Nd and La (Fig.13). The PL spectra for doped ZnO NPs were observed to be lower than prestine ZnO NPs, as shown in Fig. 13. Moreover, the doping of ZnO NPs with Nd resulted in the highest quenching of PL emission. This indicates that the type of metal ion used for doping has a great influence on the PL.

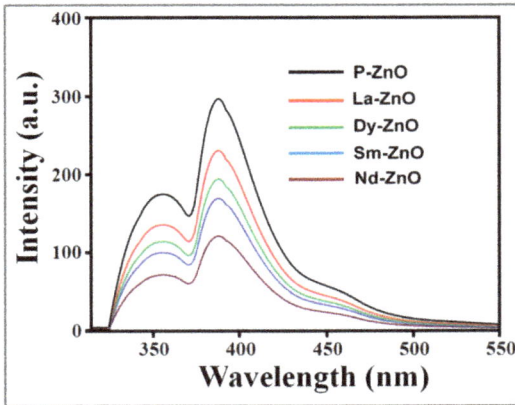

Figure 13: PL spectrum of ZnO doped with different rare earth metals, viz. P, Dy, Sm, Nd and La [94].

Effects of form of ZnO: The two most widely studied forms of ZnO are thin films and powder. It has been observed that for the incident photon having energy greater than E_{gap}, the excitonic PL signal in NPs (Powder) may easily be seen; however, there may be cases where band to band transitions are absent in powder but such transitions are mostly present in films [73].

Effects of post synthesis annealing of ZnO: The impact of post synthesis annealing on the PL of ZnO NRs formed on GaN was examined by Quang at el. [97]. They annealed the generated samples for 30 minutes at 800 degrees Celsius in oxygen (O_2) and nitrogen (N_2) environment. By comparing RT-PL all these ZnO samples, they found that annealing ZnO in N_2 ambient considerably increased UV light emission (Fig.14(a)) while defect related emissions were high ZnO NRs annealed in O_2 ambient (insets of Fig.14(a)). A similar trend of increased PL intensity by annealing ZnO in N_2 ambient was seen for LT-PL recorded at 80K (Fig.14 (b)). These findings demonstrated that the annealing environment had a significant impact on the emission characteristic of ZnO.

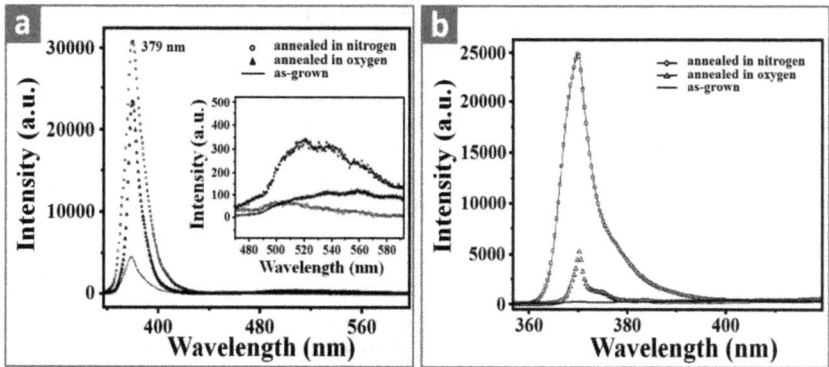

Figure 14: PL of ZnO NRs annealed in different environment: (a) at 298, and (b) at 80 K [97].

It must be noted that though PL is an extremely sensitive approach for identifying surface and bulk defects, but for this semi-quantitative analysis of defects in ZnO via PL excessive care is needed. In practice, utilizing PL intensities as a metric of impurity concentration is difficult as in addition to the defect concentrations the recombination velocities as well as matrix medium used also alter the radiative quantum yield from sample to sample.

Conclusion

This demonstrates that point defects in ZnO crystals can modulate their optical properties considerably. As a result, a complete grasp of defect engineering is required for improving the controllability of optical characteristic modulation, which might lead to useful optoelectronics applications in the future. By different methods such as annealing or introducing impurities of transition metal ions, the optical and magnetic characteristics of ZnO may be adjusted which results in its different technological applications. As a result, comprehending the function of defects in ZnO is crucial, and it is a current emphasis since it may broaden the domain of ZnO in terms of real-world applications. Keeping this in mind, we have gone through numerous intrinsic point defects found in ZnO, defect generation procedures, defect characterization, defect formation factors, and luminous characteristics of various defects in ZnO in this chapter.

References

[1] Ü. Özgür, Y.I. Alivov, C. Liu, A. Teke, M.A. Reshchikov, S. Dog˘an, V. Avrutin, S.-J. Cho, H. Morkoç, A comprehensive review of ZnO materials and devices, J. Appl. Phys. 98:4 (2005) 041301, https://doi.org/10.1063/1.1992666

[2] J. Theerthagiri, S. Salla, R.A. Senthil, P. Nithyadharseni, A. Madankumar, P. Arunachalam, T Maiyalagan, H.S. Kim, A review on ZnO nanostructured materials: energy, environmental and biological applications, Nanotechnology 30 (2019) 392001, https://doi.org/10.1088/1361-6528/ab268a

[3] H. Dong, B. Zhou, J. Li, J. Zhan, L. Zhang, Ultraviolet lasing behavior in ZnO optical microcavities, Journal of Materiomics 3:4 (2017) 255-266, https://doi.org/10.1016/j.jmat.2017.06.001

[4] F. Rahman, Zinc oxide light-emitting diodes: a review, Optical Engineering 58:1 (2019) 010901, https://doi.org/10.1117/1.OE.58.1.010901

[5] C. Torres-Torres, J.H. Castro-Chacón, L. Castañeda, R.R. Rojo, R. Torres-Martínez, L. Tamayo-Rivera, A.V. Khomenko, Ultrafast nonlinear optical response of photoconductive ZnO films with fluorine nanoparticles, Optics Express 19:17 (2011) 16346-16355, https://doi.org/10.1364/OE.19.016346

[6] Y. Liu, Y. Li, H. Zeng, ZnO-based transparent conductive thin films: doping, performance, and processing, 2013 (2013) 196521, https://doi.org/10.1155/2013/196521

[7] Y. Lu, N. Emanetoglu, Y. Chen, ZnO Piezoelectric Devices; In book: Zinc oxide bulk, thin films and nanostructures (2006) 443-489, https://doi.org/10.1016/B978-008044722-3/50013-0

[8] Ü. Özgür, D. Hofstetter, H. Morkoç, ZnO devices and applications: A review of current status and future prospects, Proceedings of the IEEE 98:7 (2010) 1255–1268, https://doi.org/10.1109/jproc.2010.2044550

[9] K. Koike, T. Aoki, R. Fujimoto, S. Sasa, M. Yano, S. Gonda, R. Ishigami, K. Kume, Radiation hardness of single-crystalline zinc oxide films, Physica Status Solidi (c) 9:7 (2012) 1577–1579, https://doi.org/10.1002/pssc.201100566

[10] D.C. Look, Recent advances in ZnO materials and devices, Materials Science and Engineering: B 80 (2001) 383–387, https://doi.org/10.1016/s0921-5107(00)00604-8

[11] P.J.P. Espitia, N. de F.F. Soares, J. S. dos R. Coimbra, N.J. de Andrade, R.S. Cruz, E.A.A. Medeiros, Zinc oxide nanoparticles: Synthesis, antimicrobial activity and food packaging applications, Food and Bioprocess Technology 5:5 (2012) 1447–1464. https://doi.org/10.1007/s11947-012-0797-6

[12] M.A. Borysiewicz, ZnO as a functional material, a review, Crystals 9 (2019) 505-533, https://doi.org/10.1109/10.3390/cryst9100505

[13] A. Kołodziejczak-Radzimska, T. Jesionowski, Zinc oxide—from synthesis to application: A review, Materials (Basel) 7:4 (2014) 2833–2881, https://doi.org/10.3390/ma7042833

[14] S. Sabir, M. Arshad, S.K. Chaudhari, Zinc oxide nanoparticles for revolutionizing agriculture: Synthesis and applications, The Scientific World Journal 2014 (2014) 925494, https://doi.org/10.1155/2014/925494

[15] S. Choi, M.R. Phillips, I. Aharonovich, S. Pornsuwan, B.C.C. Cowie, C. Ton-That, Photophysics of point defects in ZnO nanoparticles, Advanced Optical Materials 3:6 (2015) 821-827, https://doi.org/10.1002/adom.201400592

[16] P. A. Rodnyi, I. V. Khodyuk, Optical and luminescence properties of zinc oxide (Review), Optics and Spectroscopy 111 (2011) 776–785, https://doi.org/10.1134/S0030400X11120216

[17] M.A. Reshchikov, H. Morkoç, B. Nemeth, J. Nause, J. Xie, B. Hertog, A. Osinsky, Luminescence properties of defects in ZnO, Physica B: Condensed Matter 401–402 (2007) 358-361, https://doi.org/10.1016/j.physb.2007.08.187

[18] A. Galdámez-Martinez, G. Santana, F. Güell, P.R. Martínez-Alanis, A. Dutt, Photoluminescence of ZnO nanowires: A review, Nanomaterials 10:5 (2020) 857, https://doi.org/10.3390/nano10050857

[19] A. Janotti, C.G. Van de Walle, Native point defects in ZnO, Physical Review B 76 (2007) 165202, https://doi.org/10.1103/PhysRevB.76.165202

[20] T.-L. Phan, Y.D. Zhang, D.S. Yang, N.X. Nghia, T.D. Thanh, S.C. Yu, Defect-induced ferromagnetism in ZnO nanoparticles prepared by mechanical milling, Appl. Phys. Lett. 102 (2013) 072408, https://doi.org/10.1063/1.4793428

[21] K. Rainey, J. Chess, J. Eixenberger, D.A. Tenne, C.B. Hanna, A. Punnoose, Defect induced ferromagnetism in undoped ZnO nanoparticles, Journal of Applied Physics 115:17 (2014) 17D727, https://doi.org/10.1063/1.4867596

[22] Q. Xu, H. Schmidt, S. Zhou, K. Potzger, M. Helm, H. Hochmuth, M. Lorenz, A. Setzer, P. Esquinazi, C. Meinecke, M. Grundmann, Room temperature ferromagnetism in ZnO films due to defects, Appl. Phys. Lett. 92 (2008) 082508, https://doi.org/10.1063/1.2885730

[23] D.C. Look, T.C. Droubay, S.A. Chambers, Stable highly conductive ZnO via reduction of Zn vacancies, Appl. Phys. Lett. 101 (2012) 102101, https://doi.org/10.1063/1.4748869

Materials Research Forum LLC
https://doi.org/10.21741/9781644902394-2

[24] V. Dhiman, N. Kondal, MoS_2–ZnO nanocomposites for photocatalytic energy conversion and solar applications, Physica B 628 (2022) 413569, https://doi.org/10.1016/j.physb.2021.413569

[25] K. Momma, F. Izumi, VESTA 3 for three-dimensional visualization of crystal, volumetric and morphology data, J. Appl. Crystallogr. 44 (2011) 1272–1276, https://doi.org/10.1107/S0021889811038970

[26] P.K. Samanta, P.R. Chaudhuri, Growth and optical properties of chemically grown ZnO nanobelts, Sci. Adv. Mater. 3:1 (2011) 107–112, https://doi.org/10.1166/sam.2011.1141

[27] A. Eftekhari, F. Molaei, H. Arami, Flower-like bundles of ZnO nanosheets as an intermediate between hollow nanosphere and nanoparticles, Mater. Sci. Eng. A 437 (2) (2006) 446–450, https://doi.org/10.1016/j.msea.2006.08.033

[28] T.M. de B. Farias, S. Watanabe, A comparative study of the thermo luminescence properties of several varieties of Brazilian natural quartz, J. Lumin. 132:10 (2012) 2684–2692, https://doi.org/10.1016/j.jlumin.2012.04.047

[29] J. Geurts, Crystal structure, chemical binding, and lattice properties, Springer Series in Materials Science (2010) 7–37, https://doi.org/10.1007/978-3-642-10577-7_2

[30] A. Segura, J.A. Sans, F.J. Manjón, A. Muñoz, M.J. Herrera-Cabrera, Optical properties and electronic structure of rock-salt ZnO under pressure, Applied Physics Letters 83:2 (2003) 278–280, https://doi.org/10.1063/1.1591995

[31] M. Khuili, N. Fazouan, H.A.E. Makarim, DFT study of physical properties of wurtzite, zinc blende, and rocksalt phases of zinc oxide using GGA and TB-mBJ potential, 3rd International Renewable and Sustainable Energy Conference (IRSEC) (2015), https://doi.org/10.1109/irsec.2015.7454962

[32] S.S. Li, Classification of solids and crystal structure. In: Semiconductor Physical Electronics. Microdevices (Physics and Fabrication Technologies). Springer, Boston, MA (1993), https://doi.org/10.1007/978-1-4613-0489-0_1

[33] W.B. Fowler, Point defects. In: Encyclopedia of Condensed Matter Physics, Elsevier (2005) 318-323, https://doi.org/10.1016/B0-12-369401-9/00412-5

[34] J. Zhang, G. Hong, Synthetic chemistry of nonstoichiometric compounds. In: Modern Inorganic Synthetic Chemistry, Elsevier (2011) 321–338, https://doi.org/10.1016/b978-0-444-53599-3.10015-0

[35] H.L. Tuller, S.R. Bishop, Point defects in oxides: Tailoring materials through defect engineering, Annual Review of Materials Research 41 (2011) 369-398, https://doi.org/10.1146/annurev-matsci-062910-100442

[36] B. Pieraggi, Defects and transport in oxides and oxide scales. Shreir's Corrosion (2010) 101–131, https://doi.org/10.1016/b978-044452787-5.00009-3

[37] M. Jiang, D.D. Wang, B. Zou, Z.Q. Chen, A. Kawasuso, T. Sekiguchi, Effect of high temperature annealing on defects and optical properties of ZnO single crystals, Phys. Status Solidi A 209:11 (2012) 2126–2130, https://doi.org/10.1002/pssa.201127527

[38] L.E. Halliburton, N.C. Giles, N.Y. Garces, M. Luo, C. Xu, L. Bai. Production of native donors in ZnO by annealing at high temperature in Zn vapor, Appl. Phys. Lett. 87 (2005) 172108, https://doi.org/10.1063/1.2117630

[39] C.-H. Hsu, X.-P. Geng, W.-Y. Wu, M.-J. Zhao, X.-Y. Zhang, P.-H. Huang, S.-Y. Lien, Air annealing effect on oxygen vacancy defects in al-doped zno films grown by high-speed atmospheric atomic layer deposition, Molecules 25:21 (2020) 5043, https://doi.org/10.3390/molecules25215043

[40] T.P. Yadav, R.M. Yadav, D.P. Singh, Mechanical milling: A top down approach for the synthesis of nanomaterials and nanocomposites, Nanoscience and Nanotechnology 2:3 (2012) 22-48, https://doi.org/10.5923/j.nn.20120203.01

[41] S. Ghose, A. Sarkar, S. Chattopadhyay, M. Chakrabarti, D. Das, T. Rakshit, S.K. Ray, D. Jana, Surface defects induced ferromagnetism in mechanically milled nanocrystalline ZnO, Journal of Applied Physics 114 (2013) 073516, https://doi.org/10.1063/1.4818802

[42] S. Ghose, T. Rakshit, R. Ranganathan, D. Jana, Role of Zn-interstitial defect states on d^0 ferromagnetism of mechanically milled ZnO nanoparticles, RSC Adv. 5 (2015) 99766-99774, https://doi.org/10.1039/C5RA13846A

[43] H.V. Wenckstern, H. Schmidt, M. Brandt, A. Lajn, R. Pickenhain, et al., Anionic and cationic substitution in ZnO, Progress in Solid State Chemistry 37:2-3 (2009) 153–172, https://doi.org/10.1016/j.progsolidstchem.2009.11.008

[44] J.C. Fan, K.M. Sreekanth, Z. Xie, S.L. Chang, K.V. Rao, p-Type ZnO materials: Theory, growth, properties and devices, Progress in Materials Science 58:6 (2013) 874-985, https://doi.org/10.1016/j.pmatsci.2013.03.002

[45] R.G. Elliman, J.S. Williams, Advances in ion beam modification of semiconductors, Current Opinion in Solid State and Materials Science 19:1 (2015) 49-67, https://doi.org/10.1016/j.cossms.2014.11.007

[46] S. Dey, J. Mardinly, Y. Wang, J.A. Valdez, T.G. Holesinger, B.P. Uberuaga, J.J. Ditto, J.W. Drazina, R.H.R. Castro, Irradiation-induced grain growth and defect evolution in nanocrystalline zirconia with doped grain boundaries, Physical Chemistry Chemical Physics 18:25 (2016) 16921–16929, https://doi.org/10.1039/c6cp01763k

[46] F. Oba, M. Choi, A. Togo, I. Tanaka, Point defects in ZnO: An approach from first principles, Sci. Technol. Adv. Mater. 12:3 (2011) 034302, http://dx.doi.org/10.1088/1468-6996/12/3/034302

[47] S.E. Harrison, Conductivity and Hall Effect of ZnO at Low Temperatures, Phys. Rev. 93:1 (1954) 52-62, https://doi.org/10.1103/PhysRev.93.52

[48] F. Oba, S.R. Nishitani, S. Isotani, H. Adachi, I. Tanaka, Energetics of native defects in ZnO, J. Appl. Phys. 90 (2001) 824, http://dx.doi.org/10.1063/1.1380994

[49] Y. Guan, Q. Hou, D. Xia, Effect of intrinsic point defects on ZnO electronic structure and absorption spectra, International Journal of Modern Physics B 34:17 (2020) 2050147 http://dx.doi.org/10.1142/S0217979220501477

[50] A.A. Sokol, S.A. French, S.T. Bromley, C.R.A. Catlow, H.J.J. van Dam, P. Sherwood, Point defects in ZnO, Faraday Discuss. 134 (2007) 267-282, http://dx.doi.org/10.1039/b607406e

[51] S.B. Zhang, S.-H. Wei, A. Zunger, Intrinsic n-type versus p-type doping asymmetry and the defect physics of ZnO, Physical Review B 63:7 (2001) 075205, http://dx.doi.org/10.1103/physrevb.63.075205

[52] D.C. Look, J.W. Hemsky, J.R. Sizelove, Residual Native Shallow Donor in ZnO, Phys. Rev. Lett. 82:12 (1999) 2552-2555, https://doi.org/10.1103/PhysRevLett.82.2552

[53] A.F. Kohan, G. Ceder, D. Morgan, C.G. Van de Walle, First-principles study of native point defects in ZnO, Phys. Rev. B 61 (2000) 15019, https://doi.org/10.1103/PhysRevB.61.15019

[54] D.G. Thomas, Interstitial zinc in zinc oxide, J. Phys. Chem. Solids 3:3-4 (1957) 229-237, https://doi.org/10.1016/0022-3697(57)90027-6

[55] F. Oba, A. Togo, I. Tanaka, J. Paier, G. Kresse, Defect energetics in ZnO: A hybrid Hartree-Fock density functional study, Phys. Rev. B 77 (24) (2008) 245202, https://doi.org/10.1103/PhysRevB.77.245202

[56] Y.S. Kim, C.H. Park, Rich variety of defects in ZnO via an attractive interaction between O vacancies and Zn interstitials: Origin of n-type doping, Physical Review Letters 102:8 (2009) 1-4, https://doi.org/10.1103/PhysRevLett.102.086403

[57] F. Tuomisto, V. Ranki, K. Saarinen, D.C. Look, Evidence of the Zn vacancy acting as the dominant acceptor in n-Type ZnO, Phys. Rev. Lett. 91 (2003) 205502, https://doi.org/10.1103/PhysRevLett.91.205502

[58] P. Erhart, K. Albe, A. Klein, First-principles study of intrinsic point defects in ZnO: Role of band structure, volume relaxation, and finite-size effects, Phys. Rev. B 73:20 (2006) 205203, https://doi.org/10.1103/PhysRevB.73.205203

[59] P. Erhart, A. Klein, K. Albe, First-principles study of the structure and stability of oxygen defects in zinc oxide, Phys. Rev. B 72:8 (2005) 085213, https://doi.org/10.1103/PhysRevB.72.085213

[60] K. Vanheusden, W.L. Warren, C.H. Seager, D.R. Tallant, J.A. Voigt, B.E. Gnade, Mechanisms behind green photoluminescence in ZnO phosphor powders, J. Appl. Phys. 79 (10) (1996) 7983–7990, https://doi.org/10.1063/1.362349

[61] F.H. Leiter, H.R. Alves, A. Hofstaetter, D.M. Hofmann, B.K. Meyer, The oxygen vacancy as the origin of a green emission in undoped ZnO, Phys. Status Solidi Basic Res. 226:1 (2001) R4–R5, https://doi.org/ https://doi.org/10.1002/1521-3951(200107)226:1<R4::AID-PSSB99994>3.0.CO;2-F

[62] H. Akazawa, Identification of defect species in ZnO thin films through process modification and monitoring of photoluminescent properties, Journal of Vacuum Science & Technology A 37 (2019) 061514, https://doi.org/10.1116/1.5121439

[63] P.S. Venkatesh, V. Ramakrishnan, K. Jeganathan, Investigations on the growth of manifold morphologies and optical properties of ZnO nanostructures grown by radio frequency magnetron sputtering, AIP Advances 3:8 (2013), https://doi.org/10.1063/1.4820386.

[64] L.J. Brillson, W.T. Ruane, H. Gao, Y. Zhang, S. Diego, Spatially-resolved cathodoluminescence spectroscopy of ZnO defects, Materials Science in Semiconductor Processing 57:1 (2017) 197-209, https://doi.org/10.1016/j.mssp.2016.10.032

[65] N. Bano, I. Hussain, O. Nour, M. Willander, Q. Wahab, A. Henry, H.S. Kwack, D.L.S. Dang, Depth-resolved cathodoluminescence study of zinc oxide nanorods catalytically grown on p-type 4H-SiC, , Journal of Luminescence 130 (2010) 963-968, http://dx.doi.org/10.1016/j.jlumin.2010.01.006

[66] M. Ellguth, M. Schmidt, R. Pickenhain, H.V. Wenckstern, M. Grundmann, Characterization of point defects in ZnO thin films by optical deep level transient spectroscopy, Phys. Status Solidi B 248:4 (2011) 941-949, https://doi.org/10.1002/pssb.201046244

[67] L.S. Vlasenko, Point defects in ZnO: Electron paramagnetic resonance study, Physica B Condensed Matter 404:23-24 (2009) 4774-4778, https://doi.org/10.1016/j.physb.2009.08.149

[68] A. Maqsood, M. Islam, M. Ikram, S. Salam, S. Ameer, Synthesis, characterization and hall effect measurements of nanocrystalline ZnO thin films, Key Engineering Materials 510-511(2012) 186-193, https://doi.org/10.4028/www.scientific.net/KEM.510-511.186

[69] M. Brocklebank, J.J. Noël, L.V. Goncharova, In situ Rutherford backscattering spectrometry for electrochemical studies, Journal of The Electrochemical Society166:11 (2019) C3290, https://doi.org/10.1149/2.0301911jes

[70] F. Tuomisto, I. Makkonen, Defect identification in semiconductors with positron annihilation: Experiment and theory, Rev. Mod. Phys. 85 (2013) 1583, https://doi.org/10.1103/RevModPhys.85.1583.

[71] V.V. Strelchuk, V.P. Kladko, E.A. Avramenko, O.F. Kolomys, et al., X-ray diffraction analysis and scanning micro-Raman spectroscopy of structural irregularities and strains deep inside the multilayered InGaN/GaN heterostructure, Semiconductors 44:9 (2010) 1199-1210, https://doi.org/10.1134/S1063782610090174

[72] U. Ross, A. Lotnyk, E. Thelander, B. Rauschenbach, Direct imaging of crystal structure and defects in metastable $Ge_2Sb_2Te_5$ by quantitative aberration-corrected scanning transmission electron microscopy, Applied Physics Letters 104 (2014) 121904, http://dx.doi.org/10.1063/1.4869471

[73] J. Liqiang, Q. Yichun, W. Baiqi, L. Shudan, J. Baojiang, Y. Libin, F. Wei, F. Honggang, S. Jiazhong, Review of photoluminescence performance of nano-sized semiconductor materials and its relationships with photocatalytic activity, 90:12 (2006) 1773–1787, https://doi.org/10.1016/j.solmat.2005.11.007

[74] H. Rai, Prashant, N. Kondal, A review on defect related emissions in undoped ZnO nanostructures, Materials Today: Proceedings 48: 5 (2022) 1320-1324, https://doi.org/10.1016/j.matpr.2021.08.343

[75] D.M. Hofmann, D. Pfisterer, J. Sann, B.K. Meyer, R. Tena-Zaera, V. Munoz-Sanjose, T. Frank, G. Pensl, Properties of the oxygen vacancy in ZnO, Appl. Phys. A 88:1 (2007) 147-151, https://doi.org/10.1007/s00339-007-3956-2

[76] J.J. Wu, S.C. Liu, Catalyst-Free Growth and Characterization of ZnO Nanorods, J. Phys. Chem. B 106:37 (2002) 9546-9551, https://doi.org/10.1021/jp025969j

[77] H. Zeng, G. Duan, Y. Li, S. Yang, X. Xu, W. Cai, Blue luminescence of ZnO nanoparticles based on non-equilibrium processes: Defect origins and emission controls, Adv. Funct. Mater. 20 (4) (2010) 561-572, https://doi.org/10.1002/adfm.200901884

[78] H. Zeng, W. Cai, Y. Li, J. Hu, P. Liu, Composition/structural evolution and optical properties of ZnO/Zn nanoparticles by laser ablation in liquid media, J. Phys. Chem. B 109:39 (2005) 18260-18266, https://doi.org/10.1021/jp052258n

[79] D.C. Look, G.C. Farlow, P. Reunchan, S. Limpijumnong, S.B. Zhang, K. Nordlund, Evidence for native-defect donors in n-type ZnO, Phys. Rev. Lett. 95:22 (2005) 1-4, https://doi.org/10.1103/PhysRevLett.95.225502

[80] V. Ischenko, S. Polarz, D. Grote, V. Stavarache, K. Fink, M. Driess, Zinc oxide nanoparticles with defects, Adv. Funct. Mater. 15:12 (2005) 1945-1954, https://doi.org/10.1002/adfm.200500087

[81] Z. Zang, M. Wen, W. Chen, Y. Zeng, Z. Zu, X. Zeng, X. Tang, Strong yellow emission of ZnO hollow nanospheres fabricated using polystyrene spheres as templates, Materials & Design 84 (2015) 418-421, https://doi.org /10.1016/j.matdes.2015.06.141

[82] W.Q. Peng, S.C. Qu, G.W. Cong, Z.G. Wang, Structure and visible luminescence of ZnO nanoparticles, Mater. Sci. Semicond. Process. 9:1-3 (2006) 156-159, https://doi.org /10.1016/j.mssp.2006.01.038

[83] D. Zwingel, Trapping and recombination processes in the thermoluminescence of Li-doped ZnO single crystals, J. Lumin. 5:6 (1972) 385-405, https://doi.org/10.1016/0022-2313(72)90001-4

[84] F.K. Shan, G.X. Liu, W.J. Lee, G.H. Lee, I.S. Kim, B.C. Shin, Aging effect and origin of deep-level emission in ZnO thin film deposited by pulsed laser deposition, Appl. Phys. Lett. 86:22 (2005) 221910, https://doi.org/10.1063/1.1939078

[85] Q.X. Zhao, P. Klason, M. Willander, H.M. Zhong, W. Lu, J.H. Yang, Deep-level emissions influenced by O and Zn implantations in ZnO, Appl. Phys. Lett. 87:21 (2005) 211912, https://doi.org/10.1063/1.2135880

[86] M. Liu, A.H. Kitai, P. Mascher, Point defects and luminescence centres in zinc oxide and zinc oxide doped with manganese, J. Lumin. 54:1 (1992) 35-42, https://doi.org/10.1016/0022-2313(92)90047-d.

[87] P.S. Xu, Y.M. Sun, C.S. Shi, F.Q. Xu, H.B. Pan, The electronic structure and spectral properties of ZnO and its defects, Nucl. Instruments Methods Phys. Res. Sect. B Beam Interact. with Mater. Atoms 199 (2003) 286–290, https://doi.org/10.1016/S0168-583X(02)01425-8

[88] N.Y. Garces, L. Wang, L. Bai, N.C. Giles, L.E. Halliburton, G. Cantwell, Role of copper in the green luminescence from ZnO crystals, Appl. Phys. Lett. 81:4 (2002) 622-624, https://doi.org/10.1063/1.1494125

[89] T.H. Gfroerer, Photoluminescence in analysis of surface and interfaces, Encyclopedia of Analytical Chemistry, R. A. Meyers, John Wiley & Sons Ltd (2000) 9209-9231

[90] L. Jing, F. Yuan, H. Hou, B. Xin, W. Cai, H. Fu, Relationships of surface oxygen vacancies with photoluminescence and photocatalytic performance of ZnO nanoparticles, Science in China Ser. B Chemistry 48 (2005) 25-30, https://doi.org/10.1007/BF02990909

[91] G.J. Lee, Y.P. Lee, H.-H. Lim, M. Cha, S.S. Kim, H. Cheong, S.-K. Min, S.-H. Han, Photoluminescence and lasing properties of ZnO nanorods, Journal of the Korean Physical Society 57:6 (2010) 1624-1629, https://doi.org/10.3938/jkps.57.1624

[92] S.C. Choi, D.K. Lee, S.H. Sohn, Morphological and optical properties of cobalt ion-modified ZnO nanowires, Catalysts 10 (2020) 614, https://doi.org/10.3390/catal10060614

[93] K.P. Raj, K. Sadaiyandi, A. Kennedy, et al., Influence of Mg doping on ZnO nanoparticles for enhanced photocatalytic evaluation and antibacterial analysis, Nanoscale Research Letters 13 (2018) 229-241, https://doi.org/10.1186/s11671-018-2643-x

[94] U. Alam, A. Khan, D. Ali, D. Bahnemann, M. Muneer, Comparative photocatalytic activity of sol–gel derived rare earth metal (La, Nd, Sm and Dy)-doped ZnO photocatalysts for degradation of dyes, RSC Adv. 8 (2018) 17582, https://doi.org/10.1039/c8ra01638k

[95] P.R. Talakonda, Excitation-intensity (EI) effect on photoluminescence of ZnO materials with various morphologies. In: Luminescence - An outlook on the phenomena and their applications, IntechOpen (2016) 91-105, http://dx.doi.org/10.5772/64937

[96] B. Das, P. Kumar, C.N.R. Rao, Factors affecting laser-excited photoluminescence from ZnO nanostructures, J. Clust. Sci. 23 (2012) 649–659, http://dx.doi.org/10.1007/s10876-012-0453-3

[97] L.H. Quang, S.J. Chua, K.P. Loh, E. Fitzgerald, The effect of post-annealing treatment on photoluminescence of ZnO nanorods prepared by hydrothermal synthesis, Journal of Crystal Growth 287:1 (2006), 157–161, http://dx.doi.org/10.1016/j.jcrysgro.2005.10.060

Materials Research Forum LLC
https://doi.org/10.21741/9781644902394-3

Chapter 3

ZnO: A Potential Candidate as a Host Material for Diluted Magnetic Semiconductors

Pooja Dhiman[1*], Amit Kumar[1], Gaurav Sharma[1], Garima Rana[1], Pawan Kumar[2]

[1] International Research Centre of Nanotechnology for Himalayan Sustainability (IRCNHS), Shoolini University, India

[2] School of Physics and Materials Sciences, Shoolini University, India

* dhimanpooja85@gmail.com

Abstract

Researchers have investigated GaN and ZnO-based diluted magnetic semiconductors since the discovery of the first DMS in 2000. The investigation is progressing with both theoretical and empirical discoveries. For a material to function as DMS in actual applications, researchers are intensively engaged in attaining two key conditions. One is ferromagnetism at ambient temperature, whereas the other has a higher Curie temperature. This chapter discusses the advancements in the realm of diluted magnetic semiconductors based on ZnO. The issues, difficulties, and competing causes of magnetism continue to be the focus of this chapter. Diverse doping techniques, such as transition metals, rare earth metals, and co-doped systems, have been investigated. In this chapter, we focused mostly on recent breakthroughs in experimental and theoretical ZnO nanostructures DMS investigations.

Keywords

ZnO, Spintronics, Diluted Magnetic Semiconductors

Contents

1.   Introduction

Today's Nano science relies on the mass, charge, and spin of electrons in a solid state. Due to human capacity to create, design, and adjust the characteristics of nanoparticles, this field has gained a great deal of attention in recent years. Due of this potential, nanotechnology research and its applications have accelerated. In general, nanoparticles have diameters smaller than 100 nm and contain 20 to 15000 atoms. The characteristics of materials vary proportionally as the particle size of bulk materials is reduced to the nanoscale. Nano-materials include semiconductor metal oxides, metal nanoparticles, polymeric materials, ceramics etc. One of the most important reasons why nanoscale has different characteristics (optical, electrical, magnetic, chemical, and mechanical) than bulk material is because in this range of sizes, quantum effects begin to prevail due to an increase in the surface-to-volume ratio. IC's , part of semiconductor electronics industries relies on the charge of the carriers for processing of information while mass storage purely depends upon the magnetic properties utilizing the spin of the electronics. The semiconductors utilised in electronics and integrated circuits lack magnetic ions, are nonmagnetic, and their magnetic Lande – g factors are often relatively low. A typical semiconductor may be represented using the one-electron approximation, however a magnetic semiconductor containing localised magnetic moments and necessitating many-particle interaction cannot. It is difficult to establish a relationship between magnetic and semiconducting phenomena in the absence of electron-electron interactions, several particle descriptions, and correlation effects. Magnetic semiconductors comprise semiconductor materials that display both ferromagnetism and the desirable features of a semiconductor. Based on control of quantum spin and charge, these materials give a novel sort of conduction control. This provides potentially close to 100% spin polarisation, in contrast to iron and other metals that produce just 50% polarisation, and is a crucial characteristic for spintronic applications. The schematic for spintronics is shown in Figure 1.

1.1   Spintronics

The term "spintronics" (spin-based electronics) refers to technologies that utilise both the "spin" quantum property of electrons and their "charge" fundamental characteristic [1]. The initial generation of spintronics systems were centred on passive magneto-resistive sensors and memory components with electrodes comprised of alloys of 3D ferromagnetic metals. With the discovery of enormous magneto-resistance in $(Fe/ Cr)_n$ multilayers and tunnelling magneto-resistance, these devices underwent rapid development. Active spin-based devices that require the synthesis and manipulation of spin-polarized electrons in a host semiconductor matrix are anticipated to comprise the future generation of electronic components. The most important point in the roadmap of spintronics was the discovery of the diluted magnetic semiconductors popularly known as DMS's [2] [3].

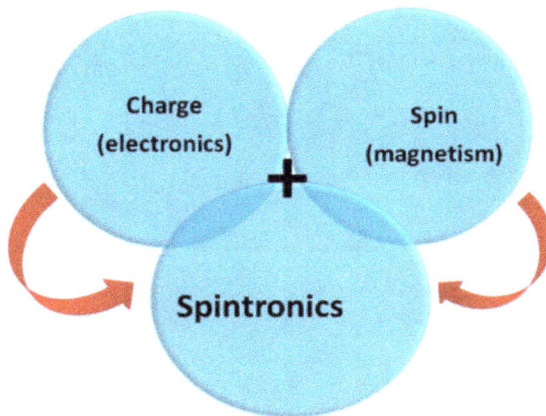

Figure 1: Schematic diagram showing spintronics relies on charge and spin.

1.2 Introduction to diluted magnetic semiconductors

Diluted magnetic semiconductors can be well understood by differentiating them from conventional and magnetic semiconductors. Figure 2 represents the schematics of three types of semiconductors. Conventional semiconductors purely rely on the type and amount of impurity elements for applications. The addition of the impurity element is capable of altering only the transport properties of the semiconductors. While the scenario is quite different in magnetic semiconductors, where an abundant array of magnetic ions can results into both semiconducting and magnetic properties. However, magnetic semiconductors suffer from variety of growth related issue due to non-compatibility with some of the semiconductors. The second drawback is associated low Curie temperature which limits the utilization of these materials for spintronic applications [4].

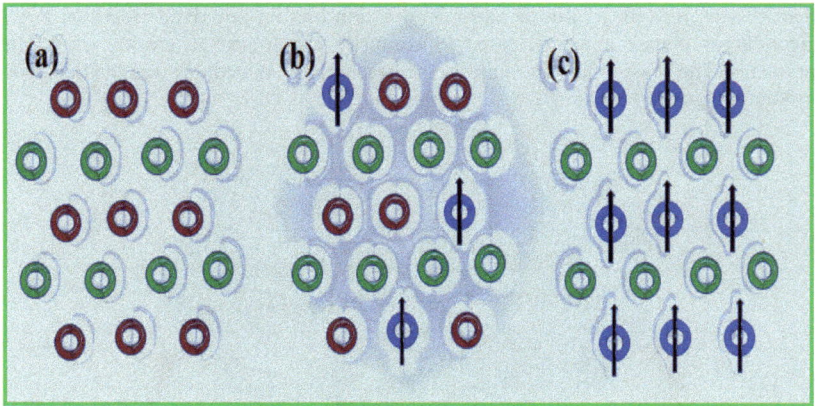

Figure 2: Schematics for Non Magnetic Semiconductor, (b) Diluted Magnetic Semiconductor (c) Magnetic semiconductor.

Diluted magnetic semiconductors have emerged as potential candidate for spintronics applications because they fulfil both criteria which are mandate for practical applications. One is room temperature ferromagnetism and other is high curie temperature (above room temperature) [5]. A number of reported DMS's are capable of exhibiting room temperature ferromagnetism and high curie temperature [6,7,8,9]. $Zn_{1-x}Mn_xTe$ and $Cd_{1-x}Mn_xTe$, which are II-VI semiconductor alloys, were the first identified DMSs [2]. Actual progress in DMS emerges with the reported results by T. Dietl in the year '2000'. In the work, Curie temperature of various p types semiconductors were reported for 5at. % of Mn (3.5×10^{20} holes per cm^3). In the work, Curie temperature for various wide band gap semiconductors were predicted. The results shows the most promising results are for ZnO and GaN which are capable of exhibiting higher Curie temperature. These theoretical predictions are dependent on the anisotropy of the carrier-mediated exchange interaction linked with the spin-orbit coupling in the host material, as predicted by the mean field model. As per this model, Curie temperature was directly linked to the doping of Mn ions concentration and carrier density (holes). It was predicted that room temperature ferromagnetism is hard to achieve in n types of materials (attributed to less s-d exchange interaction) and can be achieved only by raising sufficiently higher holes concentration in the both ZnO and GaN. Sato and Katayama-Yoshida computed the total energy difference between ferromagnetic, antiferromagnetic, and spin glass states using ab initio calculations using the KKR-CPA approach [10,11]. These results revealed that Cr and V doping is promising for ZnO based DMS while Fe, Ni, Co doping may results into spin glass state for material.

A number of models are there to support the room temperature ferromagnetism in diluted magnetic semiconductors. Some of the most prominent models are direct exchange, super-exchange, carrier mediated mechanism, and bound magnetic polaron model. Direct exchange interaction is identified by the direct connection of magnetic ions through magnetic orbital overlap, which determines the residual magnetization of the material. Super exchange interaction generally describes the exchange interaction in insulators type of materials. This interaction type of interaction requires an intermediate ligand for coupling between nearest neighbours. This model describes the interactions in transition metal complexes since their d orbitals are localized and hopping is merely a possibility. The carrier-mediated exchange includes interactions between the localized moments of 'd' or 'f' orbitals that are mediated by the system's unbound carrier concentration [12] [13]. Large concentrations of free carriers are essential for this process to explain ferromagnetism at ambient temperature. This belongs to the special case of carrier mediated magnetism model which are RKKY, Zener model of exchange interaction and double exchange interactions. RKKY is the prominent interaction for higher carrier density systems while Zener model is applicable to the systems with carrier density less than metals. And the double exchange describes the systems with double valence states of the dopant [14]. Most important model of magnetism which is popular in explaining magnetism inside DMS is bound magnetic polarons which explains the system with low carrier densities. In this concept, the generation of a bound magnetic polaron is explained by exchange interactions between the localized spins of transition metal ions, mediated by a far smaller number of weakly localized carriers [15]. The overlapping of neighbouring polarons leads to long-range ferromagnetic order, and ferromagnetic transition arises when the polaron size approaches the size of the sample.

2.   Suitability of ZnO as host for DMS

With the prediction of above results, ZnO is being continuously explored for desired results. ZnO belongs to the II-VI type of DMS classification. As a semiconductor with a broad band gap (3.4 eV) and a high exciton binding energy, it may be employed in blue and ultraviolet optical systems. ZnO do exists in three structures, wurtzite, Zinc blende and rocksalt. The iconicity of ZnO sits at the boundary between covalent and ionic semiconductors, giving wurtzite structure the greatest stability under ambient surroundings. The piezoelectric nature of ZnO also comes from the ionic bonding of the structure. ZnO is a well-known diamagnetic material in bulk form in nature, however, due to defects, interactions, it is also reported to have paramagnetic or even weakly ferromagnetic in literature [16]. Even when doped with non-magnetic ions, ferromagnetism can be achieved [17]. However, there are enormous reports on transition metals, rare earth metals and mixed doped ZnO based diluted magnetic semiconductors. There are large numbers of reports observing room temperature ferromagnetism for doped ZnO nanostructures, however there is small section focussing on systematic approach interested in finding out the exact cause of magnetism. Major portion of the experimental results are based on defects induced magnetism following the bound magnetic polaron approach in DMS. The cause of magnetism in DMS should be tackled with systematic characterizations supporting the room temperature magnetism and also theoretical studies confirming the same. In the following sections, we have discussed some of highlighted works reported on TM doped ZnO as DMS.

3.   TM (transition metal) doped ZnO

Earlier theoretical studies have predicted that few percent (~5%) transition metals doping and co-doping into ZnO matrix may results into room temperature ferromagnetism. However the role of oxygen vacancies are crucial for this target [18]. In 2013, Zhang et. al. have performed DFT calculations and observed that doping at Zn sites in ZnO monolayers with Cr, Mn, Fe, Co, Ni, and Cu could result into room temperature magnetization while doping with Sc, Ti, and V atoms were not preferred [19]. In other reported work, Fe, Ni, and Mn (5-21 at.%) doped ZnO nanoparticles were synthesized using co-precipitation method and it was found that most prominent magnetic properties were observed for Fe doped samples [20]. The role of hydrogen induced magnetism was observed; later on ESR+ magnetization results confirmed the role of dual valence state of Fe responsible for magnetism. In the same year Dhiman et. al. have synthesized Ni −Fe co-doped ZnO nanoparticles through solution combustion method and maximum room temperature magnetization was found for $Fe_{0.05}Ni_{0.01}Zn_{0.98}O$ nanoparticles [21]. In 2014, other group of scientists have confirmed the high saturation magnetization and large coercivity for Ni, Mn, Co doped ZnO thin films fabricated by sol-gel spin coating method [22]. Cr and Fe co-doping has also been reported for high saturation magnetization for ZnO nanoparticles synthesized through sol-gel auto combustion method [23]. In a recent report, Mn doped, Mn-In co-doped ZnO nanoflowers were fabricated in the presence of nitrogen gas environment [24]. Both Mn doped and Mn-In co-doped samples were found to exhibit room temperature ferromagnetism. The cause of magnetism was estimated and supported with systematically characterizations in the work. In the case of RTFM observed for Mn doped ZnO nanoflowers, Zn interstitials were found to be responsible for generation of BMP's in the presence of spin-up Mn ions. However, the induction of ions in Mn doped ZnO samples lead to an abrupt 18% increase in magnetization and almost 10 times increase in coercive field value. Figure 3(a-b) represents the susceptibility vs. temperature curves for Mn doped ZnO and Mn-In doped ZnO nano flowers. The curves were fitted for segregation of ferromagnetic component from data by eliminating the paramagnetic component. The residual ferromagnetic contribution can be visualized in Figure 3c. One can clearly observe the subtle increased magnetization in Mn-In doped ZnO nano flowers. The reason was explained in view of delocalized conduction band electrons which are capable of introducing ferromagnetic interaction between Mn ions. The new type of coupling was mentioned as "delocalized charge carrier-mediated ferromagnetic coupling".

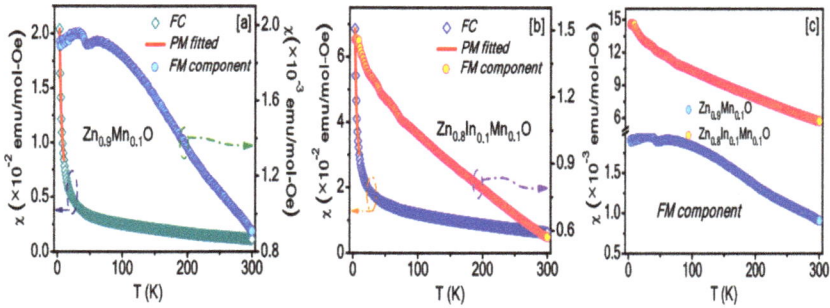

Figure 3: Susceptibility versus temperature curve under the application of 100 Oe field strength [24].

Besides the great work carried by scientists, the magnetism in transition metal doping in ZnO matrix is still controversial. The same compositions and same fabrication route may results into different magnetic properties. Moreover, number of theories is the backbone of the reported magnetism, like hydrogen mediated, oxygen vacancies mediated, defects induced, and higher carrier density approach. Table 1 summarizes the latest reported results of transition metals doped ZnO nanostructures. Table clearly revealed that varied synthesis routes, slightly varied composition, and different experimental conditions can be opted for acquiring room temperature ferromagnetism in ZnO based diluted magnetic semiconductors. Researchers are also exploring rare earth metal doping on Zn sites to achieve and stabilize the room temperature magnetism in diluted magnetic conductors. In the further section we have discussed some of rare earth metal doped ZnO nano- structures already explored for magnetic properties.

Table 1: Some of reported transition metal doped ZnO materials as diluted magnetic semiconductors.

Composition	Method of synthesis	Nature	Magnetization	Reference
$Zn_{0.95}Co_{0.05}O{:}H$ (hydrogenated)	solid state method	ferro	M_s-2.9 emu/g	[25]
$Zn_{0.96}Mn_{0.04}O$ $Zn_{0.95}Mn_{0.04}Co_{0.01}O$ $Zn_{0.94}Mn_{0.04}Co_{0.02}O$ $Zn_{0.92}Mn_{0.04}Co_{0.04}O$	co-precipitation method	ferro	M_R -0.2×10^{-2} emu/g 0.25×10^{-2} emu/g 0.1×10^{-2} emu/g 0.09×10^{-2} emu/g	[26]
$Zn_{0.95}Co_{0.05}O$	Solid state method	para	-	[25]
$Zn_{0.99}Co_{0.01}O$	hydrothermal	ferro	-	[27]

$Zn_{0.98}Ni_{0.02}O$ $Zn_{0.97}Ni_{0.02}Mn_{0.01}O$ $Zn_{0.96}Ni_{0.02}Mn_{0.02}O$	Co-precipitation	ferro	M_s-270.12x10^{-6} emu/g 493.50x10^{-6} 530.24 x10^{-6}	[28]
$Zn_{0.99}Co_{0.01}O$	hydrothermal method.	ferro	M_s-0.002 emu/g	[29]
$Zn_{1-x}Co_xO$ (x = 0.03, 0.05, and 0.10)	Derived from MOF	ferro+ dia	-	[30]
$Zn_{0.8}Co_{0.10}Ga_{0.10}O$	solvo-thermal	ferro	M_s-4.88 emu/g	[31]
$Zn_{0.97}Fe_{0.03}O$	Sol-gel	ferro	M_s-3.45 emu/g	[32]
$Zn_{0.95}Fe_{0.03}Cu_{0.02}O$	sol-gel	ferro	M_s-0.46 emu/g	[32]
Fe doped ZnO	Sono-chemical	ferro	M_R- 1.9489X10^{-3} emu/g	[33]
$Zn_{0.95}Mn_{0.05}O$ nanorods	hydrothermal	ferro	M_s-34.5 memu/g	[34]
1D $Zn_{0.95}Mn_{0.05}O$ nanorods(vaccum annealed at 100 ˚C	hydrothermal	ferro	M_s-35.5 memu/g	
$Zn_{0.95}Mn_{0.05}O$ nanorods(vaccum annealed at 300 ˚C	hydrothermal	ferro	M_s-37.5 memu/g	
$Zn_{0.95}Mn_{0.05}O$ nanorods(vaccum annealed at 700 ˚C	hydrothermal	ferro	M_s-71.6 memu/g,	
$Fe_{0.01}Mn_{0.01}Zn_{0.98}O$ $Mn_{0.01}Ni_{0.01}Zn_{0.98}O$ $Ni_{0.01}Co_{0.01}Zn_{0.98}O$ $Ni_{0.01}Fe_{0.01}Zn_{0.98}O$	sol-gel	ferro	M_s-3.15 emu/g M_s-2.25 emu/g M_s-1.32 emu/g M_s-2.64 emu/g	[35]
$Zn_{0.97}Fe_{0.03}O$ $Zn_{0.95}Fe_{0.05}O$	Sol-gel	ferro	M_s- 0.0075 emu/g M_s- 0.0065 emu/g	[36]
$Fe_{0.01}Ni_{0.01}Zn_{0.98}O$ $Fe_{0.03}Ni_{0.01}Zn_{0.98}O$ $Fe_{0.05}Ni_{0.01}Zn_{0.98}O$	Solution combustion method	ferro	M_s-0.02 emu/g M_s-0.08 emu/g M_s-0.09 emu/g	[21]

Magnetization field values may or may not be same for reported values.

4.   RE (rare earth) doped ZnO

Rare earth doping strategy has resulted into quite fruitful insights in ZnO based DMS's. Due to the presence of an unpaired electron in their outermost shell, i.e., 4f, which has a large magnetic moment and is responsible for ferromagnetism at ambient temperature, rare earth metals were deployed as doping magnetic ions in the host semiconducting oxide. A number of researchers have observed room temperature ferromagnetism in rare earth doped ZnO nanostructures. We have listed some of them in Table 2. A clear variation in the magnetic moment values can be observed for same or different rare earth dopant. However, in general, magnetic moments observed in rare earth elements doped ZnO nanostructures are slight less than the TM doping. However, cause of magnetism is still controversial, whether intrinsic or extrinsic. In a recent report, theoretical calculations have been performed to calculate the partial density of states utilizing spin polarized calculations to impart more knowledge on cause of magnetism in Er doped ZnO system [37]. Figure 4 shows the partial density of the states observed for all compositions. The symmetrical spin up-down calculations for pure ZnO indicate the non-magnetic nature for ZnO, while for all Er doped samples (Er-ZnO), asymmetry was observed for spin up and down states revealing the magnetic nature of all doped samples. For Er doped ZnO, spin polarization can be visually seen by observing the partial density of states near Fermi level. It is very clear from the results that generated magnetic moment is due to spin polarization making the composition ferromagnetic. The 4f electrons of the Er have resulted into magnetic moment value. Moreover, in the presence of oxygen vacancies and Zn vacancies, the net magnetic moment increases to high value indicating the defect mediated magnetism also. In summary, the proper knowledge of defects chemistry, density of states information, valence states of the dopant element, or any other phase formation may give a deep understanding of the cause of magnetism inside ZnO based DMS.

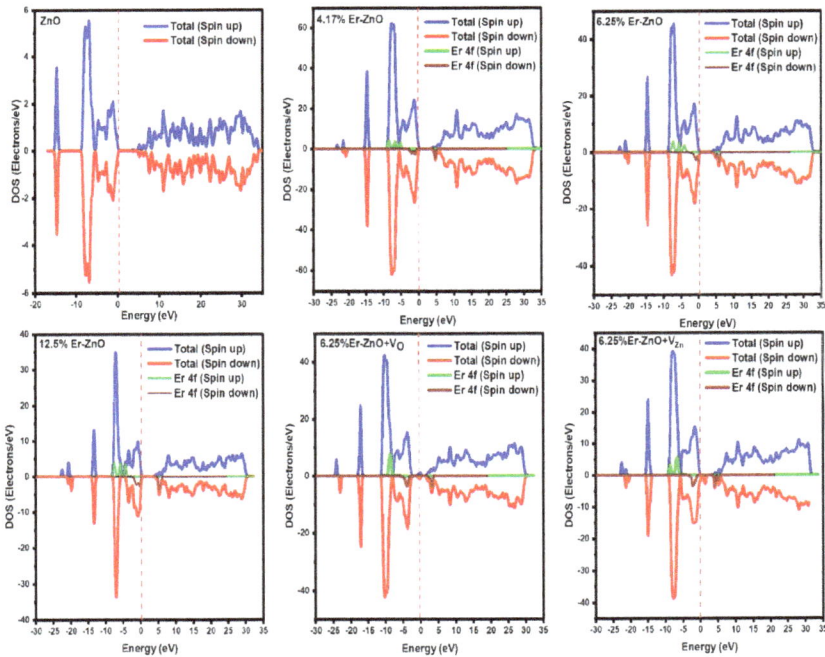

Figure 4: The total density of states and the partial density of Er-4f of pure and Er-doped ZnO with varying Er concentrations and the presence of native defects (V_O: oxygen vacancy and V_{Zn}: Zn vacancy). Spin polarization [37].

Table 2: Comparison of reported room temperature magnetization observed for rare-earth doped ZnO nanostructures.

Composition	Method of synthesis	Magnetization	Reference
ZnO	co-precipitation route	M_s-0.434x10-3 emu/g	[38]
$Zn_{0.095}Sm_{0.005}O$		0.171 x10-3 emu/g	
$Zn_{0.090}Sm._{010}O$		0.136 x10-3 emu/g	
$Zn_{0.085}Sm_{0.015}O$		0.064 x10-3 emu/g	
$Zn_{0.080}Sm_{0.020}O$		0.066 x10-3 emu/g	
$Zn_{0.075}Sm_{0.025}O$		0.041 x10-3 emu/g	
ZnO (thin films)	sol gel dip coating technique	M_s-7.25±0.01x10^3A/m	[39]
$Zn_{0.099}Sm_{0.001}O$		9.77 ± 0.07 x10^3A/m	
$Zn_{0.098}Sm_{0.002}O$		8.48 ± 0.06 x10^3A/m	
$Zn_{0.097}Sm_{0.003}O$		7.75 ± 0.05 x10^3A/m	
$Zn_{0.096}Sm_{0.004}O$		7.56 ± 0.03 x10^3A/m	
$Zn_{0.095}Sm_{0.005}O$		7.36 ± 0.02 x10^3A/m	
$Zn_{0.99}Sm_{0.01}O$	chemical co-precipitation method	M_s- 0.2970 emu/g	[40]
$Zn_{0.98}Sm_{0.02}O$		1.2602 emu/g	
$Zn_{0.96}Sm_{0.04}O$		0.3243 emu/g	
$Zn_{0.94}Sm_{0.06}O$		1.3665 emu/g	
$Zn_{0.92}Sm_{0.08}O$		0.0540 emu/g	
$Zn_{0.90}Sm_{0.10}O$		1.8370 emu/g	
$Zn_{0.97}Nd_{0.03}O$	Co- precipitation method	M_s- 0.041emu/g	
$Zn_{0.94}Nd_{0.06}O$		0.051 emu/g	[41]
$Zn_{0.91}Nd_{0.09}O$		0.069 emu/g	
$Zn_{0.97}Nd_{0.03}O$	Sol-gel method	M_s- 0.022 emu/g	[42]
ZnO	Sol-gel method	M_s- 0.001emu/g	
$Zn_{0.98}Gd_{0.02}O$		0.006 emu/g	[43]
$Zn_{0.95}Gd_{0.05}O$		0.007 emu/g	
$Zn_{0.90}Gd_{0.10}O$		0.075 emu/g	
$Zn_{0.85}Gd_{0.15}O$		0.176 emu/g	
$Zn_{0.989}Gd_{0.011}O$	Thermal decomposition	M_s- 0.0001 emu/g	[44]
$Zn_{0.965}Gd_{0.035}O$		0.05 emu/g	
$Zn_{0.949}Gd_{0.051}O$		0.0032 emu/g	
$Zn_{0.99}Ce_{0.01}O$	Co-precipitation method	M_s- 3.93×10^{-4} emu/g	
$Zn_{0.97}Ce_{0.03}O$		3.81×10^{-3} emu/g	[45]
$Zn_{0.95}Ce_{0.05}O$		1.08×10^{-4} emu/g	
ZnO	Solid state reaction route	M_s- 0.764×10^{-3} emu/g	
$Zn_{0.99}Sm_{0.010}O$		2.090×10^{-3} emu/g	[46]
$Zn_{0.97}Sm_{0.030}O$		3.205×10^{-3} emu/g	
ZnO	solid-state reaction method	M_s- 2.92x10^{-3} emu/g	
$Zn_{0.99}Er_{0.01}O$		4.44 x10^{-3} emu/g	[47]
$Zn_{0.97}Er_{0.03}O$		8.11 x10^{-3} emu/g	
$Zn_{0.95}Er_{0.05}O$		3.64 x10^{-3} emu/g	
ZnO	wet chemical method	M_s- 0.013 emu/g	
$Zn_{0.99}Nd_{0.01}O$		0.018 emu/g	[48]
$Zn_{0.98}Nd_{0.02}O$		0.025 emu/g	
$Zn_{0.97}Nd_{0.03}O$		0.028 emu/g	
$Zn_{0.096}Nd_{0.04}O$		0.033 emu/g	

$Zn_{0.95}Nd_{0.05}O$		0.025 emu/g	
$Zn_{0.97}Gd_{0.03}O$	hydrothermal method	M_s- 0.27emu/g	[49]
$Zn_{0.95}Gd_{0.05}O$		0.91 emu/g	
$Zn_{0.93}Gd_{0.07}O$		1.03 emu/g	

* *Magnetization field values may or may not be same for reported values.*

5.   TM+RE doped ZnO

In a recent report, Ni and Er doped ZnO nanoparticles were synthesized by chemical precipitation technique [50]. XRD results revealed the pure phase formation of samples without the presence of any secondary phase. Figure 5a shows the hysteresis loops for samples synthesized recorded at room temperature. The ZnO exhibit paramagnetic like curve, while for 2 at. % Ni doped ZnO sample was found to possess retentive filed of 0.0032 emu/g. However, $Zn_{0.97}Ni_{0.02}Er_{0.01}O$ nanoparticles showed well defined saturated hysteresis loop with M_r-0.048 emu/g and M_s-0.0586 emu/g respectively indicating the stable ferromagnetic nature for the samples. Figure 5b demonstrates the PL spectra for all samples which confirm the presence of oxygen vacancies, deep donor sites, and interstitial occupation of dopant and co-dopant ions. Moreover, incorporating rare earth ions with transition metal ions in nano materials can also increase luminescence. This is supported by the stretching of electron–phonon combinations or polarons and exchange interaction disturbances caused by the difference in ionic radius between Ni and Er ions. The robust ferromagnetism in the mentioned samples were explained in view of more number of oxygen vacancies on Er ions doping resulting into creation of free electrons and holes for the charge neutrality which thereby favours the formation of bound magnetic polarons. Table 3 summarizes the recently reported different TM+RE doped ZnO nanostructures as DMS for spintronic applications.

In a recent experiment, bare Ni, Ce, Cu and Ce-Cu co-doping in ZnO matrix is opted for magnetic properties exploration [52]. Figure 6a demonstrates the observed the hysteresis loops for synthesized samples at room temperature (300K) and low temperature (10K). The room temperature saturation magnetization observed follow the trend as, Ce-Cu co-doped ZnO> Ce, Ni codoped > Ni doped> Cu doped ZnO nanostructures. The magnetism observed for Ni or Cu doped samples were explained in view of oxygen vacancies which promotes the formation of magnetic polarons, while in case of Ce-Cu and Ce Ni co-doped ZnO, magnetism is explained on the basis of double exchange model because of presence of Ce ions in different valence states as confirmed by XPS analysis. Moreover, at low temperature (10K), abrupt coercive field values were observed which are explained on the basis of presence of ferromagnetic clusters in the samples. The ferromagnetic clusters may have increased on Ni, Cu Ce doping. The magnetization process takes place through interconnected chains of elements and collinear arrangement of magnetic moments. To know the exact cause of magnetism, FC-ZFC was performed for all samples. Figure 6b shows the FC-ZFC curves for all samples. From results, the separation between the FC-ZFC curves was found to be more for co-doped samples then others. The concave type curves for the Ce doped samples confirm the weightage of bound magnetic polarons in magnetization process. However, dip near low temperature is a signature of various factors like spin glass transition, super paramagnetic relaxation or Neels temperature. Inset curves (Figure 6b: inset) for susceptibility vs. temperature curves have negative slopes for all samples indicating the predominance of antiferromagnetic interactions in the samples. The room temperature and low temperature magnetism are explained on the basis of magnetic polarons formation and the predominant model of magnetism was bound magnetic polaron model.

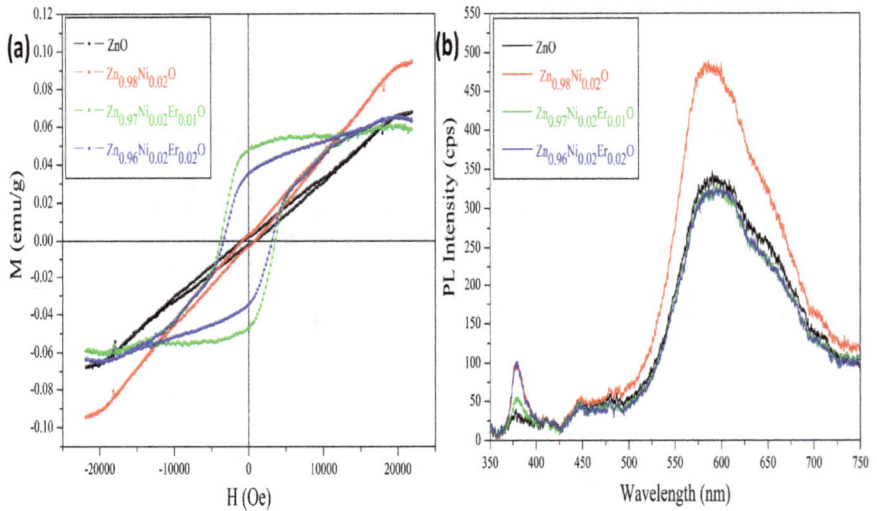

Figure:5(a) Room temperature M-H loops for bare ZnO, $Zn_{0.98}Ni_{0.02}O$, $Zn_{0.97}Ni_{0.02}Er_{0.01}O$, and $Zn_{0.96}Ni_{0.02}Er_{0.02}O$ nanoparticles [51].

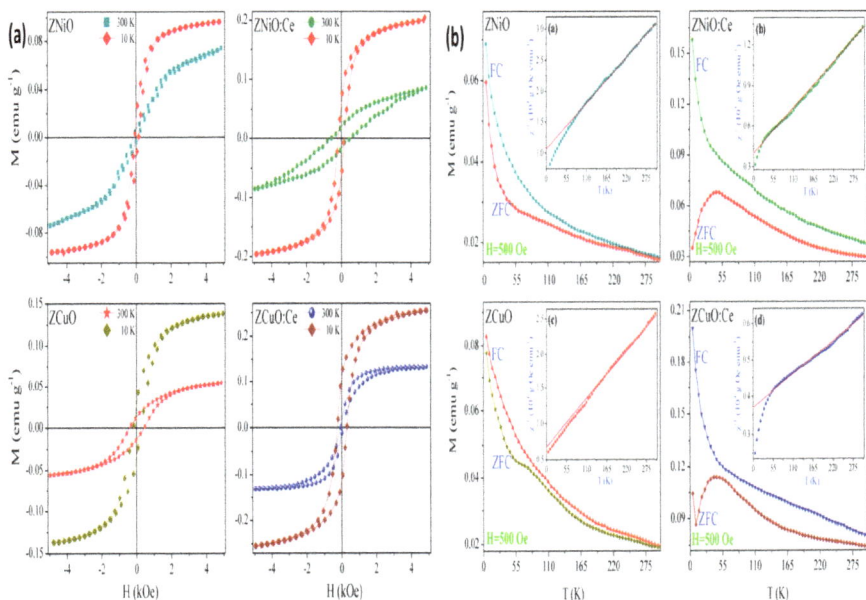

Figure 6: (a) Room temperature and low temperature (10K) M-H loops for Ni, Ce,Cu, and Cu, Ce co-dopedZnO , (b) FC-ZFC curves Ni, Ce,Cu, and Cu, Ce co-doped ZnO (inset:dc magnetic susceptibility inverse vs. temperature curve [52].

Table 3: Rare earth+transition metal doped approach for ZnO based DMS.

Sr. no.	Composition	Method of synthesis	Nature	Magnetization	Reference
1.	Pr, Fe co-doped ZnO	co-precipitation method	ferro	-	[53]
2.	Gd, Mn co-doped ZnO	hydrothermal method	ferro	M_s- 0.00011 emu/g	[54]
3.	La–Eu co-doped ZnO (1 wt. %)	co-precipitation technique	ferro	M_s-0.05198 emu/g	[55]
4.	$Zn_{0.98}Ni_{0.02}O$	chemical precipitation method	ferro	M_r-0.0032 emu/g	[50]
5.	$Zn_{0.97}Ni_{0.02}Er_{0.01}O$			M_s-0.0586 emu/g	

* *Magnetization field values may or may not be same for reported values.*

6.   Challenges and Future prospective

It has been seen and claimed that ZnO DMS materials exhibit ferromagnetic behaviour above room temperature. Current ZnO DMS research focuses mostly on elucidating the source and mechanism of ferromagnetism in ZnO DMS materials. In order to find the exact cause of magnetism in ZnO DMS, it is advisable to keep in mind that ZnO is semiconductor oxide which contains number of native defects at ambient conditions. From the literature, it is also evident that hydrogen can also play important role in inducing magnetism in ZnO system. The role of native defects in magnetism needs to be assured with variety of characterization. Large carrier strategies should be opted to induce the carrier mediated magnetism in DMS. For the doped ZnO nanostructures, dopant choice should be based on the fact whether the dopant can add hole or electrons in the system or not. The cause of magnetism is still highly controversial for ZnO based DMS. It is a fact that researchers have explored this field in many ways; there are less systematic experimental reports on ZnO as DMS. The exact cause of magnetism whether it is intrinsic or extrinsic needs to be more searched. The experimental evidences should contain the complete picture starting from synthesis route, structural characterization, defects information, valence state confirmation and magnetic properties at room temperature, low temperature and even at high temperature. The reproducibility of the results can boost the current research field. In summary, the above-room-temperature ferromagnetism observed in ZnO DMS materials has a long way to go before it can be utilised in practical devices, despite massive efforts.

References

[1] F. Pulizzi, Spintronics, Nature Materials, 11 (2012) 367-367. https://doi.org/10.1038/nmat3327

[2] J.K. Furdyna, Diluted magnetic semiconductors, Journal of Applied Physics, 64 (1988) R29-R64. https://doi.org/10.1063/1.341700

[3] A. Kumar, P. Dhiman, M. Singh, Effect of Fe-doping on the structural, optical and magnetic properties of ZnO thin films prepared by RF magnetron sputtering, Ceramics International, 42 (2016) 7918-7923. https://doi.org/10.1016/j.ceramint.2016.01.136

[4] T. Dietl, Magnetic semiconductors, Diluted Magnetic Semiconductors, (1991) 141. https://doi.org/10.1142/9789814368216_0005

[5] M.K. Jain, Diluted magnetic semiconductors, World Scientific, 1991. https://doi.org/10.1142/1065

[6] Y. Li, S. Ding, Y. Luo, P. Yu, Y. Cui, X. Wang, Z. Cheng, Z. Wu, Room temperature intrinsic diluted magnetic semiconductor Li(Cd,Mn)As, Journal of Materials Chemistry C, 10 (2022) 3217-3223. https://doi.org/10.1039/D1TC05482A

[7] A. Sarikhani, L. Avazpour, W. Liyanage, R. Florez, E. Bohannan, D. Satterfield, M. Nath, J.E. Medvedeva, Y.S. Hor, Transparency and room temperature ferromagnetism in diluted magnetic polycrystalline Zn1−xCrxTe non-oxide II-VI semiconductor

Materials Research Foundations 146 (2023) 67-85   https://doi.org/10.21741/9781644902394-3

compounds, Journal of Alloys and Compounds, 924 (2022) 166478.
https://doi.org/10.1016/j.jallcom.2022.166478

[8] V.N. Jafarova, N.T. Mamedov, M.A. Musaev, High Curie Temperature and Half-Metallic Ferromagnetism in ZnSe:Co,Ni with Wurtzite Structure: First-Principles Study, physica status solidi (b), n/a (2022) 2200360.
https://doi.org/10.1002/pssb.202200360

[9] Y. Liu, Y. Yang, J. Yang, Q. Guan, H. Liu, L. Yang, Y. Zhang, Y. Wang, M. Wei, X. Liu, L. Fei, X. Cheng, Intrinsic ferromagnetic properties in Cr-doped ZnO diluted magnetic semiconductors, Journal of Solid State Chemistry, 184 (2011) 1273-1278.
https://doi.org/10.1016/j.jssc.2011.03.049

[10] K.S.K. Sato, H.K.-Y.H. Katayama-Yoshida, Material design for transparent ferromagnets with ZnO-based magnetic semiconductors, Japanese Journal of Applied Physics, 39 (2000) L555. https://doi.org/10.1143/JJAP.39.L555

[11] K. Sato, H. Katayama-Yoshida, First principles materials design for semiconductor spintronics, Semiconductor Science and Technology, 17 (2002) 367.
https://doi.org/10.1088/0268-1242/17/4/309

[12] A. Walsh, J.L.F. Da Silva, S.-H. Wei, Theoretical Description of Carrier Mediated Magnetism in Cobalt Doped ZnO, Physical Review Letters, 100 (2008) 256401.
https://doi.org/10.1103/PhysRevLett.100.256401

[13] G. Bouzerar, R. Bouzerar, Unraveling the nature of carrier-mediated ferromagnetism in diluted magnetic semiconductors, Comptes Rendus Physique, 16 (2015) 731-738.
https://doi.org/10.1016/j.crhy.2015.09.003

[14] K. Sato, P.H. Dederichs, H. Katayama-Yoshida, J. Kudrnovský, Exchange interactions in diluted magnetic semiconductors, Journal of Physics: Condensed Matter, 16 (2004) S5491. https://doi.org/10.1088/0953-8984/16/48/003

[15] A.C. Durst, R.N. Bhatt, P.A. Wolff, Bound magnetic polaron interactions in insulating doped diluted magnetic semiconductors, Physical Review B, 65 (2002) 235205. https://doi.org/10.1103/PhysRevB.65.235205

[16] J.J. Beltrán, C.A. Barrero, A. Punnoose, Understanding the role of iron in the magnetism of Fe doped ZnO nanoparticles, Physical Chemistry Chemical Physics, 17 (2015) 15284-15296. https://doi.org/10.1039/C5CP01408E

[17] H. Guo, Y. Zhao, N. Lu, E. Kan, X.C. Zeng, X. Wu, J. Yang, Tunable Magnetism in a Nonmetal-Substituted ZnO Monolayer: A First-Principles Study, The Journal of Physical Chemistry C, 116 (2012) 11336-11342. https://doi.org/10.1021/jp2125069

[18] P. Gopal, N.A. Spaldin, Magnetic interactions in transition-metal-doped ZnO: An ab initio study, Physical Review B, 74 (2006) 094418.
https://doi.org/10.1103/PhysRevB.74.094418

Materials Research Forum LLC
https://doi.org/10.21741/9781644902394-3

[19] J. Ren, H. Zhang, X. Cheng, Electronic and magnetic properties of all 3d transition-metal-doped ZnO monolayers, International Journal of Quantum Chemistry, 113 (2013) 2243-2250. https://doi.org/10.1002/qua.24442

[20] R. Saleh, N.F. Djaja, S.P. Prakoso, The correlation between magnetic and structural properties of nanocrystalline transition metal-doped ZnO particles prepared by the co-precipitation method, Journal of Alloys and Compounds, 546 (2013) 48-56. https://doi.org/10.1016/j.jallcom.2012.08.056

[21] P. Dhiman, K.M. Batoo, R.K. Kotnala, J. Chand, M. Singh, Room temperature ferromagnetism and structural characterization of Fe,Ni co-doped ZnO nanocrystals, Applied Surface Science, 287 (2013) 287-292. https://doi.org/10.1016/j.apsusc.2013.09.144

[22] G. Vijayaprasath, R. Murugan, G. Ravi, T. Mahalingam, Y. Hayakawa, Characterization of dilute magnetic semiconducting transition metal doped ZnO thin films by sol-gel spin coating method, Applied Surface Science, 313 (2014) 870-876. https://doi.org/10.1016/j.apsusc.2014.06.093

[23] K. Irshad, M.T. Khan, A. Murtaza, Synthesis and characterization of transition-metals-doped ZnO nanoparticles by sol-gel auto-combustion method, Physica B: Condensed Matter, 543 (2018) 1-6. https://doi.org/10.1016/j.physb.2018.05.006

[24] S. Paul, B. Dalal, M. Das, P. Mandal, S.K. De, Enhanced Magnetic Properties of In-Mn-Codoped Plasmonic ZnO Nanoflowers: Evidence of Delocalized Charge Carrier-Mediated Ferromagnetic Coupling, Chemistry of Materials, 31 (2019) 8191-8204. https://doi.org/10.1021/acs.chemmater.9b03059

[25] R.K. Singhal, A. Samariya, Y.T. Xing, S. Kumar, S.N. Dolia, U.P. Deshpande, T. Shripathi, E.B. Saitovitch, Electronic and magnetic properties of Co-doped ZnO diluted magnetic semiconductor, Journal of Alloys and Compounds, 496 (2010) 324-330. https://doi.org/10.1016/j.jallcom.2010.02.005

[26] R. Khan, Zulfiqar, S. Fashu, Z.U. Rehman, A. Khan, M.U. Rahman, Structure and magnetic properties of (Co, Mn) co-doped ZnO diluted magnetic semiconductor nanoparticles, Journal of Materials Science: Materials in Electronics, 29 (2018) 32-37. https://doi.org/10.1007/s10854-017-7884-4

[27] H. Li, Y. Qiao, J. Li, H. Fang, D. Fan, W. Wang, A sensitive and label-free photoelectrochemical aptasensor using Co-doped ZnO diluted magnetic semiconductor nanoparticles, Biosensors and Bioelectronics, 77 (2016) 378-384. https://doi.org/10.1016/j.bios.2015.09.066

[28] R. Gopalakrishnan, R. Kabilan, M. Ashokkumar, Investigations of Mn introduced structural modifications on Ni-doped ZnO diluted magnetic semiconductors and improved magnetic and antibacterial properties, Journal of Molecular Structure, 1251 (2022) 132060. https://doi.org/10.1016/j.molstruc.2021.132060

[29] M. Zhong, W. Wu, H. Wu, S. Guo, A facile way to regulating room-temperature ferromagnetic interaction in Co-doped ZnO diluted magnetic semiconductor by reduced graphene oxide coating, Journal of Alloys and Compounds, 765 (2018) 69-74. https://doi.org/10.1016/j.jallcom.2018.06.228

[30] Y. Lü, Q. Zhou, L. Chen, W. Zhan, Z. Xie, Q. Kuang, L. Zheng, Templated synthesis of diluted magnetic semiconductors using transition metal ion-doped metal-organic frameworks: the case of Co-doped ZnO, CrystEngComm, 18 (2016) 4121-4126. https://doi.org/10.1039/C5CE02488A

[31] A. Šutka, T. Käämbre, U. Joost, K. Kooser, M. Kook, R.F. Duarte, V. Kisand, M. Maiorov, N. Döbelin, K. Smits, Solvothermal synthesis derived Co-Ga codoped ZnO diluted magnetic degenerated semiconductor nanocrystals, Journal of Alloys and Compounds, 763 (2018) 164-172. https://doi.org/10.1016/j.jallcom.2018.05.036

[32] H. Liu, J. Yang, Z. Hua, Y. Liu, L. Yang, Y. Zhang, J. Cao, Cu-doping effect on structure and magnetic properties of Fe-doped ZnO powders, Materials Chemistry and Physics, 125 (2011) 656-659. https://doi.org/10.1016/j.matchemphys.2010.10.002

[33] B. Babu, G. Thirumala Rao, V. Pushpa Manjari, K. Ravindranadh, R. Joyce Stella, R.V.S.S.N. Ravikumar, Sonochemical assisted synthesis and spectroscopic characterization of Fe3+ doped ZnO diluted magnetic semiconductor, Journal of Materials Science: Materials in Electronics, 25 (2014) 4179-4186. https://doi.org/10.1007/s10854-014-2146-1

[34] L. Dong, C. Liu, Z. Shen, B. Zhou, T. Zheng, Q. Li, Y. Zhong, Influence of Annealing Temperature on the Magnetic Properties of One-dimensional Diluted Magnetic Semiconductor Zn0.95Mn0.05O Tuning with Vacuum Atmospheric Annealing, Journal of Superconductivity and Novel Magnetism, (2022). https://doi.org/10.1007/s10948-022-06451-x

[35] L.K. Sharma, D. Mandal, R.K. Choubey, S. Mukherjee, On the correlation of the effect of defects on the microstructural, optical and magnetic properties of doped ZnO, Physica E: Low-dimensional Systems and Nanostructures, 144 (2022) 115370. https://doi.org/10.1016/j.physe.2022.115370

[36] P. Dhiman, J. Chand, A. Kumar, R.K. Kotnala, K.M. Batoo, M. Singh, Synthesis and characterization of novel Fe@ZnO nanosystem, Journal of Alloys and Compounds, 578 (2013) 235-241. https://doi.org/10.1016/j.jallcom.2013.05.015

[37] M. Achehboune, M. Khenfouch, I. Boukhoubza, I. Derkaoui, B.M. Mothudi, I. Zorkani, A. Jorio, A DFT study on the electronic structure, magnetic and optical properties of Er doped ZnO: Effect of Er concentration and native defects, Computational Condensed Matter, 31 (2022) e00627. https://doi.org/10.1016/j.cocom.2021.e00627

[38] P. Kaur, S. Chalotra, H. Kaur, A. Kandasami, D.P. Singh, Role of Bound Magnetic Polaron Model in Sm Doped ZnO: Evidence from Magnetic and Electronic Structures,

Applied Surface Science Advances, 5 (2021) 100100.
https://doi.org/10.1016/j.apsadv.2021.100100

[39] Z.N. Kayani, M. Sahar, S. Riaz, S. Naseem, Z. Saddiqe, Enhanced magnetic, antibacterial and optical properties of Sm doped ZnO thin films: role of Sm doping, Optical Materials, 108 (2020) 110457. https://doi.org/10.1016/j.optmat.2020.110457

[40] K. Badreddine, I. Kazah, M. Rekaby, R. Awad, Structural, morphological, optical, and room temperature magnetic characterization on pure and Sm-doped ZnO nanoparticles, Journal of Nanomaterials, 2018 (2018). https://doi.org/10.1155/2018/7096195

[41] G. Vijayaprasath, R. Murugan, T. Mahalingam, Y. Hayakawa, G. Ravi, Enhancement of ferromagnetic property in rare earth neodymium doped ZnO nanoparticles, Ceramics International, 41 (2015) 10607-10615. https://doi.org/10.1016/j.ceramint.2015.04.160

[42] J. Zheng, J. Song, Z. Zhao, Q. Jiang, J. Lian, Optical and magnetic properties of Nd-doped ZnO nanoparticles, Crystal Research and Technology, 47 (2012) 713-718. https://doi.org/10.1002/crat.201200026

[43] N. Aggarwal, K. Kaur, A. Vasishth, N. Verma, Structural, optical and magnetic properties of Gadolinium-doped ZnO nanoparticles, Journal of Materials Science: Materials in Electronics, 27 (2016) 13006-13011. https://doi.org/10.1007/s10854-016-5440-2

[44] A. Dakhel, M. El-Hilo, Ferromagnetic nanocrystalline Gd-doped ZnO powder synthesized by coprecipitation, Journal of Applied Physics, 107 (2010) 123905. https://doi.org/10.1063/1.3448026

[45] N. Fifere, A. Airinei, D. Timpu, A. Rotaru, L. Sacarescu, L. Ursu, New insights into structural and magnetic properties of Ce doped ZnO nanoparticles, Journal of Alloys and Compounds, 757 (2018) 60-69. https://doi.org/10.1016/j.jallcom.2018.05.031

[46] D. Arora, K. Asokan, A. Mahajan, H. Kaur, D. Singh, Structural, optical and magnetic properties of Sm doped ZnO at dilute concentrations, RSC advances, 6 (2016) 78122-78131. https://doi.org/10.1039/C6RA12905F

[47] P. Kumar, V. Sharma, A. Sarwa, A. Kumar, R. Goyal, K. Sachdev, S. Annapoorni, K. Asokan, D. Kanjilal, Understanding the origin of ferromagnetism in Er-doped ZnO system, RSC advances, 6 (2016) 89242-89249. https://doi.org/10.1039/C6RA17761A

[48] S. Kumar, P. Sahare, Nd-doped ZnO as a multifunctional nanomaterial, Journal of rare earths, 30 (2012) 761-768. https://doi.org/10.1016/S1002-0721(12)60126-4

[49] S. Das, S. Das, A. Roychowdhury, D. Das, S. Sutradhar, Effect of Gd doping concentration and sintering temperature on structural, optical, dielectric and magnetic properties of hydrothermally synthesized ZnO nanostructure, Journal of Alloys and Compounds, 708 (2017) 231-246. https://doi.org/10.1016/j.jallcom.2017.02.216

[50] B. Poornaprakash, S. Ramu, K. Subramanyam, Y.L. Kim, M. Kumar, M.S. Pratap Reddy, Robust ferromagnetism of ZnO:(Ni+Er) diluted magnetic semiconductor nanoparticles for spintronic applications, Ceramics International, 47 (2021) 18557-18564. https://doi.org/10.1016/j.ceramint.2021.03.181

[51] B. Poornaprakash, S. Ramu, K. Subramanyam, Y. Kim, M. Kumar, M.S.P. Reddy, Robust ferromagnetism of ZnO:(Ni+ Er) diluted magnetic semiconductor nanoparticles for spintronic applications, Ceramics International, 47 (2021) 18557-18564. https://doi.org/10.1016/j.ceramint.2021.03.181

[52] K.C. Verma, R.K. Kotnala, Understanding lattice defects to influence ferromagnetic order of ZnO nanoparticles by Ni, Cu, Ce ions, Journal of Solid State Chemistry, 246 (2017) 150-159. https://doi.org/10.1016/j.jssc.2016.11.018

[53] F. Kabir, A. Murtaza, A. Saeed, A. Ghani, A. Ali, S. Khan, K. Li, Q. Zhao, K.K. Yao, Y. Zhang, S. Yang, Structural, optical and magnetic behavior of (Pr, Fe) co-doped ZnO based dilute magnetic semiconducting nanocrystals, Ceramics International, 48 (2022) 19606-19617. https://doi.org/10.1016/j.ceramint.2022.03.096

[54] B. Poornaprakash, U. Chalapathi, S. Babu, S.-H. Park, Structural, morphological, optical, and magnetic properties of Gd-doped and (Gd, Mn) co-doped ZnO nanoparticles, Physica E: Low-dimensional Systems and Nanostructures, 93 (2017) 111-115. https://doi.org/10.1016/j.physe.2017.06.007

[55] V. Parthasaradi, M. Kavitha, A. Sridevi, J.J. Rubia, Novel rare-earth Eu and La co-doped ZnO nanoparticles synthesized via co-precipitation method: optical, electrical, and magnetic properties, Journal of Materials Science: Materials in Electronics, 33 (2022) 25805-25819. https://doi.org/10.1007/s10854-022-09272-9

Materials Research Forum LLC
https://doi.org/10.21741/9781644902394-4

Chapter 4

Recent Advances and Trends in ZnO Hybrid Nanostructures

Rahul Kalia[1], Ritesh Verma[2]*, Ankush Chauhan[3], Anand Sharma[1], Rajesh Kumar[4]

[1]School of Physics and Materials Science, Shoolini University, Bajhol, Solan, H.P., India-173212

[2]Department of Physics, Amity University Haryana, Gurugram, India-122413.

[3]Chettinad Hospital and Research Institute, Chettinad Academy of Research and Education, Kelambakkam, Tamil Nadu, India-603103

[4]Department of Physics, Sardar Patel University, Mandi, H.P., India-175001

*vermaritesh.rv40@gmail.com

Abstract

ZnO nanostructures are excellent candidates for use in the production of functional devices because to their low toxicity, robust thermal stability, excellent corrosion resistance, biocompatibility, high specific surface area, and high conductivity. In this chapter we have discussed the various nanostructures of ZnO and their kind, synthesis of hybrid ZnO nanostructure, modification in nanostructure of ZnO, Hybrid nanostructure of ZnO. In addition, we have discussed the various methods like Sol-gel method, Hydrothermal method and Green Synthesis method in detail for the synthesis of ZnO nanostructures. The effect of these methods on variation of nanostructures have been discussed in detail. It has been observed that because of its great sensitivity to the chemical environment, ZnO nanostructures have been extensively used in sensing applications. Also, the ZnO nano structures have been widely used in light emitting diodes and in solar cells because of its semiconducting nature. The ZnO is a n-type semiconductors and has a perfect bandgap of 3.37eV for its use in solar cell applications. Thus, this chapter provides a detailed discussion about the various nano structures of ZnO, the synthesis methods and various applications.

Keywords

ZnO Nanostructures, Hybrid structures, Sensors, Solar Cells, Light Emitting Diode

Contents

1.   Introduction

Nanomaterials have been extensively studied for use in various types of nanoscale functional devices, which are widely used in the chemical industry, medical diagnostics, food technology, ultraviolet testing, national defence, and our everyday lives [1–5]. Among them, semiconductor nanoparticles such as ZnO, SnO_2, TiO_2, and ZnS get the most interest owing to the fascinating nanoscale impacts on their physical and chemical characteristics [6–9]. Because of their low toxicity, strong thermal stability, good oxidation resistance, good biocompatibility, large specific surface area, and high electron mobility, ZnO nanostructures are promising candidates for application in the fabrication of functional devices [10]. With a straight band gap ($E_g = 3.37$ eV) and a big exciton binding energy (60 meV), ZnO is a transparent semiconductor with strong conductivity and piezoelectricity [11]. It exhibits near UV emission and has a substantial exciton binding energy (60 meV). A great deal of research has been done on the morphology-controlled synthesis of ZnO nanostructures. The morphology of ZnO can be changed from nanorods to nanotubes to nanoneedles and nanocomb to nanoinjectors, nanohelixes, and nanodisks by simply adjusting the preparation method and preparation conditions [12–16]. The availability of a wide range of morphologies and established growing processes allows for the rapid fabrication of ZnO-based devices. ZnO nanostructures have been extensively employed in a variety of applications, including field-effect transistors, light emitters, lasers, solar cells, and sensors [17–19]. The use of ZnO nanostructures for sensing applications has gotten a lot of interest lately. Several features of ZnO are being exploited; one of them is the fact that conductance varies with the reversible chemisorption process of reactive gases on the surface of ZnO [20]. Pressure sensors are based on the piezoelectric property of ZnO, which was discovered by Wang et al. [11] and used to detect pressure. In part, this is due to the biocompatibility and nontoxicity of ZnO [21], which allows it to be used as a biosensor. Different devices with high sensing performances have been described; nevertheless, achieving a high selective response has remained a significant difficulty to date.

In this chapter we will discussed on nanostructure and their kind, synthesis of hybrid ZnO nanostructure, modification in nanostructure of ZnO, Hybrid nanostructure of ZnO and also discussed on the application of ZnO nanostructure-based sensors such as gas sensors, chemical sensors, biosensors, UV sensors, and pH sensors.

2.   Nanostructures and their kind

2.1   Nanorods and nanowires

In order to attain certain qualities for high performance materials, many researchers have produced metal oxide nanorods and nanowires at different scales. In this section, some of the most notable pieces of work are examined. Deka et al., [22] successfully generated CuO nanowires enclosed in WCF polyester resin composites using the VARTM process for hydrothermal ZnO nanostructure enhancement. Hydrothermal synthesis was used to generate nanowires on a section of WCF that had been seeded with CuO. The growth of nanowires is entirely dependent on the seeding cycles, and the concentration of development solution and development duration have no effect on the growth of nanowires. The strength and tensile modulus of WCF will rise by 42.8 percent and 33.1 percent, respectively, as a result of the formation of nanowires on the material. It was shown by Ko et al., [23] that a simple selected hierarchical development may significantly improve the efficiency of a DSSC power converter. Tree-like ZnO photoanodes with varying levels of crystalline ZnO have been developed in a nanoforest. Stretched nanowire DSSCs outperformed upstanding ZnO nanowires in terms of short circuit current density and total light transformation efficiency. Due to a larger surface area, which allows for more dye stacking and light collection as well as less charge recombination along ZnO nanotree branches, increased productivity is possible. ZnO nanoforests are built on a base of long ZnO nanowires that have been vertically adjusted (Figure 1). "Seed effect" is shown in Figure 1b, 1c, whereas "polymer removal effect" is shown in Fig. 2d, 2e.

Figure 1. Shows the SEM image of ZnO NWs (a) Growth of length (b) with no seeds (c) with seeds after polymer removal (d) without polymer removal and (e) with polymer removal after seed NP deposition [23].

Hydrothermal treatment by Hazarika et al. was used to successfully generate ZnO nanorods on the woven Kevlar fibre WKF [24] for the purpose of improving the interfacial strength of aramid fibre composites. First, the WKF was hydrolyzed on the surface and then subjected to an ion exchange method in order to improve the attachment of ZnO to the major strands of WKF. Results from scanning electron microscopy (SEM) revealed that the growth of ZnO nanorods was entirely dependent on the amount of seeding, treatment duration, and convergence of ZnO used. There has been research on the effects of pH and temperature on ZnO nanostructure morphology by Amin et al., who examined the formation of ZnO nanorods over a period of 10 hours [25]. Hydrothermal technique was used by Kong et al., [26] to produce ZnO NRs on woven carbon fibre. The ZnO/WCFs were thoroughly mixed with polyester resin using the VARTM method. Cross-connected networks transmit energy over interfaces, enhancing load exchange and interfacial strength, according to the findings of the tests. Piezoelectric nano-generators were created by generating ZnO NRs on the inner wall of a horizontal quartz tube by Ruqeishi et al., [27]. Using nanorods as an alternating electric current generator was a success. 3-5g of ZnO NRs were grown every cycle using the tube-in-tube CVD process, which is more than enough to produce current in response to mechanical stress. Using a hydrothermal method, Salahuddin et al., [28] successfully synthesised ZnO nanotubes with mean exterior diameters and radii of 200 nm and 2.4 m. The typical absorption bands at 508 and 404 cm⁻

[1] were clearly seen in the FTIR. The mass To-phonon frequency and the mass Lo-phonon frequency are connected by these two retention peaks. The hexagonal wurtzite structure of ZnO nanotubes was confirmed using XRD analysis. Measurements were made using UV spectroscopy. The resulting tubes have a length of 2.4 nm and a wall thickness of 200 nm. ZnO needle-like and plate-like morphologies were achieved by Ghasaban et al., [29] using hydrothermal responses at low temperatures of 115 °C and short reaction times of 2 or 6 hours. To achieve ZnO nanostructures of the desired size and shape, they proposed using the hydrothermal reaction as a method that could be easily scaled up. A needle-like morphology with a diameter of 240 nm is formed when ammonia (pH = 9) is used as an anionic antecedent. Plate-like nanostructures with a mean diameter of less than 50 nm are formed when the anionic antecedent is replaced by sodium hydroxide (pH = 13). Figure 2 shows the XRD results for ZnO nanoparticles. Because of their uneven orientation, both ZnO nanostructures show comparable XRD patterns, save from relative pinnacle intensities. Hexagonal Wurtzite ZnO period XRD patterns were obtained (a = 3.249, c = 5.206; JCPDS card no. 36-1451) using XRD. XRD designs provide clear and strong evidence that the object is made of an extremely crystalline substance. The crystallite size (L) of the particles may be determined using the Scherrer condition as follows:

$$L = \frac{K\lambda}{\beta s} cos\theta \rightarrow cos\theta = \left(\frac{K\lambda}{L}\right)\left(\frac{1}{\beta s}\right) \tag{1}$$

Where, L is crystallite size in nm, λ is the radiation wavelength (1.54056 Å for CuKα) in nm, βs is the full width (at half-maximum) of ZnO diffraction crest profile in radian, θ is the diffraction angle (top position) and K is the shape factor (0.89 < K < 1) which is viewed as 0.94 here.

It is important to decide the instrumental expanding and correct the deliberate β as following:

$$(\beta_s)^2 = (\beta_{measured})^2 - (\beta_{instrumental})^2 \tag{2}$$

As indicated by Scherrer condition, the crystallite size can be acquired by plotting Cosθ versus 1/βs and figuring L parameter from the slant of regression line going through the origin facilitates. The other technique is directed by making logarithm on both sides of equation (1) as following [30]:

$$\beta_s = \left(\frac{K\lambda}{L}\right)\left(\frac{1}{cos\theta}\right) \rightarrow ln\beta_s = ln\left(\frac{K\lambda}{L}\right) + ln\left(\frac{1}{cos\theta}\right) \tag{3}$$

By drawing lnβs versus ln(1/Cosθ), a straight line with a slant of around one and an intercept ca. lnK/L must be found. Yilmaz et al., [31] viably applied hydrothermal technique and chemical spray pyrolysis to successfully create nanocubes and nanorods. Amount of Zn2+ ion assumes a critical part for different shapes of the nanoparticles.

Presence of nanoparticles of ZnO indicates great photo-iridescence so that synthesized product can have opto-electronic utilities. Grain size (D) and dislocation density (δ) have been estimated with the help of mathematical relations given by Scherer [32] as mentioned below.

$$D = \frac{0.9\lambda}{\beta cos\theta} \tag{4}$$

$$\delta = \frac{1}{D^2} \tag{5}$$

Where, 'λ' represent the incident x-beam's wavelength

'β' represent the FWHM

'θ' represent the Bragg's angle

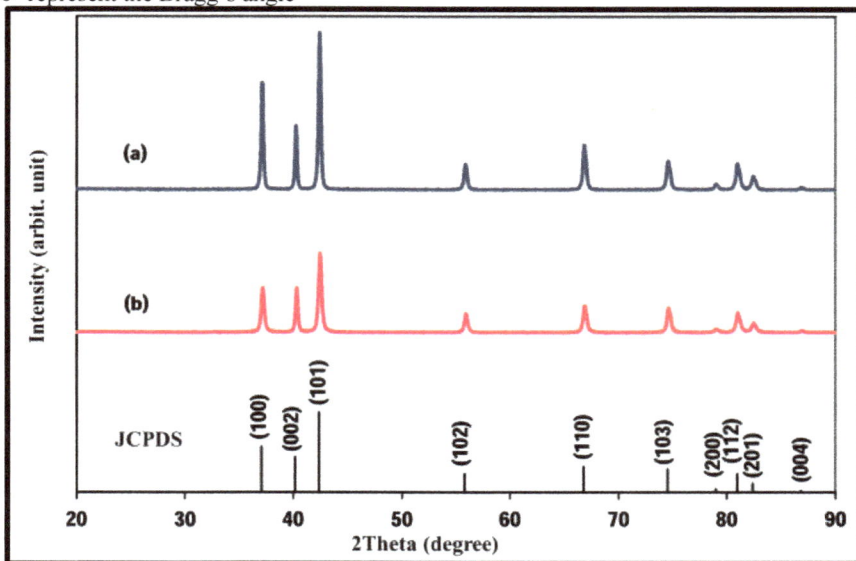

Figure 2. Shows the XRD result of (a) needle-like (b) plate-like ZnO NPs. [31]

2.2   Blossoms and cabbage-like nanostructures

Zinc oxide nanostructures resembling "flowers" may be generated using hydrothermal methods. Yoo et al., [33] exhibited the manufacture of nanopowders of CuO on the high surface area of ZnO flowers using hydrothermal method. DMMP gas detection is made possible because to the materials created as a result of this process. PXRD and TEM analysis confirmed the CuO/ZnO hetero-junction configuration. Figure 4 shows the findings of time-dependent SEM examinations of ZnO and CuO/ZnO morphologies. ZnO structures were framed, and by increasing ZnO production time, the bloom's diameter grew from 3-3-10 nm to a diameter of 10 nm or more. CuO nanoparticles were continuously preserved at the apex of the flowers when ZnO blooms were mixed with CuO, as seen in Figure 3 (d-f). No extra seeds were needed to produce hierarchical ZnO crystals on polyimide film under hydrothermal conditions, according to Xu et al. Just by altering the concentration of zinc ion solution, ZnO crystal morphologies may be fine-tuned. ZnO-modified PI film may self-assemble into nanoflowers when it is submerged in a solution of stearic acid (SA). SA's self-assembly influenced the film's capacity to cling to the surface. However, zinc concentration has a considerable impact on ZnO morphology on PI film, as shown in this study. PI(ZnO-SA) films' surface morphologies were further studied as a means of better understanding the surface wettability transition. A unique belt- or sheet-like morphology was discovered in PI(ZnO) films. Original ZnO flower shapes on PI film are somewhat altered after immersing them in stearic acid (SA) solution, which may be ascribed to the wet chemical manufacture of ZnO flowers with SA in the presence of SA. [34]. A new approach for the selective creation of 3D ZnO flowerlike nanostructures was suggested by Guo et al., [35] changing the hydrothermal reaction conditions and laser irradiation parameters may alter the flower-like structure. Synthetic flexibility in film architecture, coating texture and crystallite size may be achieved with this method. Another key feature of spatial organisation is the management of blossom density. In addition, no nanostructures were found in the region that was not exposed to ionising radiation. Hydrothermal ZnO cluster synthesis on FTO substrates was described in detail by Abdelfatah et al., [36]. It was KOH's hilarity that dictated the development process's course of action. According to the results, ZnO exhibits along the [0 0 2] axis. Optoelectronic devices based on these ZnO rod clusters with the smallest diameter will be the most efficient ones. Using ZnO nanorods and nanoflowers, which have a large surface volume and a respectably high roughness, is both feasible and transformative for sensor implementation in innovative devices [37]. By using an aqueous approach, Fan et al., [38] successfully produced vast amounts of zinc-oxide nanoflowers on graphene/SiO2/Si substrates. The results of the XRD test show that nanoflowers contain pure wurtzite. ZnO nanoflowers were found to have a star-like morphology. The existence of an O-vacancy in ZnO nanoflowers resulted in a narrower band gap than in ZnO powders.

Figure 3. Shows the SEM images of the surface morphology of both (a–c) ZnO and (d–f) ZnO/CuO structures as a function of ZnO synthesis time [33].

2.3 Other shapes

Nanodumbell, nanoflakes, nano disk, twinned dumbbells, and double disc are some of the various types of nanostructures that have been produced by renowned researchers. Guo et al., [39] used CTAB-assisted aqueous method to successfully make twinned ZnO discs at reduced temperatures. According to research, the size and intensity of the UV NBE peak decrease with increasing development time. The apparent discharges of imperfection increase in number as development proceeds. CTAB and improved temperature are indicated as the primary components for the fabrication of twinned plates moulded ZnO in light of the early results. Using an aqueous approach, Wang et al., [40] were able to insert ZnO microstructures in the shape of dumbbells. ZnO has a length of 5-20 nm and a diameter of 1-5 nm for the two closures and 0.5-3 nm for the middle section. SEM images of ZnO generated in the form of dumbbells are shown in Figure 4. CTAB assisted low temperature aqueous approach generated twinned flower like ZnO structure by Sun et al., [41]. The temperature-dependent PL spectra show that the main emanation peaks are associated with the acceptor's bound exciton recombination, and the crystal nature of the specimens is further established by these spectra. According to the first results, the enhancement of stream-like structures may be attributed to self-etching and renewal. Figure 5 depicts the impact of twinned bloom-like ZnO structures as currently constructed (A). Figure 5 shows the twinned bloom-like ZnO structures enlarged (B). ZnO particles in the twinned bloom-like shape show the hexagonal profile on the exterior, unlike the double-disk moulded ZnO in Figure 5 (A). Small wafers encased in a ZnO hexagon provide the basis of the bloom shape. Figure 5 (C) depicts a hexagonal nut structure that may be used to form bloom-like formations. Using hydrothermal aided electrochemical discharge deposition, Kumar et al., [42] devised a new concept to generate ultrathin ZnO nanoflakes to build a tip-based tool. Engineers are creating a new hybrid technique in conjunction with hydrothermal processes in order to create superior nanostructures.

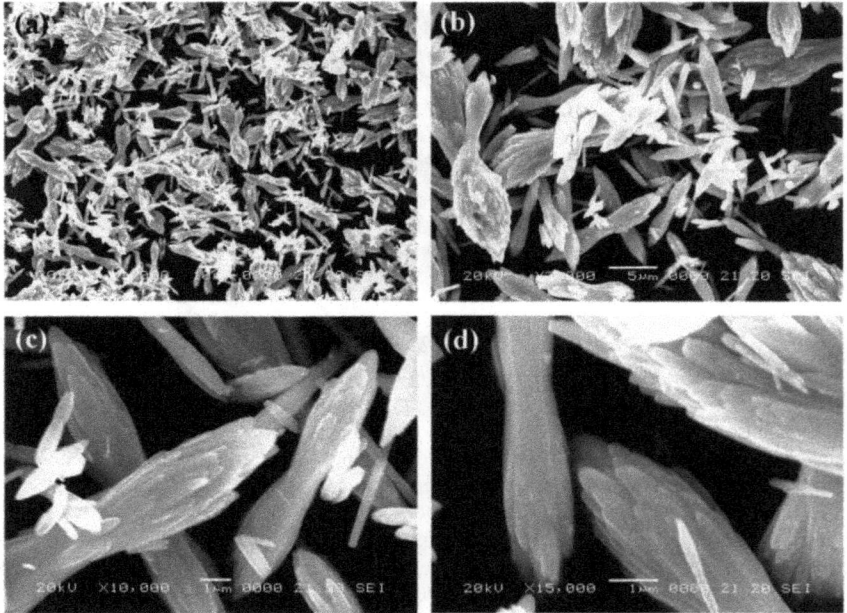

Figure 4. Shows the SEM analysis of dumbbell-shaped ZnO at various magnifications: (a & b) low magnification; (c & d) high magnification [40].

Figure 5. Shows the (A) SEM image of ZnO NSs; (B) twinned flower-like ZnO; (C) hexagonal nut structure [41].

3. Synthesis of hybrid ZnO nanostructure

This technique details the fabrication routes for ZONSs used in photocatalytic, photovoltaic, biological, and sensing applications. The overall flow chart in Figure 6 illustrates the numerous synthesis techniques that have been successfully employed to fabricate several kinds of ZONSs and other nanostructures [43]. Our goal is to broaden the use of wet processing technologies, particularly sol–gel, hydrothermal, and biological processes, via current research advancements.

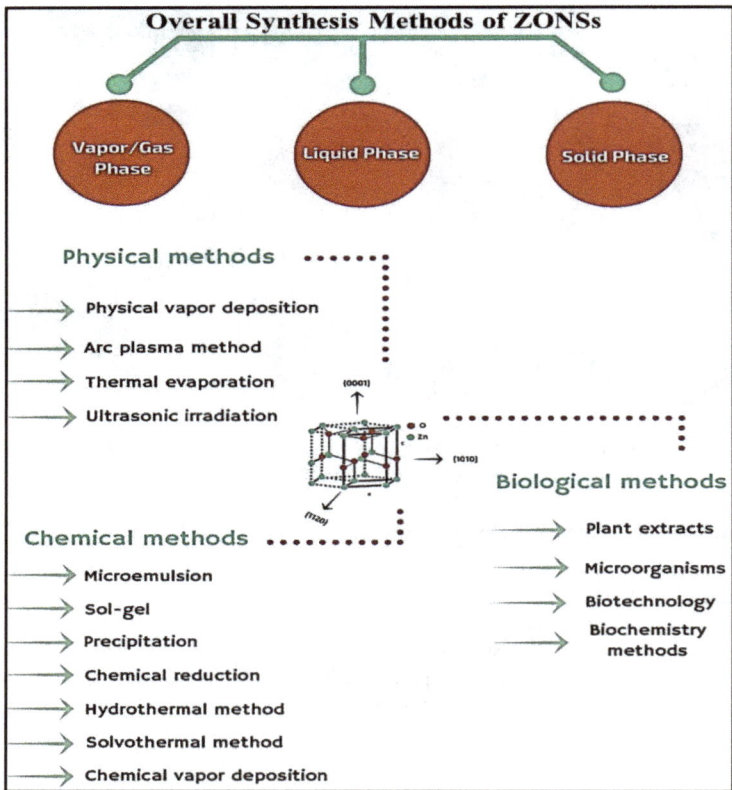

Figure 6. General flow chart of synthesis routes used in the fabrication of ZONSs. Reprinted with permission from Ref. [43]. Copyright 2017, with permission from Elsevier.

3.1   Sol-gel synthesis

It is one of the most adaptable technologies for nanofabrication of various materials, and it is employed in a variety of applications. [44–47]. The hydrolysis of precursors results in the formation of a colloidal suspension known as sol in this approach. As a result, liquid sols turn into solid gels as a result of polymerization. Furthermore, a consistent, homogeneous, and ultrafine powder is created at the end of the process. To better understand the mechanism of metal oxide sol–gel synthesis for supercritical fluid applications, Sui and Charpentier offered a complete description in their paper. They discussed the advantages of supercritical fluids and supercritical drying in the synthesis of solid goods, as well as the limitations of these technologies. It is well known that, at room temperature and pressure, some inherent shrinkage of solids causes the microstructure to collapse, resulting in a low specific surface area for the solid. When sols are formed in liquid phase, they have an unlimited amount of viscosity when converted to gel form. They also reported the creation of xerogel and aerogel with ambient drying and supercritical drying, respectively; and the formation of xerogel and aerogel with supercritical drying. Also included was an overview of the chemistry of metal oxide nanostructures created using the sol–gel technique, as well as recommendations for the use of supercritical fluids such as water, CO_2, and organic solvents in the process of nanofabrication. Alias and colleagues described the sol–gel production of ZnO nanoparticles under a variety of pH settings. When the pH was between 6 and 7, agglomeration was seen in the resultant nano powder (acidic and neutral conditions). The alkaline pH settings (pH = 9) were found to be more conducive to the production of fine powder. They indicated that under alkaline circumstances, the optical characteristics of the nanoparticles produced were better than in neutral settings. As shown in Figure 7, the findings of field emission scanning electron microscopy (FESEM) and the analysis of ZnO NPs generated at various pH settings are shown. Their findings further clarified the method through which ZnO NPs proliferate. Hydrolysis and polycondensation (nucleation) are the essential processes in the preparation of particles for sol–gel synthesis, whereas a significant quantity of OH⁻ ions are required for the formation of ZnO NPs with high crystallinity [48].

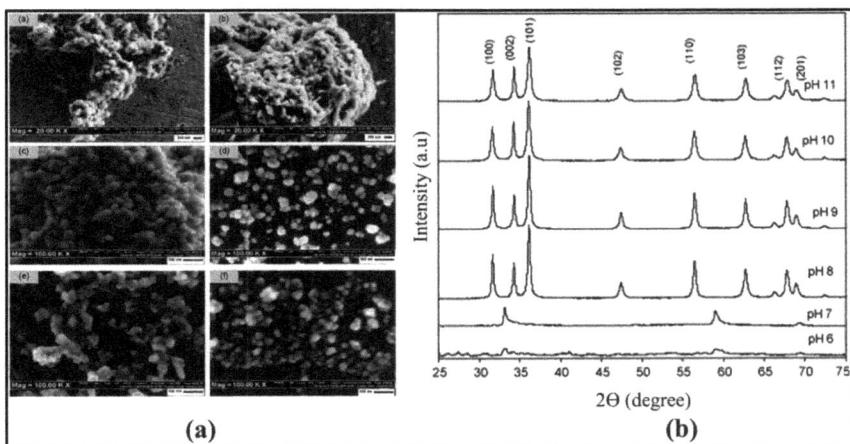

Figure 7. (a) FESEM analysis of sol–gel synthesized ZnO nano powder at varies pH from 6 to 11 in a–f, respectively and (b) XRD analysis of sol–gel synthesized ZnO nano powder at different pH conditions. Reprinted with permission from Ref. [48]. Copyright 2010, with permission from Elsevier.

This approach was utilised by Zak et al., to generate ZnO NPs in both pure and hybrid forms. Magnesium-doped ZnO NPs were synthesised at a high calcination temperature, i.e. 650 C. Adding magnesium to a material has a positive impact on its structural integrity. The synthetic NPs had a hexagonal wurtzite crystal phase and were spherical in form. They were extremely crystalline in nature. Magnesium doping modifies the normal ZnO NP characteristics and reduces structural flaws. Figure 8 depicts the particle size distribution and TEM examination findings of magnesium-doped ZnO NPs [49]. Plate-shaped ZnO NPs were synthesised by Zak et al. using a modified sol–gel combustion technique. According to the findings of this research, annealing temperature had a significant impact on optical and structural characteristics of ZnO nanoparticles [50]. At low temperatures and in starch medium, the optical characteristics of ZnO NPs were investigated in a different work by Zak et al. The ZnO NPs were stabilised using starch to regulate their mobility and development. XRD examination indicated that the NPs have a hexagonal wurtzite structure. They suggested employing starch as a stabiliser for the large-scale manufacture of nano ZnO.

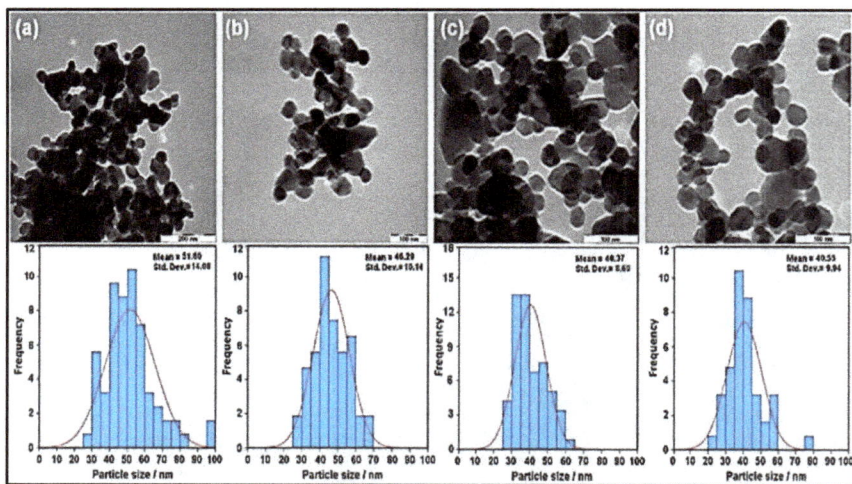

Figure 8. (a–d) TEM analysis of ZnO, Zn0.99Mg0.01O, Zn0.97Mg0.03O and Zn0.95Mg0.05O, respectively. Below TEM is the size distribution. Reprinted with permission from Ref. [49]. Copyright 2012, with permission from Elsevier.

The chemical reaction shown in Figure 9 hydrolyses starch and produces a binding connection for Zn^{+2} ions, which are attracted to oxygen [51]. Habibi and Karimi also created hybrid ZnO/CuO nanostructures by the sol–gel method with thermal breakdown. They employed both materials' carbonates as major precursors and discovered that the resultant NPs are naturally spherical [52]. Other researchers previously described a modified sol–gel synthesis of ZONSs. Konne and Christopher published a paper on the production of hydrogenated zinc oxide (ZnO:H). They utilised starch as a stabiliser and observed the formation of a new peak inside the ZnO lattice using XRD [53]. Jurablu et al. demonstrated the glycolated sol–gel production of zinc oxide nano powder utilising zinc sulphate heptahydrate as the principal reagent. They discovered that the generated samples have a hexagonal wurtzite crystal structure [54]. Humanez et al. prepared ZnO NPs through the citrate sol–gel technique and then calcined them at various temperatures to improve their crystalline structure. Additionally, they revealed the wurtzite structure of ZnO NPs. When samples were treated to electron paramagnetic resonance, they detected flaws in the crystal structure of ZnO NPs (EPR). They determined that these flaws were caused by oxygen and zinc vacancies [55].

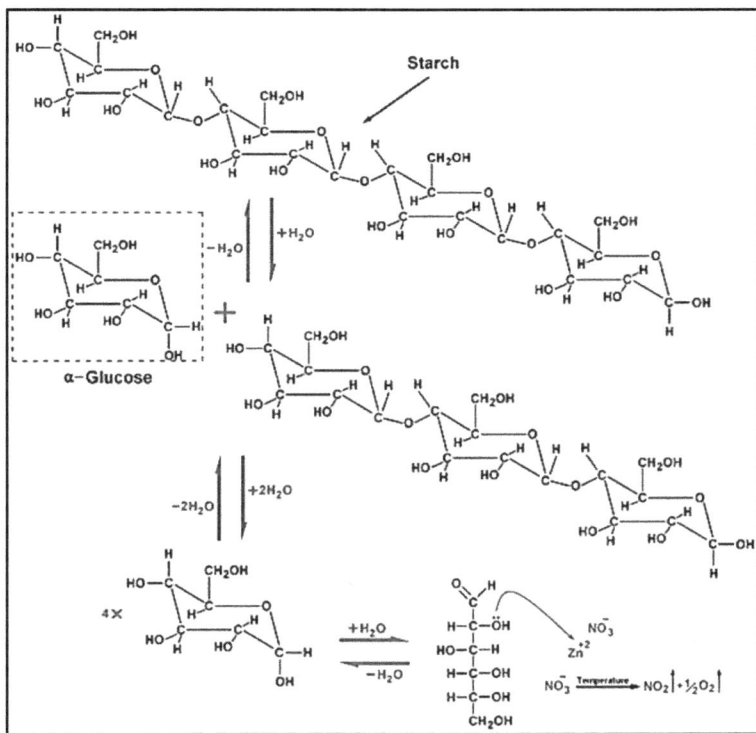

Figure 9. The process shows the hydrolysis of Starch and the binding side of Znþ2 ions. Reprinted with permission from Ref. [51]. Copyright 2013, with permission from Elsevier.

3.2   Hydrothermal synthesis

This procedure is carried out in autoclaves under carefully regulated circumstances. The temperature exceeded 100 degrees Celsius and attained saturation vapour pressure. This approach is often used to synthesise nanoparticles. Lu et al. investigated the ferromagnetism at room temperature in hybrid ZnO/CuO nanocomposites synthesised through a facile one-step hydrothermal process. The nanocomposites created have a range of phase ratios. They discovered that oxygen vacancies are responsible for the phase transition and development of ferromagnetism in nanocomposites [56]. Marlinda et al. used a single hydrothermal procedure to manufacture hybrid zinc oxide nanorods coated with reduced graphene oxide (ZnO/rGO) nanocomposites. They demonstrated that the density distribution, optical characteristics, and shape of ZnO nanorods are all affected by the GO content. They recommended using the nanocomposites they created for gas sensing applications [57]. The typical photocatalytic effect of nano ZnO for photodegradation of organic pollutants on its surface in the presence of sunlight, as well as ball and stick models of three different ZnO crystals, namely cubic rocksalt, cubic zincblende, and hexagonal wurtzite, with their characteristic planes, are graphically illustrated in Figure 10 [58]. Zak et al. published multiple research publications on the synthesis of ZONSs, in which they synthesised nano ZnO in a variety of dimensions using a variety of various synthesis techniques. Zak et al. employed a solvothermal approach to synthesise ZnO NPs in triethanolamine (TEA) medium to regulate the development of ZnO NPs in a research. The NPs formed had a hexagonal wurtzite structure with an average particle size of 48 nm. As previously reported, the use of TEA as a polymer agent increased the homogeneity of manufactured nano ZnO [59]. Saranya et al. investigated electrochemical supercapacitor applications and developed hybrid graphene–ZnO nanocomposites using a straightforward solvothermal process. Their findings indicated that ZnO nanoparticles were consistently deposited and disseminated on graphene. The electrochemical characteristics were determined using the galvanostatic charge/discharge technique. The created hybrid nanocomposite outperformed pure graphene in terms of performance. Figure 11 [60] illustrates the fabrication of hybrid graphene–ZnO nanocomposites schematically. Wu et al. described a straightforward solvothermal approach for fabricating graphene–ZnO nanocomposites with a sandwich-like structure. They employed zinc acetylacetonate and graphene oxide as precursors for zinc oxide and graphene, respectively, in a medium of ethylene glycol (EG). Additionally, they stated that the resultant NPs with a diameter of 5 nm were deposited homogeneously, evenly, and extremely thickly on the surface of graphene to create a sandwich-like structure. Additionally, the hybrid graphene–ZnO nanocomposite produced considerably enhanced ZnO's photocatalytic and sensing characteristics [61]. Lian et al., [62] demonstrated a simple template-free solvothermal ZnO NPs production in a regulated sized mixed solvent environment in a separate investigation. The findings indicate that the generated NPs display a range of optical characteristics depending on their size, with samples with a smaller size, a greater surface area, and more oxygen vacancies exhibiting a stronger visible emission spectrum peak.

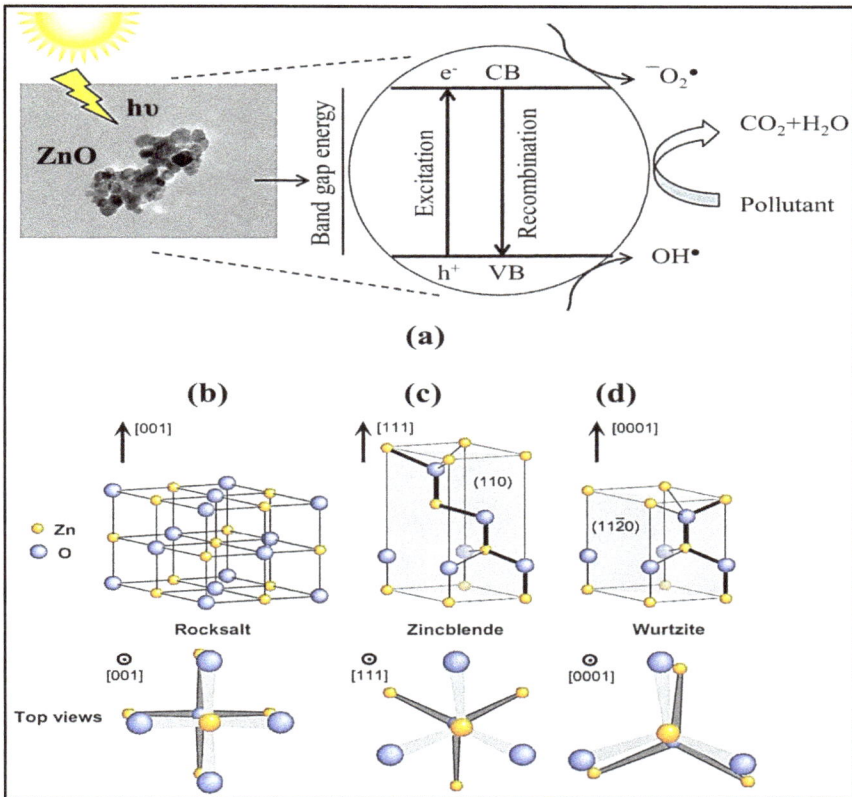

Figure 10. (a) Photodegradation of organic pollutants in the presence of sunlight by ZnO, and graphical representation of ball and stick models of different crystal structures of ZnO (b) cubic rocksalt, (c) cubic zincblende and (d) hexagonal wurtzite, respectively. Reprinted with permission from Ref. [58]. Copyright 2018, with permission from Elsevier.

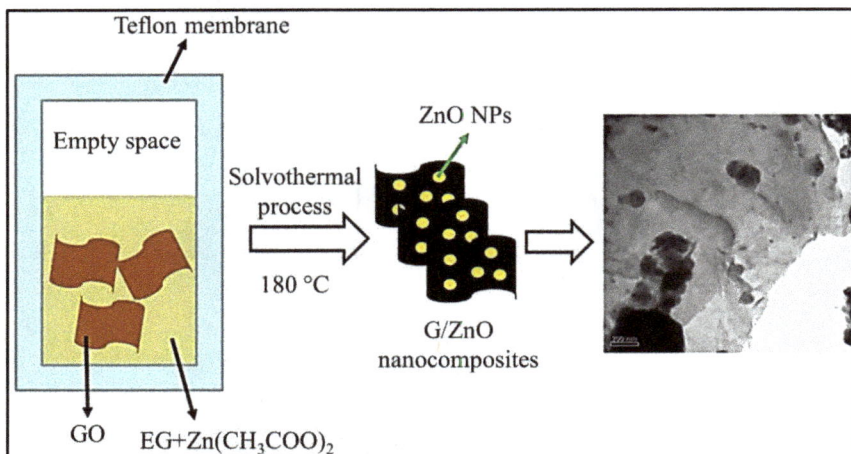

Figure 11. Graphical illustration of the synthesis of graphene–ZnO nanocomposites. Reprinted with permission from Ref. [60]. Copyright 2016, with permission from Elsevier.

Figure 12 [62] shows the SEM and TEM investigation of the produced ZnO NPs. Matei et al. produced nano ZnO in a polymer matrix utilising a straightforward coprecipitation approach. The polymer matrix employed was poly vinyl alcohol (PVA). The SEM analysis revealed that the NPs formed were homogeneous and evenly dispersed throughout the polymer matrix. SEM analysis indicated that the resultant NPs ranged in size from 20 to 150nm [63]. Kumar et al. used a straightforward precipitation approach to manufacture ZnO NPs using zinc sulphate and sodium hydroxide (NaOH). Additionally, they calcined ZnO NPs at various temperatures to improve their crystallinity. They discovered that increasing the calcination temperature lowered the band gap of the produced ZnO NPs [64]. Singhal et al. employed the co-precipitation approach to manufacture pure and copper-doped nano ZnO at different copper concentrations (Cu). They noticed a quasi-spherical shape in the generated nano ZnO, which was validated by transmission electron microscopy investigation. Additionally, they indicated that raising the Cu concentration to 15% first lowered and then enhanced the Zn–O bond length [65]. ZONSs were manufactured in 3D with better gas sensing characteristics using a microwave aided technique, as reported by Gu et al.,

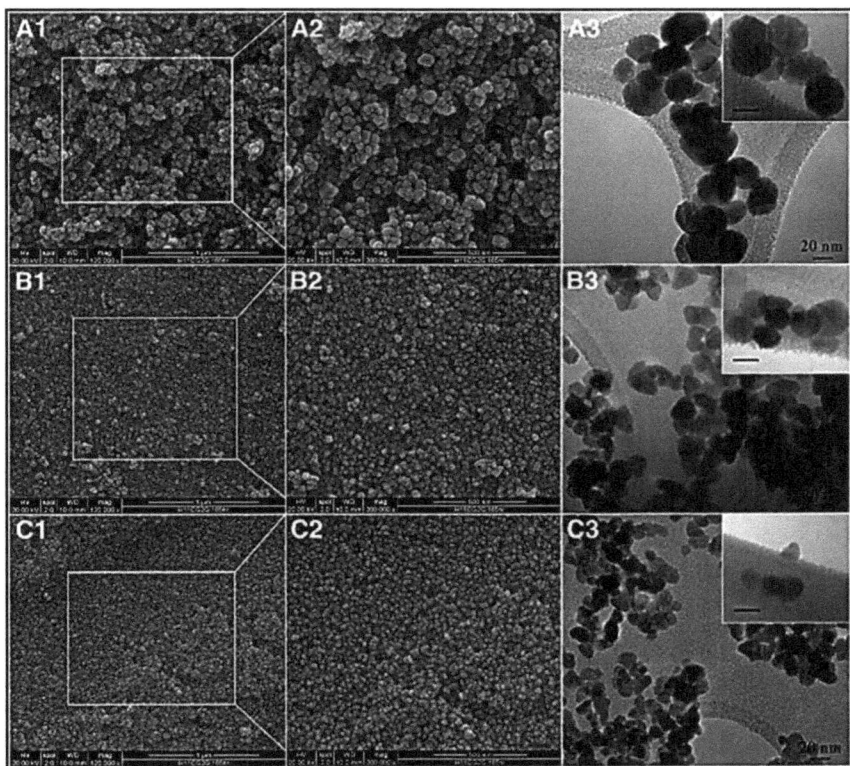

Figure 12. (A1, B1, C1) low-magnification SEM, (A2, B2, C2) high-magnification SEM, respectively. (A3, B3, C3) low-magnification TEM, inset of (A3, B3, C3) high-magnification TEM results, respectively for ZnO NPs. Reprinted with permission from Ref. [62]. Copyright 2012, with permission from Elsevier.

To create three-dimensional nanostructures, urea and zinc nitrates were utilised as precursors. SEM analysis verified the presence of flower-like 3D ZONSs. The amount of urea used in the synthesis dictates the morphology of ZONSs. Calcination temperature increases the gas sensing effectiveness of ZONSs even more. Figure 13 [66] illustrates the findings of a SEM investigation of the three-dimensional flower-like ZONSs. Bhatte et al. developed nanocrystalline ZnO utilising an additive-free single-step microwave-assisted technique in novel research. Zinc acetate was employed as the primary precursor and butanediol as the solvent, respectively. They claimed that the process employed to synthesise ZnO is more cost effective, quicker, and cleaner than standard methods, and that the resultant NPs exhibit a high photocatalytic effectiveness [67]. Soumya et al. employed the reflux microwave heating technique to synthesise nanocrystalline zinc oxide utilising a variety of capping agents, including polyethylene glycol (PEG), cetyl trimethyl ammonium bromide (CTAB), and poly vinyl pyrrolidone (PVP) (PVP). They created 1D nanostructures with a variety of morphologies using this approach, depending on the quantity and kind of capping agents used. Additionally, they synthesised PMMA/ZnO nanocomposites using PMMA as the polymer matrix. The produced hybrid nanocomposites were utilised to shelter glass and other polymer-based sandwich structures from solar thermal radiation [68]. In a separate work, He et al. described a facile co-precipitation approach for the production and characterization of cobalt-doped ZnO NPs.

Figure 13. SEM micrographs of 3D flower-like ZONSs calcinated at (a, d) 400 °C, (b, e) 500 °C and (c, f) 600 °C. Reprinted with permission from Ref. [66]. Copyright 2014, with permission from Elsevier.

They investigated the valence state and position of cobalt (Co) ions in the structure of ZnO NPs and demonstrated in a systematic manner that Co doping occupies Zn2 sites in the hexagonal wurtzite crystal structure and exhibits a two valence state [69]. Sharma et al. described a simple microwave-assisted approach for fabricating ZnO NPs under a variety of circumstances. They found that altering the circumstances of synthesis may alter the shape and size of the resultant NPs. They next tested the antibacterial and photocatalytic capabilities of the produced ZnO NPs and discovered excellent findings. We conducted photocatalytic tests using the dye methylene blue (MB). Figure 14 [70] depicts a graphical depiction of the shape-controlled synthesis of ZnO NPs.

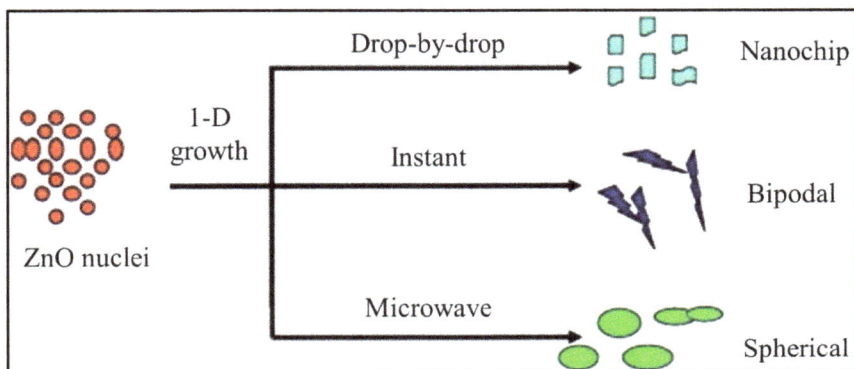

Figure 14. A schematic illustration for the shape-controlled synthesis of ZnO NPs. Reprinted with permission from Ref. [70]. Copyright 2011, with permission from Elsevier.

3.3   Green synthesis

Green synthesis, also known as biological synthesis, is gaining popularity as a means of synthesising ZONSs using plants or plant materials (flowers, stems, seeds, roots, fruits, peel, bark, and leaves). Until recently, nanostructures were created via chemical and physical processes, but the newest green chemistry discovery looks to be an approach to the synthesis of harmless nanostructures. Scientists may manage the morphology of their products without the use of harmful and costly solvents utilising the green synthesis approach, for example. ZONSs were successfully synthesised using green synthesis techniques by researchers from many disciplines [71–75]. Biomedical applications rely heavily on ZONSs, which Mirzaei and Darroudi shown may be synthesised from a variety of natural sources, including polymers (gelatin and starch), algae (Hypnea valencia and marine macroalgae), and moulds (Aspergillus fumigates and Aspergillus aeneus). Plants, rather than other green sources, were recommended as the best choice for synthesising ZONSs. Many plants, such as the genus Aloe barbadensis, Calotropis procera, Acalypha indica, Calotropis gigantean, Cassandra sp., Citrus paradisi and the species Camellia sinensis have previously been employed in studies for green synthesis. As shown in Figure 15, some of the plants and their components employed in the synthesis of ZONSs are shown. Using Hibiscus subdariffa plant leaf extracts, Bala et al. demonstrated a temperature-dependent green production technique for ZnO NPs. NPs were tested against Escherichia coli and Staphylococcus aureus for antibacterial efficacy. The anti-diabetic activity of ZnO was further examined and substantial reductions in blood glucose levels in mice were discovered [76]. Green synthesis techniques for ZONSs were reviewed and discussed by Lakshmi et al., who also noted current advancements in biomedical uses of green produced ZONSs. ZONSs and other metal oxide nanostructures may benefit from the inclusion of certain biological species, which they described in detail. A variety of plant components employed in green synthesis processes were detailed in the report [77]. Allium sativum plant extract was employed by Slman et al. for the green production of ZnO NPs in hot water. Green synthesis was carried out using five distinct techniques, each of which yielded almost identical results, as shown by XRD and AFM analyses [78]. Figure 16 depicts the schematic diagram of the green synthesis technique utilised to create ZONSs. ZnO NPs may be made using a simple, environmentally friendly, and inexpensive process proposed by Azizi et al. In a single pot, Sargassum muticum aqueous extract from the brown marine macro algae was used for the production of ZONSs. Nano ZnO with an average particle size of 30–57 nm was obtained by calcining at 450 C to produce pure hexagonal wurtzite crystals. Cosmetics and medicine might benefit from using as-prepared NPs [79]. It was shown that the leaves of an Ocimum basilicum Benth Lamiaceae plant may be utilised as a reagent to make ZnO NPs in another research, which employed zinc nitrate and Ocimum purpurascens Benth leaves as the raw materials. As validated by XRD and TEM investigation [80], hexagonal wurtzite crystals of ZnO with an average particle size of 50 nm formed. For the manufacture of ZnO NPs using aloe barbadensis miller leaf extract, Sangeetha et al. presented a new green technique that was more cost-effective than previous chemical and physical approaches.

Figure 15. Plants used in green synthesis of ZONSs (a) Cassia auriculata (b) Parthenium hysterophorous (c) Aloe vera (d) Acalypha indica (e) Calotropis gigantean and (f) Abrus precatorius. Reprinted with permission from Ref. [81]. Copyright 2017, with permission from Elsevier.

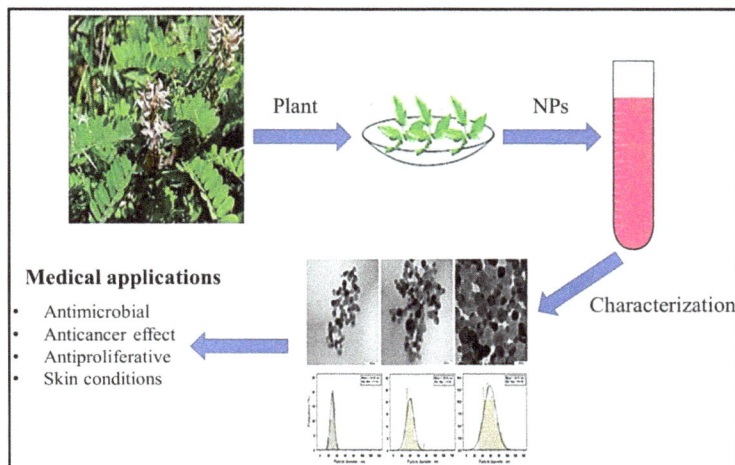

Figure 16. Graphical representation of green synthesis method for ZONSs. Reprinted with permission from Ref. [81]. Copyright 2017, with permission from Elsevier.

The use of zinc nitrate and aloe leaf extract as reagents demonstrated the synthesis of spherical and very stable ZnO NPs. The resultant NPs were poly distributed and had optical characteristics that could be tuned by varying their size, as shown in the experiment. Figure 17 shows the SEM findings of the NPs produced at various concentrations of aloe leaf extract. Aloe leaf extract, on the other hand, was used to examine the dispersion of biosynthesized ZnO nanoparticles in TEM examination (Figure 18). Increased usage of biological agents has been used to reduce the hazardous impact of chemicals on ZONSs. ZnO NPs were synthesised in a green environment by Jamdagni et al. using Nyctanthes arbor-tristis flower extract as a biological reduction agent. An average particle size of 12–32 nm was measured by TEM and validated by the improved experimental conditions [82]. Zinc nitrate and Calotropis Gigantea leaf extract were utilised in a research to produce ZnO NPs. For ZnO NPs, the average particle size was between 30 and 35 nanometers [83].

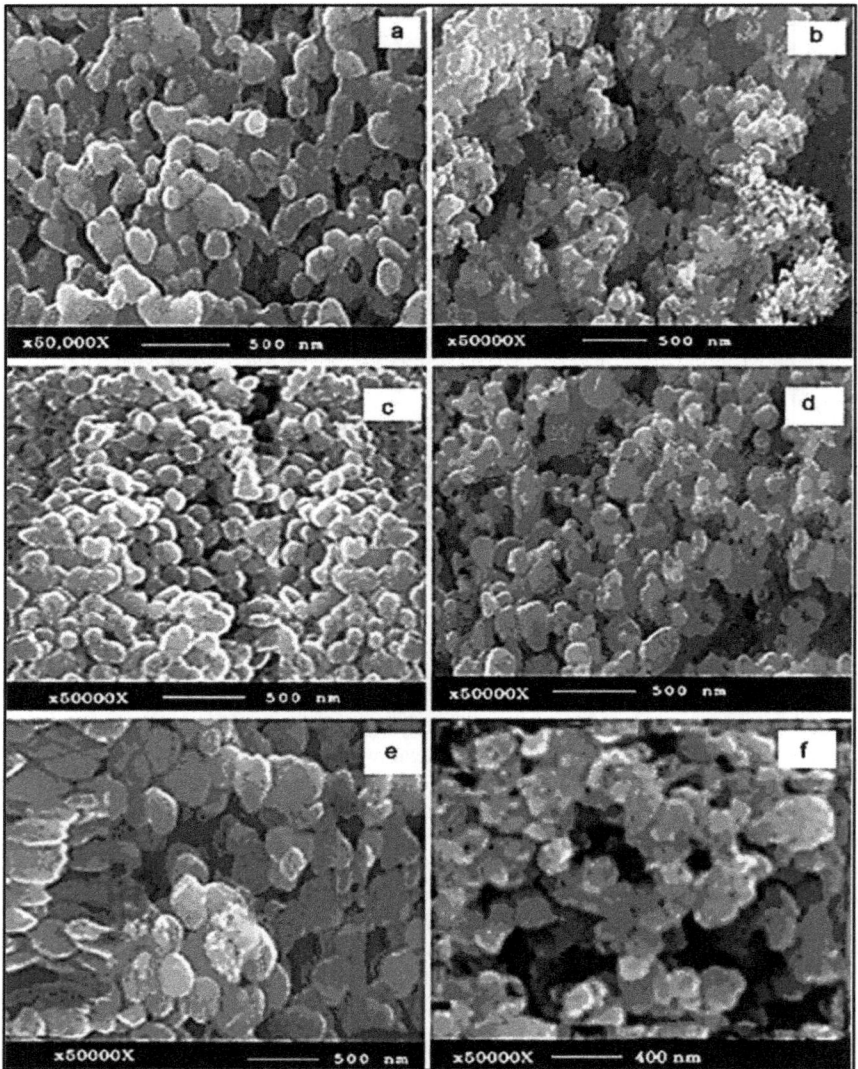

Figure 17. SEM analysis for ZnO NPs under varying concentration of aloe leaves extract (a) 0% (b) 5% (c) 10% (d) 15% (e) 25% and (f) 50%, respectively. Reprinted with permission from Ref. [84]. Copyright 2011, with permission from Elsevier.

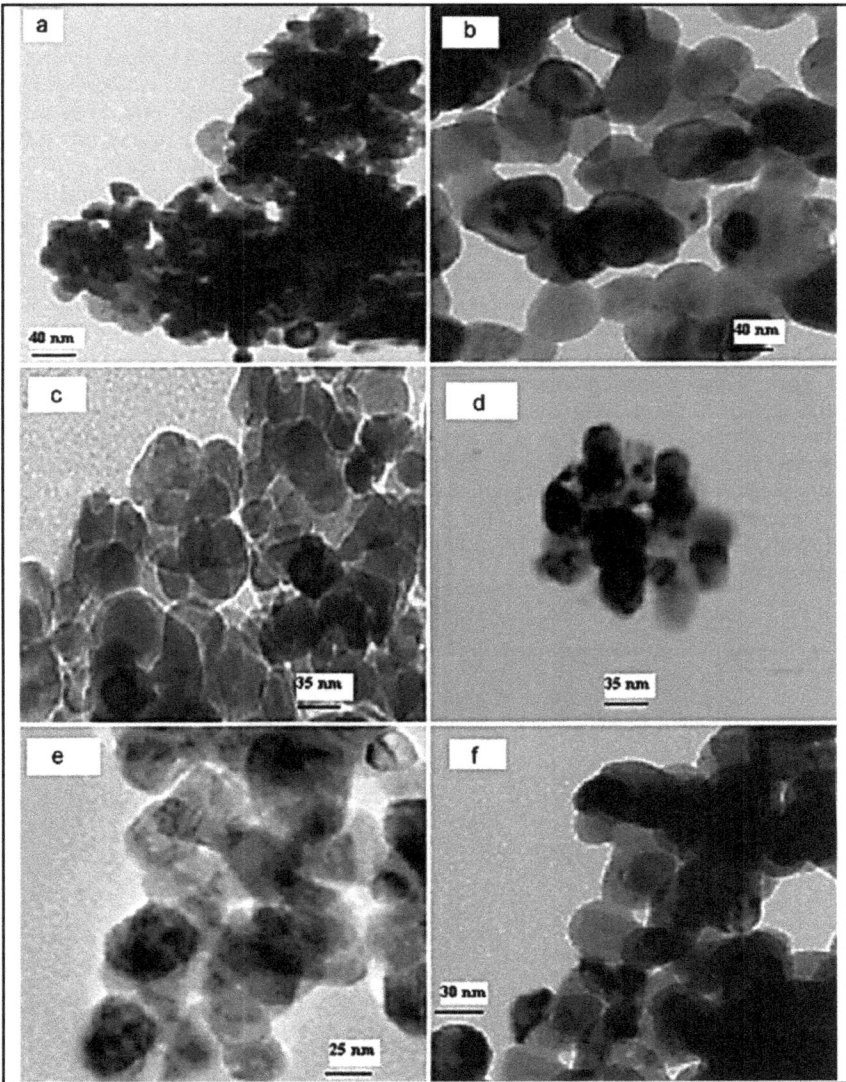

Figure 18. TEM analysis for ZnO NPs under varying concentration of aloe leaves extract (a) 0% (b) 5% (c) 10% (d) 15% (e) 25% and (f) 50%, respectively. Reprinted with permission from Ref. [84]. Copyright 2011, with permission from Elsevier.

3.4 Miscellaneous synthesis

ZONSs have been synthesised using a variety of techniques under a variety of circumstances in their pure and modified forms [85–90]. ZONSs were synthesised using the pyro-sol technique by Vasile et al. As a precursor, zinc nitrate was used. Sizes ranged from 5.7 to 21.8 nm for the ZONSs that were produced. By using this new method, ZONSs may be made without organic additives by calcination at 600 degrees Celsius [91]. ZnO and graphene–ZnO nanostructures were synthesised at various temperatures using lithium hydroxide monohydrate and zinc acetate dihydrate as reactants, according to a study by Fu et al. Results showed that ZnO NPs were present on both the top and bottom surfaces of the graphene sheet, as expected. As compared to pure ZONSs, the hybrid graphene–ZnO nanostructures demonstrated considerable results in the fight against photodegradation. Images of both ZONSs and graphene–ZnO nanostructures before and after photodegradation are shown in Figure 19 [92]. Nanostructures may be controlled in size and shape using distinct production techniques [93–95]. Using PEG as a surfactant, Vidyasagar et al. developed a unique solid-state single-step mechanochemical approach for the creation of several types of metal oxide nanostructures. Specifically for photovoltaic purposes, they created ZONSs and doped ZONSs. An efficient and simple procedure was adopted, which resulted in a high yield at a cheap cost. With increasing annealing temperatures, the band gap energy of the resultant nanostructures was shown to decrease. Fig. 20 shows the FESEM study of pure ZnO and copper doped ZnO (Cu–ZnO) at various temperatures [96].

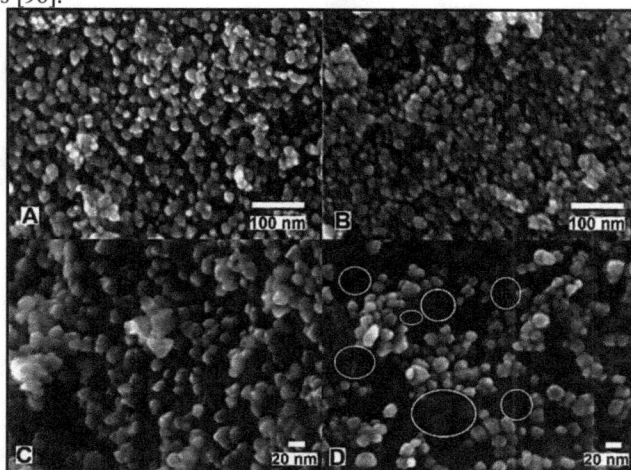

Figure 19. SEM micrographs of ZONSs and hybrid graphene–ZnO nanostructures (A) pure ZnO (B) graphene–ZnO (C) and (D) after photodegradation of MB for pure and hybrid nanostructures. Reprinted with permission from Ref. [92]. Copyright 2012, with permission from Elsevier.

Materials Research Forum LLC
https://doi.org/10.21741/9781644902394-4

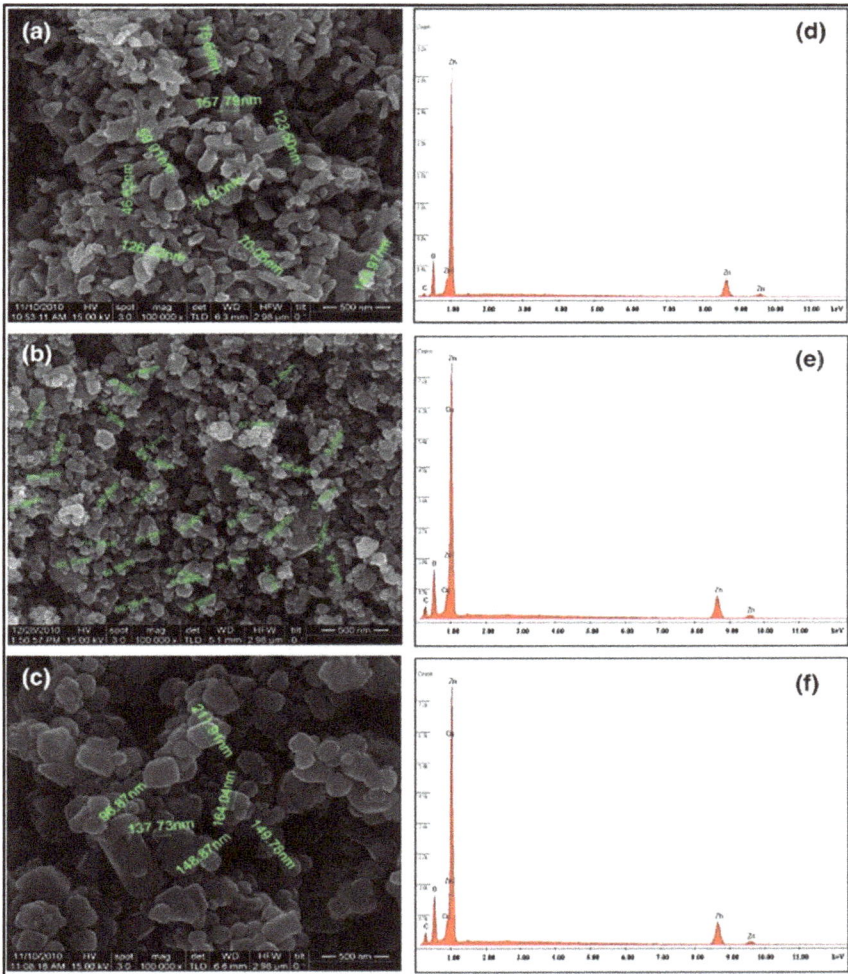

Figure 20. FESEM micrographs of pure and annealed samples (a) ZnO NPs at 350 °C, (b) Cu–ZnO at 350 °C, (c) Cu–ZnO at 650 °C, respectively. EDX spectra of pure and annealed samples (d) ZnO nanoparticles at 350 °C, (e) Cu–ZnO at 350 °C and (f) Cu–ZnO at 650 °C, respectively. Reprinted with permission from Ref. [96]. Copyright 2011, with permission from Elsevier.

Materials Research Forum LLC

https://doi.org/10.21741/9781644902394-4

Using new precursors, Akhtari et al. synthesised pure ZnO NPs and cobalt-doped ZnO (Co–ZnO) NPs by the coprecipitation approach. XRD examination verified the synthesised ZONSs to have a wurtzite hexagonal crystal structure. By increasing the cobalt content in the samples, they found that the band gap energy decreased. Nanocomposites of graphene oxide and zinc oxide (RGO-ZnO) were created utilising a simple ultrasonic process by Luo et al. The photocatalytic and photocurrent properties of the nanocomposites were greatly improved. Reduced graphene oxide–ZnO nanocomposites are shown schematically in Figure 21 [97] and are intended to improve photocatalysis.

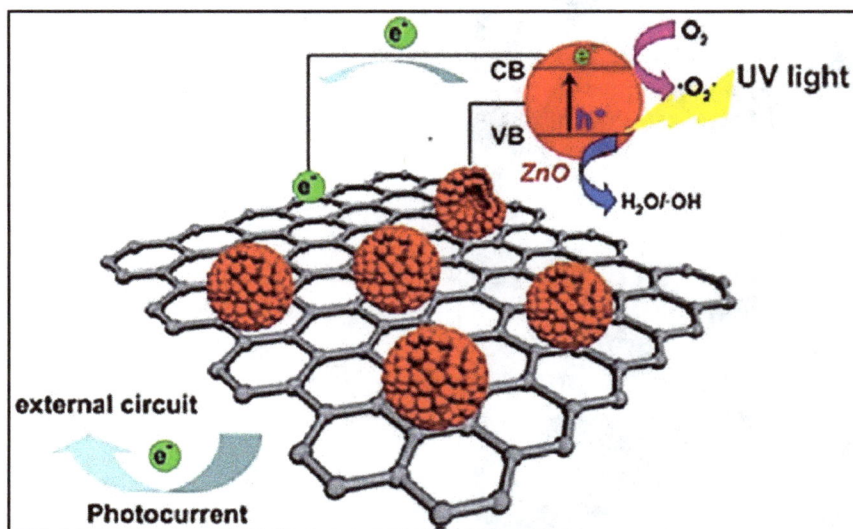

Figure 21. Graphical illustration of RGO–ZnO hierarchical nanocomposite with enhanced photocatalytic activity. Reprinted with permission from Ref. [97]. Copyright 2012, with permission from American Chemical Society.

ZnO nanoparticles were effectively produced by Noman and colleagues using sonication in a recent work to test comfort performance. We were able to synthesise and cover the ZnCl$_2$ precursor with an average particle size of 30nm using the ultrasonic approach. The thermophysiological characteristics of ZnO-coated materials were superior to those of uncoated and TiO$_2$-coated samples [98]. To get a better sense of how ZONSs are made, the synthesis techniques employed are outlined in Table 1.

Table 1. Overview of some successful methods used for the fabrication of ZONSs.

Method	Used precursors	Conditions	Structure; size	References
Sol–gel	Ethanol, $Zn(CH_3CO_2)_2$, oxalic acid $(C_2H_2O_4)$	50 °C, 60 min; drying 80 °C, 20h; calcination at 650 °C under air for 4 h	hexagonal wurtzite; NPs	[99]
Sol–gel	$Zn(CH_3CO_2)_2$, diethanolamine, ethanol	Room temperature; annealing at 500 °C for 2h	hexagonal wurtzite; nanotubes, size 70 nm	[100]
Sol–gel	$Zn(CH_3CO_2)_2$, methanol, acetyl acetone	Temperature 60 °C, continuous stirring for 8 h, heat treatment at 600 °C for 30 min	Hexagonal wurtzite, (002) orientation of the plane, ZnO thin films	[101]
Hydrothermal	$ZnCl_2$, NaOH, sodium dodecyl sulfate, Iso-butyl alcohol	Reaction: room temperature, 30 min., autoclave heating 120 °C for 12 h. drying 70 °C for 12 h	Disk shape, polar crystal, hexagonal structure	[102]
Hydrothermal	$Zn(NO_3)_2.6H_2O$, $C_6H_{12}N_4$	Temperature 90 °C, growth time 10 h	Hexagonal rings/prisms, ZnO films	[103]
Hydrothermal	$Zn(NO_3)_2.6H_2O$, $C_6H_{12}N_4$	Temperature 90 °C, growth time 3 h, UV laser power 0–120mW	ZnO nanowires, hexagonal rings	[104]
Solvothermal	Ethanol, $Zn(NO_3)_2$, $Zn(CH_3CO_2)_2$	Temp 150–180 °C, drying at 80 °C, calcined at 500 °C	hexagonal wurtzite, hollow microspheres with nanorods size 20 nm	[105]
Microwave assisted	Zinc Oximate, zinc acetylacetonate	Microwave power 800 W, 4 min; drying at 75 °C	zincite structure, size 9–31 nm; particles diameter: 40–200 nm	[106]
Precipitation	NaOH, $Zn(NO_3)_2$	Synthesis: 2 h; drying: 2 h, 100 °C	Spherical, size 40 nm	[107]
Precipitation	NH_4HCO_3, micro sized ZnO powder	25 °C, time 2 h; drying at 80 °C; calcined at 350 °C for 1 h	hexagonal wurtzite, rod like, flower-like, size 15–25 nm	[108]

Precipitation	NH_3 aqueous, $Zn(CH_3CO_2)_2$	85 °C; drying at 60 °C for 10 h	hexagonal wurtzite, rod like, flower-like, size 200 nm	[109]
Precipitation	NaOH, $Zn(CH_3CO_2)_2$	75 °C, 30 min; overnight drying at room temperature	hexagonal wurtzite, flower-like, size 800 nm	[110]
Precipitation	Ethanol, $ZnSO_4$, NH_4HCO_3	100 °C; calcined at 300–500 °C	Hexagonal wurtzite; size 9–20 nm	[111]
Ultrasonic	$Zn(CH_3CO_2)_2$, $NaNO_3$, H_2SO_4, $KMnO_4$	Sonication for 15 min, room temperature	Hierarchical hollow sphere, RGO-ZnO composites	[97]
Ultrasonic	$Zn(NO_3)_2.6H_2O$, $C_6H_8O_7 H_2O$	Sonication for 0.5–2 h, room temperature	Hexagonal, spherical particles	[112]
Biological/green	Azadirachta indica (Neem) leaves extract, $Zn(CH_3CO_2)_2.2H_2O$, NaOH	25% leaf extract, 50 ml NaOH, 2M zinc acetate	Hexagonal wurtzite, spherical, size 9.6–25 nm	[113]
Biological/green	Tabernaemontana divaricata leaves extract, $Zn(NO_3)_2.6H_2O$	80 °C temperature, continuous stirring, drying at 450 °C for 2 h	Hexagonal wurtzite, spherical, size 20–50 nm	[114]
Biological/green	Abutilon indicum leaves extract, $Zn(NO_3)_2.6H_2O$	Temperature of furnace at 200 °C for 3 min., calcination for 2 h	Hexagonal wurtzite, spheroid rod like shape, size 16–20 nm, band gap 3.36 eV.	[115]
Biological/green	Agathosma betulina plant extract, $Zn(NO_3)_2.6H_2O$	pH 5, temperature 100 °C for 2 h, annealing at 500 °C	Hexagonal wurtzite, quasi spherical, size 15.8 nm	[116]
Biological/green	Pongamia pinnata leaves extract, $Zn(NO_3)_2.6H_2O$, Citric acid	0.1 M solution for 24 h, oven dry at 80 °C, calcined at 350 °C for 3 h	Hexagonal wurtzite, spherical, size 100 nm	[117]

4.   Applications

4.1   Sensors

Because of its great sensitivity to the chemical environment, ZnO nanostructures have been extensively used in sensing applications. In addition to having a large surface area, nanostructures offer the benefit that electronic processes are substantially impacted by surface processes. When tested at ambient temperature, ZnO nanowires were shown to be very sensitive, while thin-film gas sensors are often required to work at higher temperatures. The sensing process is controlled by oxygen vacancies on the surface of ZnO, which have an impact on the electrical characteristics of the material. Electrons are removed and effectively depleted from the conduction band during oxidation, which occurs as a result of the adsorption of molecules such as NO_2 at vacancy sites that take electrons. This results in a loss in conductivity. When reducing molecules like hydrogen atoms react with surface-adsorbed oxygen, one electron is left behind, which results in a greater conductivity of the resulting solution. The problem is to detect certain gases in a selective manner. The development of a ZnO nanorod H_2 sensor has been completed [118]. The sensitivity of this sensor was increased by sputter deposition of Pd clusters on the ZnO rod surface, which was achieved by electroplating. When Pd is added to H_2, it seems to be successful in catalysing the dissociation of H_2 into atomic hydrogen, which results in an increase in the sensitivity of the sensor. At ambient temperature, the sensor can detect hydrogen concentrations as low as 10 parts per million (ppm) in N_2, but it has no reaction to O_2. Upon exposure to air or oxygen, the conductivity of the sensor recovers to 95 percent after 20 seconds. H_2 sensitivity for Pt-coated ZnO nanorods has also been shown by the same group [119]. ZnO nanoparticles were employed in a thick ZnO sheet for H_2 sensing [120]. A Pt-impregnated, 3 percent Co-doped ZnO nanoparticle film at a working temperature of 125 °C or below attained a sensitivity of 10-1000 ppm H_2 The oxidising sensors of O_2, NO_2, and NH_3 have also been developed in a field-effect transistor architecture of single nanowires [121,122]. It is possible to control the oxygen sensitivity of nanowires by varying the gate voltage. It is noticed when a significant negative gate voltage is supplied that NO_2 molecules desorb. Using this procedure, the sensor may be brought back to its original state. Pt interdigitating electrodes [123] have been used to develop an ethanol sensor with high sensitivity and rapid response at 300 °C. Ethanol may be detected quickly using tetrapod-films made in a humidified Air flow [124]. Researchers have discovered that gas molecules adsorbing on the surface of a photocurrent gas sensor composed of Ru-sensitized ZnO nanoparticles have a significant impact on its performance. The photoconductivity of ZnO is increased by the addition of a CO molecule to the ZnO surface [125]. When it comes to photoconductivity, however, O_2 may directly trap electrons on the surface and reduce the concentration of charge carriers in the ZnO condensing band. The use of ZnO nanorods as a glucose sensor has also been documented [126]. Electrostatic forces hold the negatively charged glucose oxidase (GOx) enzyme in place on ZnO. As long as the applied voltage is +0.8 V, the glucose biosensor's response ranges from 0.01-3.45 mM and its detection limit is less than 0.01 mM. The response time was less than 5 s.

4.2   Light-emitting diodes

Bao et al., [127] have developed a single nanowire light-emitting device that emits light. They first spread ZnO nanowires on a Si substrate, and then spin-coat a thin coating of poly(methyl methacrylate), or PMMA, over top of the nanowires and onto the substrate. After imaging the wire using a focus ion beam (FIB) system, a pattern for e-beam exposure of the polymethylmethacrylate (PMMA) is developed. It is necessary to remove the unexposed and partly exposed PMMA before a metallic contact may be placed onto the top surface of a single nanowire to be used. By doing so, the researchers are able to measure the current-voltage characteristics, photoluminescence characteristics, and electroluminescence characteristics of a single nanowire. The fabrication of a ZnO homojunction light-emitting diode (LED) on a single-crystal GaAs substrate using ultrasonic-assisted spray pyrolysis [128] has been demonstrated on a GaAs substrate. The diode is composed of a p-type ZnO film that has been doped with N-In and an n-type ZnO layer. Under forward bias, the homojunction LED exhibits a threshold value of around 4 V. One form of p-n junction diode, ZnO nanorods, was used to create the light-emitting diode, while the other type of diode, poly-2, 4-ethylene dioxythiophene poly(styrenesulfonate), or (PEDOT/PSS), was used to create the hole-conducting polymer. PS was utilised to separate nearby nanorods from one another. It is possible to see the ZnO band edge emission at 383 nm in the electroluminescence spectrum, as well as emission peaks at 430 and 640 nm, as well as emission peaks at 748 nm. It is 3 V when it comes to the threshold bias for UV light emission [129]. Könenkamp et al., [130] have utilised the same structure and obtained findings that are comparable to ours. Different groups have detected lasing in ZnO nanowire arrays that have been generated using a number of different ways. Researchers Huang et al. [131] have shown the usage of nanowires generated by CVD on sapphire and demonstrated an optical excitation threshold of 40 KW/cm2. Lasing of solution-grown ZnO nanorod arrays produced on F-doped SnO2 glass substrates has been reported for the first time, according to this study [126]. Solution grown ZnO on Si has also been reported to emit under the influence of a stimulator. It has been reported that nanowire arrays formed on Al2O3 substrates in a high-temperature method have comparable UV-lasing efficiency to that seen in this study [132]. A dielectrophoretic effect has been exploited to trap nanowires in a microelectrode gap, and this has been shown in experiments. These devices may therefore be employed as UV detectors, and they exhibit a remarkable sensitivity to UV light down to 10 nW/cm^2, allowing them to be utilised in a variety of applications [133].

4.3   Solar cells

Recently, zinc oxide (ZnO) has been used as an electrode material in dye-sensitized solar cells [134–136]. Observations of ultrafast electron injection into the conduction band of nanoporous dye-sensitized ZnO films have revealed that this occurs [137–139], which is comparable to the timescale of electron injection into TiO$_2$ layers, which has been the subject of extensive research for a long period of time [139]. For efficient dye-sensitization and light harvesting, semiconductors should have a wide bandgap and high charge carrier mobility, and films fabricated from the material should deliver a large surface area. This is only possible with a nanostructured film, which is only possible with a nanostructured film. As a result, zinc oxide (ZnO) seems to be a suitable material for this sort of solar cell, and it has the benefit over other metal oxides in that it is simple to synthesise controlled nanostructures, as opposed to other metal oxides. Dye-sensitized solar cells are often equipped with an electrolyte (redox system) that allows the dye to be regenerated by electron donation to its ground state after stimulation. The use of a solid organic hole-transporter material to replace the liquid electrolyte has been examined more recently [140], and it has the potential to be more environmentally friendly. This example involves the infiltration of an organic hole-transporting substance into a nanoporous framework. The issue for these solar cells is to fill the holes of the nanoporous layer [141,142], which represents a significant technical obstacle. In order to completely fill the nanoporous TiO_2 layer, only a few number of organic hole-transporter materials are available. Polymers, in general, seem to obstruct the pores and prevent the pores from being effectively filled. Thus, nanowire arrays have become attractive options for hybrid solar cell applications as a result of this research. This method is highly appealing because it allows for the use of a wide variety of organic hole-transporter materials, despite the fact that the surface area of a nanowire array is much lower than that of a nanoporous layer of the same thickness. Because of this, in addition to being a sensitizer, the dye must also function as a light-absorbing substance and contribute to the creation of charged particles. The benefit of nanostructured hybrid solar cells, as opposed to completely organic solar cells, is that the shape of the film may be regulated by the development of the ZnO structure. As an electrode material and a template for nanoscale phase separation, zinc oxide (ZnO) is used in this experiment. Recently, poly(3-hexylthiophene) (P3HT) has been employed as a hole-transporter in conjunction with ZnO nanostructures, and this has shown its effectiveness. When operated under normal solar circumstances (AM 1.5, 100 mW/cm^2) [143], these devices have an efficiency of 0.5 percent and exhibit a current density of JSC = 2.2 mA/cm2, an open circuit voltage of VOC = 440 mV, and a fill factor of 56 percent. Using a combination of P3HT and (6, 6)-phenyl C61 butyric acid methyl ester, the performance of this cell may be considerably enhanced to JSC = 10.0 mA/cm^2, VOC = 475 mV, and a fill factor of 43 percent, resulting in an efficiency of 2 percent and a fill factor of 43 percent (PCBM). The reason for the low open circuit voltage seen in hybrid solar cells that use ZnO as the electrode material is still being investigated. There is little doubt that more inquiry is required to determine the source of the leakage, and then improved cell efficiencies may be predicted. Ravirajan et al., [144] have conducted a rigorous investigation on the effect of the ZnO structure on the performance of the device. They examined flat ZnO layers, ZnO nanoporous layers, and vertically oriented nanowires to see which one was the most effective. It has been discovered that the charge recombination time in vertically aligned ZnO nanorods is exceptionally slow, with a half-life of several milliseconds, which is more than two orders of magnitude slower than that of nanoparticles, which should be advantageous for solar cell applications in the future. In a solar cell construction that contains P3HT as a hole-transporter and is sensitised with an amphiphilic Ru dye, the performance of the aligned nanorod structure is still four times greater than the performance of the aligned nanoparticle structure. According to the findings of these experiments, nanowire architectures will continue to improve the performance of hybrid solar cells in the future.

Conclusion

Because of their low toxicity, strong thermal stability, great corrosion resistance, biocompatibility, high specific surface area, and high conductivity, ZnO nanostructures are ideal candidates for application in the creation of functional devices due to the fact that they are biocompatible. In this chapter, we covered a variety of topics, including the synthesis of hybrid ZnO nanostructures, modifications in ZnO nanostructures, and ZnO nanostructures that are hybridised. In addition, we have provided an in-depth analysis of the several approaches that may be used to synthesise ZnO nanostructures, including the Sol-gel technique, the Hydrothermal method, and the Green Synthesis method. The impact that these techniques have on the variety of nanostructures has been dissected in extensive detail. It has been noted that ZnO nanostructures have been employed widely in sensing applications because of their considerable sensitivity to the chemical environment. The reason for this may be found in the previous sentence. Because ZnO is a semiconducting material, its nanostructures have found widespread use in light-emitting diodes and solar cells. ZnO is an n-type semiconductor that has a bandgap of 3.37 eV, making it ideal for use in solar cell applications. As a result, this chapter presents a comprehensive discussion on the numerous nano structures of ZnO, as well as the techniques of its manufacture and the many uses it has.

Declaration

Authors declare that they do not have any conflict of interest among them.

Acknowledgement

Authors would like to acknowledge the honourable chancellor of Shoolini University for his continuous support.

Reference

[1] X. Fang, L. Zhang, Controlled growth of one-dimensional oxide nanomaterials, Cailiao Kexue Yu JishuJournal Mater. Sci. Technol. 22 (2006) 1-18.

[2] E. Comini, C. Baratto, G. Faglia, M. Ferroni, A. Vomiero, G. Sberveglieri, Quasi-one dimensional metal oxide semiconductors: Preparation, characterization and application as chemical sensors, Prog. Mater. Sci. 54 (2009) 1-67. https://doi.org/10.1016/j.pmatsci.2008.06.003

[3] S. Barth, F. Hernandez-Ramirez, J.D. Holmes, A. Romano-Rodriguez, Synthesis and applications of one-dimensional semiconductors, Prog. Mater. Sci. 55 (2010) 563-627. https://doi.org/10.1016/j.pmatsci.2010.02.001

[4] X. Fang, T. Zhai, U.K. Gautam, L. Li, L. Wu, Y. Bando, D. Golberg, ZnS nanostructures: from synthesis to applications, Prog. Mater. Sci. 56 (2011) 175-287. https://doi.org/10.1016/j.pmatsci.2010.10.001

[5] X. Fang, L. Wu, L. Hu, ZnS nanostructure arrays: a developing material star, Adv. Mater. 23 (2011) 585-598. https://doi.org/10.1002/adma.201003624

[6] X. Fang, Y. Bando, M. Liao, U.K. Gautam, C. Zhi, B. Dierre, B. Liu, T. Zhai, T. Sekiguchi, Y. Koide, Single-crystalline ZnS nanobelts as ultraviolet-light sensors, Adv. Mater. 21 (2009) 2034-2039. https://doi.org/10.1002/adma.200802441

[7] Z. Jing, J. Zhan, Fabrication and gas-sensing properties of porous ZnO nanoplates, Adv. Mater. 20 (2008) 4547-4551. https://doi.org/10.1002/adma.200800243

[8] J. Liu, C. Roussel, G. Lagger, P. Tacchini, H.H. Girault, Antioxidant sensors based on DNA-modified electrodes, Anal. Chem. 77 (2005) 7687-7694. https://doi.org/10.1021/ac0509298

[9] Y. Wang, X. Jiang, Y. Xia, A solution-phase, precursor route to polycrystalline SnO2 nanowires that can be used for gas sensing under ambient conditions, J. Am. Chem. Soc. 125 (2003) 16176-16177. https://doi.org/10.1021/ja037743f

[10] A. Wei, C. Xu, X.W. Sun, W. Huang, G.-Q. Lo, Field emission from hydrothermally grown ZnO nanoinjectors, J. Disp. Technol. 4 (2008) 9-12. https://doi.org/10.1109/JDT.2007.901569

[11] Z.L. Wang, Towards self-powered nanosystems: from nanogenerators to nanopiezotronics, Adv. Funct. Mater. 18 (2008) 3553-3567. https://doi.org/10.1002/adfm.200800541

[12] Y.-C. Chang, W.-C. Yang, C.-M. Chang, P.-C. Hsu, L.-J. Chen, Controlled growth of ZnO nanopagoda arrays with varied lamination and apex angles, Cryst. Growth Des. 9 (2009) 3161-3167. https://doi.org/10.1021/cg801172h

[13] S. Cho, S.-H. Jung, J.-W. Jang, E. Oh, K.-H. Lee, Simultaneous synthesis of Al-doped ZnO nanoneedles and zinc aluminum hydroxides through use of a seed layer, Cryst. Growth Des. 8 (2008) 4553-4558. https://doi.org/10.1021/cg800593q

[14] X.Y. Kong, Z.L. Wang, Spontaneous polarization-induced nanohelixes, nanosprings, and nanorings of piezoelectric nanobelts, Nano Lett. 3 (2003) 1625-1631. https://doi.org/10.1021/nl034463p

[15] L. Mazeina, Y.N. Picard, S.M. Prokes, Controlled growth of parallel oriented ZnO nanostructural arrays on Ga2O3 nanowires, Cryst. Growth Des. 9 (2009) 1164-1169. https://doi.org/10.1021/cg800993b

[16] A. Wei, X.W. Sun, C.X. Xu, Z.L. Dong, Y. Yang, S.T. Tan, W. Huang, Growth mechanism of tubular ZnO formed in aqueous solution, Nanotechnology. 17 (2006) 1740. https://doi.org/10.1088/0957-4484/17/6/033

[17] H. Gullapalli, V.S. Vemuru, A. Kumar, A. Botello-Mendez, R. Vajtai, M. Terrones, S. Nagarajaiah, P.M. Ajayan, Flexible piezoelectric ZnO-paper nanocomposite strain sensor, Small. 6 (2010) 1641-1646. https://doi.org/10.1002/smll.201000254

[18] V. Pachauri, A. Vlandas, K. Kern, K. Balasubramanian, Site-Specific Self-Assembled Liquid-Gated ZnO Nanowire Transistors for Sensing Applications, Small. 6 (2010) 589-594. https://doi.org/10.1002/smll.200900876

[19] X. Fang, Y. Bando, U.K. Gautam, T. Zhai, H. Zeng, X. Xu, M. Liao, D. Golberg, ZnO and ZnS nanostructures: ultraviolet-light emitters, lasers, and sensors, Crit. Rev. Solid State Mater. Sci. 34 (2009) 190-223. https://doi.org/10.1080/10408430903245393

[20] J. Kim, K. Yong, Mechanism study of ZnO nanorod-bundle sensors for H2S gas sensing, J. Phys. Chem. C. 115 (2011) 7218-7224. https://doi.org/10.1021/jp110129f

[21] M. Ahmad, C. Pan, Z. Luo, J. Zhu, A single ZnO nanofiber-based highly sensitive amperometric glucose biosensor, J. Phys. Chem. C. 114 (2010) 9308-9313. https://doi.org/10.1021/jp102505g

[22] B.K. Deka, K. Kong, J. Seo, D. Kim, Y.-B. Park, H.W. Park, Controlled growth of CuO nanowires on woven carbon fibers and effects on the mechanical properties of woven carbon fiber/polyester composites, Compos. Part Appl. Sci. Manuf. 69 (2015) 56-63. https://doi.org/10.1016/j.compositesa.2014.11.001

[23] S.H. Ko, D. Lee, H.W. Kang, K.H. Nam, J.Y. Yeo, S.J. Hong, C.P. Grigoropoulos, H.J. Sung, Nanoforest of hydrothermally grown hierarchical ZnO nanowires for a high efficiency dye-sensitized solar cell, Nano Lett. 11 (2011) 666-671. https://doi.org/10.1021/nl1037962

[24] A. Hazarika, B.K. Deka, D. Kim, K. Kong, Y.-B. Park, H.W. Park, Growth of aligned ZnO nanorods on woven Kevlar® fiber and its performance in woven Kevlar® fiber/polyester composites Part A Applied science and manufacturing, (2015). https://doi.org/10.1016/j.compositesa.2015.08.022

[25] G. Amin, M.H. Asif, A. Zainelabdin, S. Zaman, O. Nur, M. Willander, Influence of pH, precursor concentration, growth time, and temperature on the morphology of ZnO nanostructures grown by the hydrothermal method, J. Nanomater. 2011 (2011). https://doi.org/10.1155/2011/269692

[26] K. Kong, B.K. Deka, S.K. Kwak, A. Oh, H. Kim, Y.-B. Park, H.W. Park, Processing and mechanical characterization of ZnO/polyester woven carbon-fiber composites with different ZnO concentrations, Compos. Part Appl. Sci. Manuf. 55 (2013) 152-160. https://doi.org/10.1016/j.compositesa.2013.08.013

[27] M.S. Al-Ruqeishi, T. Mohiuddin, B. Al-Habsi, F. Al-Ruqeishi, A. Al-Fahdi, A. Al-Khusaibi, Piezoelectric nanogenerator based on ZnO nanorods, Arab. J. Chem. 12 (2019) 5173-5179. https://doi.org/10.1016/j.arabjc.2016.12.010

[28] N.A. Salahuddin, M. El-Kemary, E.M. Ibrahim, Synthesis and characterization of ZnO nanotubes by hydrothermal method, Int J Sci Res Publ. 5 (2015) 3-6.

[29] S. Ghasaban, M. Atai, M. Imani, Simple mass production of zinc oxide nanostructures via low-temperature hydrothermal synthesis, Mater. Res. Express. 4 (2017) 035010. https://doi.org/10.1088/2053-1591/aa5dcc

[30] A. Monshi, M.R. Foroughi, M.R. Monshi, Modified Scherrer equation to estimate more accurately nano-crystallite size using XRD, World J Nano Sci Eng 2012 2 154. 160 (2012). https://doi.org/10.4236/wjnse.2012.23020

[31] M. Yilmaz, B. Bozkurt Cirak, C. Cirak, S. Aydogan, Hydrothermal growth of ZnO nanoparticles under different conditions, Philos. Mag. Lett. 96 (2016) 45-51. https://doi.org/10.1080/09500839.2016.1144938

[32] M. Yilmaz, Investigation of characteristics of ZnO: Ga nanocrystalline thin films with varying dopant content, Mater. Sci. Semicond. Process. 40 (2015) 99-106. https://doi.org/10.1016/j.mssp.2015.06.031

[33] R. Yoo, S. Yoo, D. Lee, J. Kim, S. Cho, W. Lee, Highly selective detection of dimethyl methylphosphonate (DMMP) using CuO nanoparticles/ZnO flowers heterojunction, Sens. Actuators B Chem. 240 (2017) 1099-1105. https://doi.org/10.1016/j.snb.2016.09.028

[34] G. Kwak, M. Seol, Y. Tak, K. Yong, Superhydrophobic ZnO nanowire surface: chemical modification and effects of UV irradiation, J. Phys. Chem. C. 113 (2009) 12085-12089. https://doi.org/10.1021/jp900072s

[35] G. Vijayaprasath, R. Murugan, T. Mahalingam, Y. Hayakawa, G. Ravi, Enhancement of ferromagnetic property in rare earth neodymium doped ZnO nanoparticles, Ceram. Int. 41 (2015) 10607-10615. https://doi.org/10.1016/j.ceramint.2015.04.160

[36] M. Abdelfatah, A. El-Shaer, One step to fabricate vertical submicron ZnO rod arrays by hydrothermal method without seed layer for optoelectronic devices, Mater. Lett. 210 (2018) 366-369. https://doi.org/10.1016/j.matlet.2017.09.064

[37] R. Sabry, O. AbdulAzeez, Hydrothermal growth of ZnO nano rods without catalysts in a single step, Manuf. Lett. 2 (2014) 69-73. https://doi.org/10.1016/j.mfglet.2014.02.001

[38] J. Fan, T. Li, H. Heng, Hydrothermal growth and optical properties of ZnO nanoflowers, Mater. Res. Express. 1 (2014) 045024. https://doi.org/10.1088/2053-1591/1/4/045024

[39] H. Guo, W. Zhang, Y. Sun, T. Zhou, Y. Qiu, K. Xu, B. Zhang, H. Yang, Double disks shaped ZnO microstructures synthesized by one-step CTAB assisted hydrothermal methods, Ceram. Int. 41 (2015) 10461-10466. https://doi.org/10.1016/j.ceramint.2015.04.122

[40] F. Wang, X. Qin, Z. Guo, Y. Meng, L. Yang, Y. Ming, Hydrothermal synthesis of dumbbell-shaped ZnO microstructures, Ceram. Int. 39 (2013) 8969-8973. https://doi.org/10.1016/j.ceramint.2013.04.096

[41] Y. Sun, H. Guo, W. Zhang, T. Zhou, Y. Qiu, K. Xu, B. Zhang, H. Yang, Synthesis and characterization of twinned flower-like ZnO structures grown by hydrothermal methods, Ceram. Int. 42 (2016) 9648-9652. https://doi.org/10.1016/j.ceramint.2016.03.051

[42] D. Kumar, R.S. Rai, N.K. Singh, An innovative approach to deposit ultrathin ZnO nanoflakes (2D) through hydrothermal assisted electrochemical discharge deposition and growth method, Ceram. Int. 46 (2020) 26216-26220. https://doi.org/10.1016/j.ceramint.2020.07.009

[43] A. Król, P. Pomastowski, K. Rafińska, V. Railean-Plugaru, B. Buszewski, Zinc oxide nanoparticles: Synthesis, antiseptic activity and toxicity mechanism, Adv. Colloid Interface Sci. 249 (2017) 37-52. https://doi.org/10.1016/j.cis.2017.07.033

[44] H. Cai, W. Mu, W. Liu, X. Zhang, Y. Deng, Sol-gel synthesis highly porous titanium dioxide microspheres with cellulose nanofibrils-based aerogel templates, Inorg. Chem. Commun. 51 (2015) 71-74. https://doi.org/10.1016/j.inoche.2014.11.013

[45] V. Caratto, F. Locardi, S. Alberti, S. Villa, E. Sanguineti, A. Martinelli, T. Balbi, L. Canesi, M. Ferretti, Different sol-gel preparations of iron-doped TiO2 nanoparticles: characterization, photocatalytic activity and cytotoxicity, J. Sol-Gel Sci. Technol. 80 (2016) 152-159. https://doi.org/10.1007/s10971-016-4057-5

[46] A.L. Chibac, V. Melinte, T. Buruiana, I. Mangalagiu, E.C. Buruiana, Preparation of photocrosslinked sol-gel composites based on urethane-acrylic matrix, silsesquioxane sequences, T iO2, and A g/A u Nanoparticles for use in photocatalytic applications, J. Polym. Sci. Part Polym. Chem. 53 (2015) 1189-1204. https://doi.org/10.1002/pola.27548

[47] D.M. Fernandes, R. Silva, A.W. Hechenleitner, E. Radovanovic, M.C. Melo, E.G. Pineda, Synthesis and characterization of ZnO, CuO and a mixed Zn and Cu oxide, Mater. Chem. Phys. 115 (2009) 110-115. https://doi.org/10.1016/j.matchemphys.2008.11.038

[48] S.S. Alias, A.B. Ismail, A.A. Mohamad, Effect of pH on ZnO nanoparticle properties synthesized by sol-gel centrifugation, J. Alloys Compd. 499 (2010) 231-237. https://doi.org/10.1016/j.jallcom.2010.03.174

[49] A.K. Zak, R. Yousefi, W.H. Abd Majid, M.R. Muhamad, Facile synthesis and X-ray peak broadening studies of Zn1- xMgxO nanoparticles, Ceram. Int. 38 (2012) 2059-2064. https://doi.org/10.1016/j.ceramint.2011.10.042

[50] A.K. Zak, M.E. Abrishami, W.A. Majid, R. Yousefi, S.M. Hosseini, Effects of annealing temperature on some structural and optical properties of ZnO nanoparticles

prepared by a modified sol-gel combustion method, Ceram. Int. 37 (2011) 393-398. https://doi.org/10.1016/j.ceramint.2010.08.017

[51] A.K. Zak, W.A. Majid, M.R. Mahmoudian, M. Darroudi, R. Yousefi, Starch-stabilized synthesis of ZnO nanopowders at low temperature and optical properties study, Adv. Powder Technol. 24 (2013) 618-624. https://doi.org/10.1016/j.apt.2012.11.008

[52] M.H. Habibi, B. Karimi, Preparation of nanostructure CuO/ZnO mixed oxide by sol-gel thermal decomposition of a CuCO3 and ZnCO3: TG, DTG, XRD, FESEM and DRS investigations, J. Ind. Eng. Chem. 20 (2014) 925-929. https://doi.org/10.1016/j.jiec.2013.06.024

[53] J.L. Konne, B.O. Christopher, Sol-gel syntheses of zinc oxide and hydrogenated zinc oxide (ZnO: H) phases, J. Nanotechnol. 2017 (2017). https://doi.org/10.1155/2017/5219850

[54] S. Jurablu, M. Farahmandjou, T.P. Firoozabadi, Sol-gel synthesis of zinc oxide (ZnO) nanoparticles: study of structural and optical properties, J. Sci. Islam. Repub. Iran. 26 (2015) 281-285.

[55] M. Acosta-Humánez, L. Montes-Vides, O. Almanza-Montero, Sol-gel synthesis of zinc oxide nanoparticle at three different temperatures and its characterization via XRD, IR and EPR, Dyna. 83 (2016) 224-228. https://doi.org/10.15446/dyna.v83n195.50833

[56] P. Lu, W. Zhou, Y. Li, J. Wang, P. Wu, Abnormal room temperature ferromagnetism in CuO/ZnO nanocomposites via hydrothermal method, Appl. Surf. Sci. 399 (2017) 396-402. https://doi.org/10.1016/j.apsusc.2016.12.113

[57] A.R. Marlinda, N.M. Huang, M.R. Muhamad, M.N. An'Amt, B.Y.S. Chang, N. Yusoff, I. Harrison, H.N. Lim, C.H. Chia, S.V. Kumar, Highly efficient preparation of ZnO nanorods decorated reduced graphene oxide nanocomposites, Mater. Lett. 80 (2012) 9-12. https://doi.org/10.1016/j.matlet.2012.04.061

[58] C.B. Ong, L.Y. Ng, A.W. Mohammad, A review of ZnO nanoparticles as solar photocatalysts: Synthesis, mechanisms and applications, Renew. Sustain. Energy Rev. 81 (2018) 536-551. https://doi.org/10.1016/j.rser.2017.08.020

[59] A.K. Zak, R. Razali, W.H. Abd Majid, M. Darroudi, Synthesis and characterization of a narrow size distribution of zinc oxide nanoparticles, Int. J. Nanomedicine. 6 (2011) 1399. https://doi.org/10.2147/IJN.S19693

[60] M. Saranya, R. Ramachandran, F. Wang, Graphene-zinc oxide (G-ZnO) nanocomposite for electrochemical supercapacitor applications, J. Sci. Adv. Mater. Devices. 1 (2016) 454-460. https://doi.org/10.1016/j.jsamd.2016.10.001

[61] J. Wu, X. Shen, L. Jiang, K. Wang, K. Chen, Solvothermal synthesis and characterization of sandwich-like graphene/ZnO nanocomposites, Appl. Surf. Sci. 256 (2010) 2826-2830. https://doi.org/10.1016/j.apsusc.2009.11.034

[62] J. Lian, Y. Liang, F. Kwong, Z. Ding, D.H. Ng, Template-free solvothermal synthesis of ZnO nanoparticles with controllable size and their size-dependent optical properties, Mater. Lett. 66 (2012) 318-320. https://doi.org/10.1016/j.matlet.2011.09.007

[63] A. Matei, I. Cernica, O. Cadar, C. Roman, V. Schiopu, Synthesis and characterization of ZnO-polymer nanocomposites, Int. J. Mater. Form. 1 (2008) 767-770. https://doi.org/10.1007/s12289-008-0288-5

[64] S.S. Kumar, P. Venkateswarlu, V.R. Rao, G.N. Rao, Synthesis, characterization and optical properties of zinc oxide nanoparticles, Int. Nano Lett. 3 (2013) 1-6. https://doi.org/10.1186/2228-5326-3-1

[65] S. Singhal, J. Kaur, T. Namgyal, R. Sharma, Cu-doped ZnO nanoparticles: synthesis, structural and electrical properties, Phys. B Condens. Matter. 407 (2012) 1223-1226. https://doi.org/10.1016/j.physb.2012.01.103

[66] F. Gu, D. You, Z. Wang, D. Han, G. Guo, Improvement of gas-sensing property by defect engineering in microwave-assisted synthesized 3D ZnO nanostructures, Sens. Actuators B Chem. 204 (2014) 342-350. https://doi.org/10.1016/j.snb.2014.07.080

[67] K.D. Bhatte, P. Tambade, S. Fujita, M. Arai, B.M. Bhanage, Microwave-assisted additive free synthesis of nanocrystalline zinc oxide, Powder Technol. 203 (2010) 415-418. https://doi.org/10.1016/j.powtec.2010.05.036

[68] S. Soumya, A.P. Mohamed, L. Paul, K. Mohan, S. Ananthakumar, Near IR reflectance characteristics of PMMA/ZnO nanocomposites for solar thermal control interface films, Sol. Energy Mater. Sol. Cells. 125 (2014) 102-112. https://doi.org/10.1016/j.solmat.2014.02.033

[69] R. He, B. Tang, C. Ton-That, M. Phillips, T. Tsuzuki, Physical structure and optical properties of Co-doped ZnO nanoparticles prepared by co-precipitation, J. Nanoparticle Res. 15 (2013) 1-8. https://doi.org/10.1007/s11051-013-2030-6

[70] D. Sharma, S. Sharma, B.S. Kaith, J. Rajput, M. Kaur, Synthesis of ZnO nanoparticles using surfactant free in-air and microwave method, Appl. Surf. Sci. 257 (2011) 9661-9672. https://doi.org/10.1016/j.apsusc.2011.06.094

[71] R. Dobrucka, J. D\lugaszewska, Biosynthesis and antibacterial activity of ZnO nanoparticles using Trifolium pratense flower extract, Saudi J. Biol. Sci. 23 (2016) 517-523. https://doi.org/10.1016/j.sjbs.2015.05.016

[72] B. Kumar, K. Smita, L. Cumbal, A. Debut, Green approach for fabrication and applications of zinc oxide nanoparticles, Bioinorg. Chem. Appl. 2014 (2014). https://doi.org/10.1155/2014/523869

[73] K. Lingaraju, H. Raja Naika, K. Manjunath, R.B. Basavaraj, H. Nagabhushana, G. Nagaraju, D. Suresh, Biogenic synthesis of zinc oxide nanoparticles using Ruta graveolens (L.) and their antibacterial and antioxidant activities, Appl. Nanosci. 6 (2016) 703-710. https://doi.org/10.1007/s13204-015-0487-6

[74] A.K. Mittal, Y. Chisti, U.C. Banerjee, Synthesis of metallic nanoparticles using plant extracts, Biotechnol. Adv. 31 (2013) 346-356. https://doi.org/10.1016/j.biotechadv.2013.01.003

[75] P. Mohanpuria, N.K. Rana, S.K. Yadav, Biosynthesis of nanoparticles: technological concepts and future applications, J. Nanoparticle Res. 10 (2008) 507-517. https://doi.org/10.1007/s11051-007-9275-x

[76] N. Bala, S. Saha, M. Chakraborty, M. Maiti, S. Das, R. Basu, P. Nandy, Green synthesis of zinc oxide nanoparticles using Hibiscus subdariffa leaf extract: effect of temperature on synthesis, anti-bacterial activity and anti-diabetic activity, RSC Adv. 5 (2015) 4993-5003. https://doi.org/10.1039/C4RA12784F

[77] S.J. Lakshmi, R.R.S. Bai, H. Sharanagouda, U.K. Nidoni, A review study of zinc oxide nanoparticles synthesis from plant extracts, Green Chem Technol Lett. 3 (2017) 26-37. https://doi.org/10.18510/gctl.2017.321

[78] D.K. Slman, R.D.A. Jalill, A.N. Abd, Biosynthesis of zinc oxide nanoparticles by hot aqueous extract of Allium sativum plants, J. Pharm. Sci. Res. 10 (2018) 1590-1596.

[79] S. Azizi, M.B. Ahmad, F. Namvar, R. Mohamad, Green biosynthesis and characterization of zinc oxide nanoparticles using brown marine macroalga Sargassum muticum aqueous extract, Mater. Lett. 116 (2014) 275-277. https://doi.org/10.1016/j.matlet.2013.11.038

[80] H.A. Salam, R. Sivaraj, R. Venckatesh, Green synthesis and characterization of zinc oxide nanoparticles from Ocimum basilicum L. var. purpurascens Benth.-Lamiaceae leaf extract, Mater. Lett. 131 (2014) 16-18. https://doi.org/10.1016/j.matlet.2014.05.033

[81] H. Mirzaei, M. Darroudi, Zinc oxide nanoparticles: Biological synthesis and biomedical applications, Ceram. Int. 43 (2017) 907-914. https://doi.org/10.1016/j.ceramint.2016.10.051

[82] P. Jamdagni, P. Khatri, J.S. Rana, Green synthesis of zinc oxide nanoparticles using flower extract of Nyctanthes arbor-tristis and their antifungal activity, J. King Saud Univ.-Sci. 30 (2018) 168-175. https://doi.org/10.1016/j.jksus.2016.10.002

[83] C. Vidya, S. Hiremath, M.N. Chandraprabha, M.L. Antonyraj, I.V. Gopal, A. Jain, K. Bansal, Green synthesis of ZnO nanoparticles by Calotropis gigantea, Int J Curr Eng Technol. 1 (2013) 118-120.

[84] G. Sangeetha, S. Rajeshwari, R. Venckatesh, Green synthesis of zinc oxide nanoparticles by aloe barbadensis miller leaf extract: Structure and optical properties, Mater. Res. Bull. 46 (2011) 2560-2566. https://doi.org/10.1016/j.materresbull.2011.07.046

[85] S. Baskoutas, Zinc oxide nanostructures: Synthesis and characterization, Materials. 11 (2018) 873. https://doi.org/10.3390/ma11060873

[86] R. Brayner, S.A. Dahoumane, C. Yéprémian, C. Djediat, M. Meyer, A. Couté, F. Fiévet, ZnO nanoparticles: synthesis, characterization, and ecotoxicological studies, Langmuir. 26 (2010) 6522-6528. https://doi.org/10.1021/la100293s

[87] T. Gordon, B. Perlstein, O. Houbara, I. Felner, E. Banin, S. Margel, Synthesis and characterization of zinc/iron oxide composite nanoparticles and their antibacterial properties, Colloids Surf. Physicochem. Eng. Asp. 374 (2011) 1-8. https://doi.org/10.1016/j.colsurfa.2010.10.015

[88] S. Sharma, R. Uttam, A. Sarika Bharti, K.N. Uttam, Interaction of zinc oxide and copper oxide nanoparticles with chlorophyll: a fluorescence quenching study, Anal. Lett. 52 (2019) 1539-1557. https://doi.org/10.1080/00032719.2018.1556277

[89] L.-H. Li, J.-C. Deng, H.-R. Deng, Z.-L. Liu, L. Xin, Synthesis and characterization of chitosan/ZnO nanoparticle composite membranes, Carbohydr. Res. 345 (2010) 994-998. https://doi.org/10.1016/j.carres.2010.03.019

[90] A.C. Mohan, B. Renjanadevi, Preparation of zinc oxide nanoparticles and its characterization using scanning electron microscopy (SEM) and X-ray diffraction (XRD), Procedia Technol. 24 (2016) 761-766. https://doi.org/10.1016/j.protcy.2016.05.078

[91] O.-R. Vasile, E. Andronescu, C. Ghitulica, B.S. Vasile, O. Oprea, E. Vasile, R. Trusca, Synthesis and characterization of nanostructured zinc oxide particles synthesized by the pyrosol method, J. Nanoparticle Res. 14 (2012) 1-13. https://doi.org/10.1007/s11051-012-1269-7

[92] D. Fu, G. Han, Y. Chang, J. Dong, The synthesis and properties of ZnO-graphene nano hybrid for photodegradation of organic pollutant in water, Mater. Chem. Phys. 132 (2012) 673-681. https://doi.org/10.1016/j.matchemphys.2011.11.085

[93] M.A. Ashraf, J. Wiener, A. Farooq, J. Saskova, M.T. Noman, Development of maghemite glass fibre nanocomposite for adsorptive removal of methylene blue, Fibers Polym. 19 (2018) 1735-1746. https://doi.org/10.1007/s12221-018-8264-2

[94] M.T. Noman, J. Militky, J. Wiener, J. Saskova, M.A. Ashraf, H. Jamshaid, M. Azeem, Sonochemical synthesis of highly crystalline photocatalyst for industrial applications, Ultrasonics. 83 (2018) 203-213. https://doi.org/10.1016/j.ultras.2017.06.012

[95] M.T. Noman, J. Wiener, J. Saskova, M.A. Ashraf, M. Vikova, H. Jamshaid, P. Kejzlar, In-situ development of highly photocatalytic multifunctional nanocomposites by ultrasonic acoustic method, Ultrason. Sonochem. 40 (2018) 41-56. https://doi.org/10.1016/j.ultsonch.2017.06.026

[96] C.C. Vidyasagar, Y.A. Naik, T.G. Venkatesh, R. Viswanatha, Solid-state synthesis and effect of temperature on optical properties of Cu-ZnO, Cu-CdO and CuO nanoparticles, Powder Technol. 214 (2011) 337-343. https://doi.org/10.1016/j.powtec.2011.08.025

[97] Q.-P. Luo, X.-Y. Yu, B.-X. Lei, H.-Y. Chen, D.-B. Kuang, C.-Y. Su, Reduced graphene oxide-hierarchical ZnO hollow sphere composites with enhanced photocurrent and photocatalytic activity, J. Phys. Chem. C. 116 (2012) 8111-8117. https://doi.org/10.1021/jp2113329

[98] M.T. Noman, M. Petru, N. Amor, P. Louda, Thermophysiological comfort of zinc oxide nanoparticles coated woven fabrics, Sci. Rep. 10 (2020) 1-12. https://doi.org/10.1038/s41598-019-56847-4

[99] H. Benhebal, M. Chaib, T. Salmon, J. Geens, A. Leonard, S.D. Lambert, M. Crine, B. Heinrichs, Photocatalytic degradation of phenol and benzoic acid using zinc oxide powders prepared by the sol-gel process, Alex. Eng. J. 52 (2013) 517-523. https://doi.org/10.1016/j.aej.2013.04.005

[100] S. Yue, Z. Yan, Y. Shi, G. Ran, Synthesis of zinc oxide nanotubes within ultrathin anodic aluminum oxide membrane by sol-gel method, Mater. Lett. 98 (2013) 246-249. https://doi.org/10.1016/j.matlet.2013.02.037

[101] E. Asikuzun, O. Ozturk, L. Arda, C. Terzioglu, Preparation, growth and characterization of nonvacuum Cu-doped ZnO thin films, J. Mol. Struct. 1165 (2018) 1-7. https://doi.org/10.1016/j.molstruc.2018.03.053

[102] K. Choi, T. Kang, S.-G. Oh, Preparation of disk shaped ZnO particles using surfactant and their PL properties, Mater. Lett. 75 (2012) 240-243. https://doi.org/10.1016/j.matlet.2012.02.031

[103] R.-M. Ko, Y.-R. Lin, C.-Y. Chen, P.-F. Tseng, S.-J. Wang, Facilitating epitaxial growth of ZnO films on patterned GaN layers: A solution-concentration-induced successive lateral growth mechanism, Curr. Appl. Phys. 18 (2018) 1-11. https://doi.org/10.1016/j.cap.2017.11.003

[104] B. Gong, T. Shi, T. Liao, X. Li, J. Huang, T. Zhou, Z. Tang, UV irradiation assisted growth of ZnO nanowires on optical fiber surface, Appl. Surf. Sci. 406 (2017) 294-300. https://doi.org/10.1016/j.apsusc.2017.02.153

[105] J. Zhang, J. Wang, S. Zhou, K. Duan, B. Feng, J. Weng, H. Tang, P. Wu, Ionic liquid-controlled synthesis of ZnO microspheres, J. Mater. Chem. 20 (2010) 9798-9804. https://doi.org/10.1039/c0jm01970d

[106] J.J. Schneider, R.C. Hoffmann, J. Engstler, A. Klyszcz, E. Erdem, P. Jakes, R.-A. Eichel, L. Pitta-Bauermann, J. Bill, Synthesis, characterization, defect chemistry, and FET properties of microwave-derived nanoscaled zinc oxide, Chem. Mater. 22 (2010) 2203-2212. https://doi.org/10.1021/cm902300q

[107] A.S. Lanje, S.J. Sharma, R.S. Ningthoujam, J.-S. Ahn, R.B. Pode, Low temperature dielectric studies of zinc oxide (ZnO) nanoparticles prepared by precipitation method, Adv. Powder Technol. 24 (2013) 331-335. https://doi.org/10.1016/j.apt.2012.08.005

[108] Z.M. Khoshhesab, M. Sarfaraz, Z. Houshyar, Influences of urea on preparation of zinc oxide nanostructures through chemical precipitation in ammonium hydrogencarbonate solution, Synth. React. Inorg. Met.-Org. Nano-Met. Chem. 42 (2012) 1363-1368. https://doi.org/10.1080/15533174.2012.680119

[109] W. Jia, S. Dang, H. Liu, Z. Zhang, C. Yu, X. Liu, B. Xu, Evidence of the formation mechanism of ZnO in aqueous solution, Mater. Lett. 82 (2012) 99-101. https://doi.org/10.1016/j.matlet.2012.05.013

[110] K.M. Kumar, B.K. Mandal, E.A. Naidu, M. Sinha, K.S. Kumar, P.S. Reddy, Synthesis and characterisation of flower shaped zinc oxide nanostructures and its antimicrobial activity, Spectrochim. Acta. A. Mol. Biomol. Spectrosc. 104 (2013) 171-174. https://doi.org/10.1016/j.saa.2012.11.025

[111] Y. Wang, C. Zhang, S. Bi, G. Luo, Preparation of ZnO nanoparticles using the direct precipitation method in a membrane dispersion micro-structured reactor, Powder Technol. 202 (2010) 130-136. https://doi.org/10.1016/j.powtec.2010.04.027

[112] C.I. Ezeh, X. Yang, J. He, C. Snape, X.M. Cheng, Correlating ultrasonic impulse and addition of ZnO promoter with CO2 conversion and methanol selectivity of CuO/ZrO2 catalysts, Ultrason. Sonochem. 42 (2018) 48-56. https://doi.org/10.1016/j.ultsonch.2017.11.013

[113] T. Bhuyan, K. Mishra, M. Khanuja, R. Prasad, A. Varma, Biosynthesis of zinc oxide nanoparticles from Azadirachta indica for antibacterial and photocatalytic applications, Mater. Sci. Semicond. Process. 32 (2015) 55-61. https://doi.org/10.1016/j.mssp.2014.12.053

[114] A. Raja, S. Ashokkumar, R.P. Marthandam, J. Jayachandiran, C.P. Khatiwada, K. Kaviyarasu, R.G. Raman, M. Swaminathan, Eco-friendly preparation of zinc oxide nanoparticles using Tabernaemontana divaricata and its photocatalytic and antimicrobial activity, J. Photochem. Photobiol. B. 181 (2018) 53-58. https://doi.org/10.1016/j.jphotobiol.2018.02.011

[115] S.A. Khan, F. Noreen, S. Kanwal, A. Iqbal, G. Hussain, Green synthesis of ZnO and Cu-doped ZnO nanoparticles from leaf extracts of Abutilon indicum, Clerodendrum infortunatum, Clerodendrum inerme and investigation of their biological and photocatalytic activities, Mater. Sci. Eng. C. 82 (2018) 46-59. https://doi.org/10.1016/j.msec.2017.08.071

Materials Research Forum LLC
https://doi.org/10.21741/9781644902394-4

[116] F.T. Thema, E. Manikandan, M.S. Dhlamini, M. Maaza, Green synthesis of ZnO nanoparticles via Agathosma betulina natural extract, Mater. Lett. 161 (2015) 124-127. https://doi.org/10.1016/j.matlet.2015.08.052

[117] M. Sundrarajan, S. Ambika, K. Bharathi, Plant-extract mediated synthesis of ZnO nanoparticles using Pongamia pinnata and their activity against pathogenic bacteria, Adv. Powder Technol. 26 (2015) 1294-1299. https://doi.org/10.1016/j.apt.2015.07.001

[118] H.-T. Wang, B.S. Kang, F. Ren, L.C. Tien, P.W. Sadik, D.P. Norton, S.J. Pearton, J. Lin, Hydrogen-selective sensing at room temperature with ZnO nanorods, Appl. Phys. Lett. 86 (2005) 243503. https://doi.org/10.1063/1.1949707

[119] L.C. Tien, S.J. Pearton, D.P. Norton, F. Ren, Synthesis and microstructure of vertically aligned ZnO nanowires grown by high-pressure-assisted pulsed-laser deposition, J. Mater. Sci. 43 (2008) 6925-6932. https://doi.org/10.1007/s10853-008-2988-0

[120] C.S. Rout, A.R. Raju, A. Govindaraj, C.N.R. Rao, Hydrogen sensors based on ZnO nanoparticles, Solid State Commun. 138 (2006) 136-138. https://doi.org/10.1016/j.ssc.2006.02.016

[121] Z. Fan, D. Wang, P.-C. Chang, W.-Y. Tseng, J.G. Lu, ZnO nanowire field-effect transistor and oxygen sensing property, Appl. Phys. Lett. 85 (2004) 5923-5925. https://doi.org/10.1063/1.1836870

[122] Z. Fan, J.G. Lu, Gate-refreshable nanowire chemical sensors, Appl. Phys. Lett. 86 (2005) 123510. https://doi.org/10.1063/1.1883715

[123] Q. Wan, Q.H. Li, Y.J. Chen, T.-H. Wang, X.L. He, J.P. Li, C.L. Lin, Fabrication and ethanol sensing characteristics of ZnO nanowire gas sensors, Appl. Phys. Lett. 84 (2004) 3654-3656. https://doi.org/10.1063/1.1738932

[124] C. Xiangfeng, J. Dongli, A.B. Djurišic, Y.H. Leung, Gas-sensing properties of thick film based on ZnO nano-tetrapods, Chem. Phys. Lett. 401 (2005) 426-429. https://doi.org/10.1016/j.cplett.2004.11.091

[125] M. Yang, D. Wang, L. Peng, Q. Zhao, Y. Lin, X. Wei, Surface photocurrent gas sensor with properties dependent on Ru (dcbpy) 2 (NCS) 2-sensitized ZnO nanoparticles, Sens. Actuators B Chem. 117 (2006) 80-85. https://doi.org/10.1016/j.snb.2005.11.014

[126] A. Wei, Z. Wang, L.-H. Pan, W.-W. Li, L. Xiong, X.-C. Dong, W. Huang, Room-temperature NH3 gas sensor based on hydrothermally grown ZnO nanorods, Chin. Phys. Lett. 28 (2011) 080702. https://doi.org/10.1088/0256-307X/28/8/080702

[127] J. Bao, M.A. Zimmler, F. Capasso, X. Wang, Z.F. Ren, Broadband ZnO single-nanowire light-emitting diode, Nano Lett. 6 (2006) 1719-1722. https://doi.org/10.1021/nl061080t

[128] X. Dong, Y. Liu, K. Huang, W. Zhao, Y. Ye, X. Xia, Y. Zhang, J. Wang, B. Zhang, G. Du, Study on the p-MgZnO/i-ZnO/n-MgZnO light-emitting diode fabricated by MOCVD, J. Phys. Appl. Phys. 42 (2009) 235101. https://doi.org/10.1088/0022-3727/42/23/235101

[129] D.G. Thomas, Interstitial zinc in zinc oxide, J. Phys. Chem. Solids. 3 (1957) 229-237. https://doi.org/10.1016/0022-3697(57)90027-6

[130] R. Könenkamp, R.C. Word, M. Godinez, Ultraviolet electroluminescence from ZnO/polymer heterojunction light-emitting diodes, Nano Lett. 5 (2005) 2005-2008. https://doi.org/10.1021/nl051501r

[131] J.F. Flores, Engineering 3D Nanostructures for a Multitude of Applications, University of California, Merced, 2015.

[132] J.-H. Choy, E.-S. Jang, J.-H. Won, J.-H. Chung, D.-J. Jang, Y.-W. Kim, Soft solution route to directionally grown ZnO nanorod arrays on Si wafer; room-temperature ultraviolet laser, Adv. Mater. 15 (2003) 1911-1914. https://doi.org/10.1002/adma.200305327

[133] J. Suehiro, N. Nakagawa, S. Hidaka, M. Ueda, K. Imasaka, M. Higashihata, T. Okada, M. Hara, Dielectrophoretic fabrication and characterization of a ZnO nanowire-based UV photosensor, Nanotechnology. 17 (2006) 2567. https://doi.org/10.1088/0957-4484/17/10/021

[134] I. Bedja, P.V. Kamat, X. Hua, A.G. Lappin, S. Hotchandani, Photosensitization of Nanocrystalline ZnO Films by Bis (2, 2 '-bipyridine)(2, 2 '-bipyridine-4, 4 '-dicarboxylic acid) ruthenium (II), Langmuir. 13 (1997) 2398-2403. https://doi.org/10.1021/la9620115

[135] K. Keis, C. Bauer, G. Boschloo, A. Hagfeldt, K. Westermark, H. Rensmo, H. Siegbahn, Nanostructured ZnO electrodes for dye-sensitized solar cell applications, J. Photochem. Photobiol. Chem. 148 (2002) 57-64. https://doi.org/10.1016/S1010-6030(02)00039-4

[136] K. Keis, E. Magnusson, H. Lindström, S.-E. Lindquist, A. Hagfeldt, A 5% efficient photoelectrochemical solar cell based on nanostructured ZnO electrodes, Sol. Energy Mater. Sol. Cells. 73 (2002) 51-58. https://doi.org/10.1016/S0927-0248(01)00110-6

[137] R. Katoh, A. Furube, Y. Tamaki, T. Yoshihara, M. Murai, K. Hara, S. Murata, H. Arakawa, M. Tachiya, Microscopic imaging of the efficiency of electron injection from excited sensitizer dye into nanocrystalline ZnO film, J. Photochem. Photobiol. Chem. 166 (2004) 69-74. https://doi.org/10.1016/j.jphotochem.2004.04.038

[138] R. Katoh, A. Furube, T. Yoshihara, K. Hara, G. Fujihashi, S. Takano, S. Murata, H. Arakawa, M. Tachiya, Efficiencies of electron injection from excited N3 dye into nanocrystalline semiconductor (ZrO2, TiO2, ZnO, Nb2O5, SnO2, In2O3) films, J. Phys. Chem. B. 108 (2004) 4818-4822. https://doi.org/10.1021/jp031260g

[139] A. Furube, R. Katoh, K. Hara, S. Murata, H. Arakawa, M. Tachiya, Ultrafast stepwise electron injection from photoexcited Ru-complex into nanocrystalline ZnO film via intermediates at the surface, J. Phys. Chem. B. 107 (2003) 4162-4166. https://doi.org/10.1021/jp034039c

[140] U. Bach, D. Lupo, P. Comte, J.-E. Moser, F. Weissörtel, J. Salbeck, H. Spreitzer, M. Grätzel, Solid-state dye-sensitized mesoporous TiO2 solar cells with high photon-to-electron conversion efficiencies, Nature. 395 (1998) 583-585. https://doi.org/10.1038/26936

[141] L. Schmidt-Mende, M. Grätzel, TiO2 pore-filling and its effect on the efficiency of solid-state dye-sensitized solar cells, Thin Solid Films. 500 (2006) 296-301. https://doi.org/10.1016/j.tsf.2005.11.020

[142] J.E. Kroeze, N. Hirata, L. Schmidt-Mende, C. Orizu, S.D. Ogier, K. Carr, M. Grätzel, J.R. Durrant, Parameters influencing charge separation in solid-state dye-sensitized solar cells using novel hole conductors, Adv. Funct. Mater. 16 (2006) 1832-1838. https://doi.org/10.1002/adfm.200500748

[143] D.C. Olson, J. Piris, R.T. Collins, S.E. Shaheen, D.S. Ginley, Hybrid photovoltaic devices of polymer and ZnO nanofiber composites, Thin Solid Films. 496 (2006) 26-29. https://doi.org/10.1016/j.tsf.2005.08.179

[144] P. Ravirajan, A.M. Peiró, M.K. Nazeeruddin, M. Graetzel, D.D. Bradley, J.R. Durrant, J. Nelson, Hybrid polymer/zinc oxide photovoltaic devices with vertically oriented ZnO nanorods and an amphiphilic molecular interface layer, J. Phys. Chem. B. 110 (2006) 7635-7639. https://doi.org/10.1021/jp0571372

Materials Research Forum LLC
https://doi.org/10.21741/9781644902394-5

Chapter 5

ZnO Hybrid Nanostructures for Solar Cell Applications

Naveen Kumar[1*,] Nupur Aggarwal[1], Payal Patial[1], Navdeep Sharma[3], Anu Kapoor[1], Ranvir Singh Panwar[2]

[1]Department of Physics, Chandigarh University, Gharuan, Mohali, India, 140413

[2]Department of Metallurgical and Materials Engineering, Punjab Engineering College (Deemed to be University), Chandigarh, India, 160012

[3]Department of Physics and Astronomical Sciences, Central University of Jammu, Jammu, India, 181143

*naveensethi99@gmail.com

Abstract

Owing to the emerging demand for energy generation by making use of efficient solar cells involving zinc oxide (ZnO) as a potential candidate due to its interesting features such as transparency, high electron mobility, conductivity, availability at a low price, and stability against light corrosion. ZnO exists with various morphologies depending upon the preparation and synthesis conditions and is then applied as the semiconducting layer on the conducting surface to act as a photo anode. Besides, ZnO nanostructure can be directly grown on the substrate to get unique electron pathways by suppressing influential surface sites. The present chapter displays recent progress and inventions of ZnO nanostructures in revolutionised and promising solar cell applications such as hybrid organic/inorganic hetero-junction solar cells, sensitized solar cells and perovskite sensitized solar cells. Extraordinary and remarkable advancements in the field of solar cell architecture are portrayed and highlighted for obtaining high efficiency for power conversion and achieving stability in operation. The role of ZnO for further improvement and future perspectives for practical application at a large scale are also addressed.

Keywords

ZnO, Perovskite Solar Cell, Dye-Sensitized Solar Cell, Hetro-Junction Solar Cell

Contents

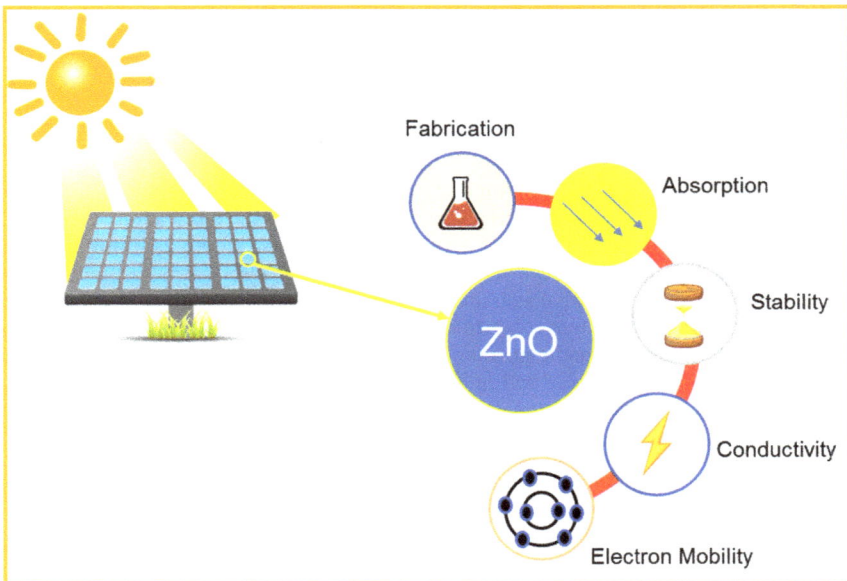

Figure 1: Interesting features of ZnO for solar cell applications.

1.   Introduction

The high rate of consumption of energy and growth in industry establishment rate has increased the demand for renewable energy resources in the entire world. In India, we still rely on coal and petroleum as a primary source of energy which fulfills about 50% demand of energy in our country. Among renewable energy resources, solar energy earns the maximum interest in the entire universe. This increase in interest in using renewable energy resources as a source of energy gain the interest of researchers to propose or find a way for the production of such kind of clean energy. In the last few years lot of discussions have been held regarding the technical as well as economic aspects of solar energy or specifically the usage of the photovoltaic cell for the generation off energy by our Indian government [1]. Power consumption around the globe has reached 16 terawatts and is expected to quadruple by 2050 [2]. The rapidly increasing demand for energy has fuelled solar cell research and innovation, which has been hastened by growing worries about carbon emissions from fossil fuel-based sources of energy. Therefore, solar cells have received a lot more attention in recent years in terms of scientific advancement. Despite the fact that silicon-based p–n junction solar cells seem to be the widely utilized solar cell system due to their high performance and dependability, their massive cost has restricted their applications. Solar cells, in addition to conventional silicon-based solar cells, have been extensively studied in the last decade. Solar cells provide a tonne of benefits, including light weight, versatility, and the prospect for low cost as designed to be compatible with a variety of solution processing processes [3].

The available range of solar cells was categorized in three generations; among which first-generation solar cells were made up of Silicon as a base layer, whereas in the second generation Si was replaced by thin films of various similar types of materials. The third generation used the complex as well as advanced materials for the fabrication of solar cells such as hetero-junction, sensitized and perovskite solar cells (Figure 2). Emergent solar cell techniques, as perovskite, dye-sensitized, organic, quantum dot, and multifunction were crested in response to the constraints of performance and flexibility. Neither of the currently available solar cell technologies has come close to meeting the power conversion limits of 90%. The energy band gap, along with transmitting and thermalization losses, is the fundamental causes of solar cell inefficiencies. This one is intimately linked to the active material's attributes. Furthermore, the intrinsic stabilization of all these materials has an impact on the solar cell systems endurance. The corresponding energy levels for excellent solar spectrum absorption efficiency, high carrier mobility, strong conductivity, and high rate removal of excited charge carriers are desirable attributes of charge carrier materials for solar cell applications.

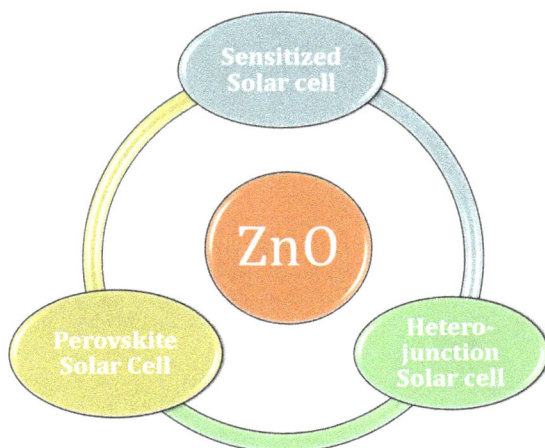

Figure 2: Pictorial representation of ZnO based solar cells.

Though first and second-generation solar cells exhibit great potential; but the fabrication of these solar cells is complex, expensive, and time-consuming. Also, Silicon is not an environment-friendly material when disposed off. So there is a requirement of using some advanced materials in the synthesis of solar cells that overcomes the drawback of Si-based solar cells [4-5].

Among II–VI group members, ZnO is a semiconducting material that shows higher conductivity, electron mobility, high efficiency, easy availability, and comparatively cheap cost making it a suitable candidate for solar cell applications. ZnO has a wide band gap of the order of 3.3eV with a large Bohr excitonic radius ($\simeq$ 2.34 nm) that absorbs light in the UV region. ZnO can be combined with materials having a small energy band gap to enhance the range of their light absorption tendency in the visible band. Further nanostructured ZnO increases the light absorption tendency because of an increase in exposed surface area that allows greater dye absorption. In nano-regime, ZnO has a variety of morphologies depending on the synthesis technique adopted such as nanoparticles, nano-spheres, nano-flowers, nano-rods, and so on.

Gratezel and Oregan [6] were the first to propose the idea of a dye-sensitized solar cell in 1991, which stands out to be a great alternative for Si-based solar cell. Ru-based dye sensitized solar cell achieve an efficiency above 10%.

The benefits and drawbacks of various ZnO nanostructures in photocatalytic applications are shown in Table 1. Nanostructure possesses a large surface area, as a result, they become a preferred choice in solar photocatalysis as more pollutants may be easily absorbed and a faster photodegradation can be accomplished. Nanowires with reduced crystallinity are proved advantageous in photocatalytic applications. This could be because hydroxyl groups are attached to vacancy and which promotes the trapping of photo-induced electron-hole pairs, hence boosting their separation [7].

Table 1: Advantages and disadvantages of varying morphology of ZnO solar cell.

Types of ZnO nanostructures	Advantages	Disadvantages
Nanowires	It is easy to carry out the growth of nanowires on mostly available substrates Have more effective surface area in comparison to thin films Post-treatment to remove catalyst is not important Have more number of defects and overall crystallinity is low	Condition required for growth mechanism is highly constrained The surface area is small in comparison
Nano-thin film	Coating is only possible for specific substrates No need of post-treatment to remove the catalyst	Small surface area limits the performance
Nanoparticles	Easy suspension in solution is possible Large surface area leads to high performance	Due to easy agglomeration in solution effective surface has been reduced. Post-treatment is mandatory for the removal of catalysts. It is hard to recover catalysts completely

1.1 Importance of ZnO in solar cell fabrication

ZnO is the most commonly used material for a wide range of applications, including sensors, surface coatings, porous ceramics, photo detectors, piezoelectrics, and in addition to solar cells, due to its exceptional properties. ZnO is desirable for solar cell applications because of its high electron mobility, which ranges from 205–300 for bulk [8-9] ZnO and 1000 cm^2 V^1 s^{-1} for nanorods ZnO [10]. ZnO is also classified as a polymorphic, with varying structures according to the technique of synthesis. Nanospheres, nanowires, nanorods, nanoflowers, nanotubes, nanocrystals, and 3D nanostructures (core–shell) are all part of ZnO's nanomorphology. Likewise, the electron diffusion coefficient for bulk ZnO is 5.2 and for nanoparticles, ZnO is 1.7×10^{-4} cm^2 s^{-1} [11]. Green synthesis employing microorganisms, hydrothermal, sol–gel, electrochemistry, inkjet printing, atomic layer deposition, and sputtering technology are among the many techniques utilized to generate

ZnO nanomaterials via various biological/physical/chemical routes. The choice of ZnO fabrication process is mostly influenced by the intended nanostructure dimensions. In the fabrication of 2D nanostructures, for example, the sol-gel technique and PECVD have been used. In comparison to the sol-gel approach, the PECVD technique has demonstrated the requirement of lower temperatures for the generation of films [12]. The impacts of various fabrication methods on zinc oxide properties are shown in Table 2. The most recent implementation of ZnO in solar cell technology is described in depth in this article. Diverse solar cell topologies, such as sensitised solar cells, will be analyzed to study the influence of synthesis processes on ZnO attributes, including all those integrated with potential perovskite materials. These methods involve the use of various ZnO nanostructures, deposition and post-treatment methods, and the incorporation of dopant compounds.

Table 2: Effect of different synthesis procedures/ precursors on surface morphology and particle size of ZnO-based solar cell.

Fabrication	Morphology	Starting Materials	Particle Size (nm)	Ref
Electrochemical	Combination of spherical and cylindrical particles	Zn electrode, oxalic acid dihydrate purified, potassium chloride, sodium hydroxide, and nitric acid.	$L_{cylindrical}$ 150–200 $D_{spherical}$ 50–100	[13]
Sonochemical	Flakes shape	Zinc nitrate hexahydrate, potassium hydroxide, and cetyltrimethylammonium bromide	200–400 wide and a few nm thick	[14]
Co-precipitation	Crystal shape	Tetrahydrated zinc nitrate, ammonium hydroxide	20–40	[15]
Microemulsion	Nanorod	Ethyl benzene acid sodium salt (EBS), dodecyl benzene sulfonic acid sodium salt (DBS), zinc acetate dihydrate, xylene, hydrazine, and ethanol	DEBS:80 DDBS:300	[16]
Sol-gel	Spherical shape	Zinc acetate dihydrate, oxalic acid dihydrate, ammonia, hydrochloric acid, and absolute ethanol	20	[17]
Microwave-assisted hydrothermal	Low microwave power: needle shape High microwave power: flower-shape	Zinc nitrate-6-hydrate, zinc acetate dehydrate, hydrazine hydrate, and ammonia	50–150	[18]

Solvothermal	Ethanol with TEA: Spherical shape Ethanol without TEA: Rod shape	Zinc acetylacetonate monohydrate, Triethanolamine, absolute ethanol, and 1-octanol	$D_{sphere} \sim 20$ $L_{rod} \sim 100$	[19]
Chemical vapor deposition	Nanorod shape	Zinc acetate di-hydrate, ethanol	Average diameter (90) and length (564)	[20]
Wet chemical	Nanodisc	Zinc chloride, sodium hydroxide	300–500	[21]

Table 3: Effect of different dopant and synthesis techniques on Photocatalytic application of Solar cells.

Type of dopants	Dopants	Fabrication method	Targeted pollutants	Photocatalytic application	Ref.
Alkaline earth	Mg	Solid state reaction	Alprazolam	Pharmaceutical compounds	[22]
		Auto combustion	MO	Dye degradation	[23]
		Co-precipitation	4-chlorophenol	Endocrine disrupting chemicals	[24]
Non-metal	S	Mechanochemical synthesis/ thermal decomposition	Resorcinol	Endocrine disrupting chemicals degradation	[25]
	C	Hydrothermal	Bisphenol A	Endocrine disrupting chemicals degradation	[26]
			RhB	Dye degradation	[27]
	N	Calcined the mixture of commercial ZnO and NH_4NO_3	Formaldehyde	Organic compounds degradation	[28]
		Sol-gel	Formaldehyde	Organic compounds degradation	[28]
			Phenol	Aromatic organic compound degradation	[29]
			MB	Dye degradation	[29]
		Solvothermal	MO	Dye degradation	[30]

		Modified non-basic solution/annealing in NH₃	RhB	Dye degradation	[31]
Other metals	Ni	Sol-gel	MG	Dye degradation	[32]
	Sn	Solid-state synthesis	MO		[33]
	Al	Plasma spraying	MB		[34]
	Bi	Parallel flaw precipitation			[35]
Transition Metals	Hf	Sol-gel	MB	Dye degradation	[36]
	Y	Microwave irradiation	MB	Dye degradation	[37]
	Pd	Sol-gel	MO	Dye degradation	[38]
		Incipient wetness impregnation	E. coli	Disinfection	[39]
	Ag	Photo reduction, chemical reduction, or polyacrylamidegel methods	4-nitrophenol	Phenolic compounds	[40]
		Spray pyrolysis	AO7	Dye degradation	[41]
		Laser-induction	MB	Dye degradation	[42]
	Fe	Sol-gel	2-chlorophenol	Aromatic organic compound degradation	[43]
	Mn	Co-precipitation	MB	Dye degradation	[44]
		Wet chemical	MB	Dye degradation	[45]
		Sol-gel	Direct Yellow 27 & Acid Blue 129	Dye degradation	[46]
		Microwave assisted hydrothermal	MB	Dye degradation	[47]
	Mo	Sol-gel	Direct Yellow 27 and Acid Blue 129	Dye degradation	[46]
	Cu	Co-precipitation	DB 71 dye, CV	Dye degradation	[48-49]
		Vapor transport	Rz	Dye degradation	[50]

		Sol-gel	MB, MO	Dye degradation	[51-52]
	Co	Co-precipitation	RhB	Dye degradation	[53]
Rare earth metal	Ce	Sonochemical	MB	Dye degradation	[54]
		Reflux method	MB	Dye degradation	[55]
		Hydrothermal	MB	Dye degradation	[56]
		Sonochemical wet impregnation	Cyanide	Pesticide degradation	[57]
	Gd	Sonochemical	AO7	Dye degradation	[58]
	Eu	Precipitation	MO	Dye degradation	[59]
	Nd	Sonochemical	MB	Dye degradation	[60]
		Wet chemical	RhB	Dye degradation	[61]
		Hydrothermal	MO	Dye degradation	[62]
	Sm	Chemical solution route	2,4-dichlorophenol	Aromatic organic compound degradation	[63]
		Solvothermal	Phenol	Aromatic organic compound degradation	[64]
	La	Precipitation/mechanical milling	MB	Dye degradation	[65]
		Co-precipitation	MCP	Pesticide degradation	[66]

Table 4. Role of different parameters on Suspended and immobilized ZnO system.

Parameters	Suspended system	Immobilized system	Ref
Active surface area	Adsorption and degradation require a large active surface area	For adsorption and degradation requires limited active surface area required	[67]
Stability of performance	Due to inconsistency towards repetitive reactions, there is a loss of nanoparticles	Comparatively more stable towards repetitive reactions	[68-69]
Aeration	Degradation has no effect	Synergistic effect on the effectiveness of system because oxygen increases the mass transfer of the system	[70]
Post-separation of catalysts	More energy steps are required	No more energy steps are required	[71]
Electric energy per order (EE/O)	To achieve the same degradation it requires lower electric energy in comparison to the immobilized system	Requires comparatively higher electric energy to achieve the same degradation	[72]

2. Hybrid organic/inorganic hetero-junction solar cells

2.1 Organic-Inorganic solar cells

The organic-inorganic hybrid solar cells have engrossed much recognition due to their mechanical flexibility, long life, and low-cost fabrication. These features made organic-inorganic hybrid solar cells more demanding and become the prime choice of the consumer inquisitiveness [73]. To achieve the allied advantages of organic-inorganic hybrid solar cells both material groups' organic and inorganic nanoparticles were combined [74]. To convert sun light in to electrical charge, two components of these hybrid solar cells combined together. In which one component act as an electron donor and a light harvester used a conjugated polymer as organic semiconductor and second component act as an electron acceptor is an inorganic semiconductor [75]. The best substitute found to be ZnO (n-type inorganic semiconductor) for the replacement of electron acceptor organic semiconductor used in fully organic solar cell. It is because; the electron mobility is high for ZnO in the inorganic components of hybrid solar cells. ZnO also had shown best physical and chemical stability in comparison to the n-type semiconductor presently available [76]. It has been found that total energy production mechanism approaching to the organic solar cells. This mechanism involves the exciton generation due to the illumination of the light, resulting in the diffusion of exciton at donor or acceptor interface up to certain diffusion length. The generation of the free positive and negative charges due to the difference in the HOMO and LUMO of the acceptor and donor material providing the driving force, which cause to reduce the binding energy of the excitons and hence separated charge carriers moves toward cathode and anode path way respectively [77]. Recently, ZnO become a suitable material because of higher electron mobility than TiO_2 and can be synthesized by different methods. For hybrid solar cell applications different nanostructure of ZnO, including nanowires, nano rods and nano particles have been examined and shown good response to enhance power efficiency in the solar cell. Consonni et al. quantified the role of abundant and non-toxic ZnO, as nanowires having good electron transporting properties for the fabrication of nanostructured solar cells. ZnO nanowires have shown excellent optical absorption through light trapping and improved the charge carrier collection and separation properties [78]. These properties make ZnO more desirable material for the fabrication of solar cell at low cost scalable [79-80]. Thus, ZnO used as a bulk hetero-junction hybrid solar cells with vertically aligned nanostructures, randomly dispersed nanocrystals and hybrid solar cells with organic-inorganic bilayers structures. Beak et al. in 2004 reported the employment of ZnO in bulk hetero-junction solar cells. The formation of organic p-type semiconductor in order to create a bulk hetero-junction hybrid solar cells were arranged individually the nano-crystalline ZnO with ~5nm diameter by hydrolysis method and condensation of zinc acetate dihydrate using KOH in methanol along with poly (2-methoxy-5-(3,7-dimethyloctyloxy)-1,4-phenylenevinylene)) [MDMO-PPV]. Such type of structural configuration has shown the power conversion efficiency of order of 1.6% at 0.71 sun equivalent intensity [81-82]. The results for the forward current density was found to be greater in ZnO:poly(2-methoxy-5-(3,7-dimethyloctyloxy)-1,4-phenylenevinylene)) configuration in comparison to the virgin poly (2- methoxy-5-(3,7-dimethyloctyloxy)-1,4-phenylenevinylene)), supporting that the presence of the nano-crystalline ZnO provide a unceasing pathways for the electron mobility [82]. Later on, the higher hole mobility were achieved by substituting the MDMO-PPV with poly (3-hexylthiophene) [P3HT]. However, the power conversion efficiency with ZnO: P3HT configuration was found to be 0.9%, which is less in comparison with nano ZnO: MDMO-PPV cell [82-83]. The low response was assumed to be due to the hole mobility of the organic phase, in organic-inorganic hybrid solar cells become the main limiting factors to performance [84]. But later it was found, the actual region of less hole mobility inside polymer was due to the existence of the hydrophilic component of pre-synthesized ZnO inside the polymer blend, which adversely affect its ability to crystallize [85]. These results were also confirmed by atomic force microscopy (AFM) study, which shows that ZnO particles distributions are not proper at the junction made with the thin layer of polymer (PEDOT/PSS). This non-uniformity and thin layer resulting to decrease in the internal quantum efficiency by reducing the rate of charge creation because it hindered the generation of excitons inside the (PEDOT/PSS) to touch the ZnO interface [85]. All these studies reveal that overall device efficiency was found to be less due to the lack of homogeneity in the particle distribution during nc –ZnO based hybrid solar cells manufacturing process, and the lack of an intimate blend between the organic and inorganic phase at the interface resulting to hinder the electron transportation [85-86]. To overcome these limitations, Beaker et al. reported the alternative route by employing ZnO in its precursor form of di-ethylzinc for the fabrication of ZnO hybrid solar cells [87]. The precursor with solvent primarily dissolved and then cast the thin film layer along with the polymer. The well distributed ZnO particles were found across the polymer film, due to the reaction of the mixture with the moisture present in the atmosphere by the hydrolysis process [88]. The hole mobility and packing in the configuration were upgraded due to the improvement in the crystallinity in the polymer by employing a thermal annealing process near the polymers glass transition temperature (TTg) [83]. However, earlier this method was employed for the inorganic n-type semiconductor TiO_2, forming the crystallinity. But, was unable to obtain desired results because TiO_2 requires a higher temperature for best results [87]. Alternatively, ZnO crystallized at a much lower temperature ~110 °C. The power conversion efficiency upgraded up to 0.5%(η=1.4%) in comparison to the earlier prepared nc-ZnO:P3HT cell due to the formation of more uniformly distributed ZnO nanoparticles on film. These improved results were obtained by employing this precursor method and the post-fabrication thermal annealing process [89]. The effect of the active thickness of the layer on the cell performance was focused on by the researcher after employing the precursor method in three-dimensional morphology of ZnO with P3HT, to obtain 167 nm active layer thickness with yielding η of 2% in the cell. The study revealed that the thin ZnO:P3HT hybrid solar cells exhibit lower performance because the abrasive phase separation lacks the continuous pathway's resulting in inadequate charge generation and hence, charge transfer rate [85].

Alternatively, for better properties of nanorods and nanowires, more aligned and controlled nanostructures are developed as a replacement for the randomly dispersed nanoparticles [90]. A highly proficient pathway for electron transportation is stated by using vertically aligned nanostructures. This kind of nanostructures arrangement provides a higher interfacial area between the organic and inorganic material to facilitate the smooth electron mobility and made them a most promising candidate [91]. It has been observed that ZnO in the vertically aligned arrangement can be grown easily on many substrates in comparison to the others mostly used inorganic materials by using the solution and hydrothermal methods. These are low-cost techniques and made ZnO a more desirable material [92]. To create hybrid solar cells devices, Olson et al. first used vertically aligned ZnO nano-fibers in combination with P3HT, the subsequent device showed η of 0.53% [93]. The various un-optimized parameters caused low performance, this happened due to the spacing between ZnO nanofibers (100 nm) causing the inefficient charge separation. Due to the large spacing in comparison to standard exciton diffusion length in P3HT (10-20 nm), yield modest results of the hybrid solar cells devices [86]. Later on, when the P3HT solution was treated with dichlorobenzene solvent instead of chloroform it showed improved results for the device. Better results of the device are due to the polymer infiltration and chain ordering [94]. The annealing process was found to be more effective to lowering the geminate recombination between two phases in nanoparticles and polymers; this is possible by the annealing process must be completed for a short period (1 min. at 225 °C)near the polymer melting point temperature. Within this short period of time polymer organised well with nanofibers, ensuing more closely interface between two phases and causing to enhance of the polymer crystallinity. It was observed that excess thermal annealing up to 15 min. showed an adverse effect on the polymer chain ordering [95]. After the completion of all optimization along with ZnO backing layer thickness, Baeten et al. studied the device performance of the ZnO nanorods arrays coupled with P3HT and reported a remarkable increase in the device performance [95]. For better performance of the hybrid solar, there is a need to improvise the surface engineering of the ZnO nanostructures, Han et al. studied the use of ZnO as a buffer layer to make a tri-laminar $ZnO/ZnS/Sb_2Se_3$ nanotube arrays inimitable surface structure with P3HT. Results showed power conversion efficiency rises by 1.32%, this occurred due to the distinctiveness of the structure resulting in to increase in the electron collection efficiency by quashing carrier recombination and providing improved energy level configuration of the tri-laminar structure with P3HT [96]. Alshanableh et al. in 2018 reported the importance of the defect quenching phenomenon because the result showed utmost perfection in the ZnO nanorods morphology when treated with KOH also chemical etching with protonic and anionic agents exhibited favorable to the ZnO morphology which exposed the positive outcomes in power conversion efficiency [97]. J.Huang et al. also studied the importance of ZnO in organic hybrid solar cell applications. This is because of ZnO structural properties which plays an important role in electrode applications and for the buffer layers. The utility of ZnO is due to its hole-blocking and electron collection properties which determine the power conversion efficiency, lifetime and stability, etc. of the device [98].

Chen et al. [99] obtained Ga-doped ZnO (GZO) films on quartz substrate RF magnetron sputtering for the application of perovskite solar cells. It was reported by the author that about a 20% of efficiency rise in power conversion can be obtained with a higher dopant concentration along with a transmittance rate of more than 87% in the range of $0.4 - 1.2$ μm. The addition of charge carriers by Ga-dopant ions made the electron fill up the vacancies at the interface and hence reduced the probability of charge carrier recombination at those boundaries which result in high charge carrier transport efficiency from perovskite to the electron transport layer (ETL). Dong et al. [100] also proposed that the efficiency of perovskite solar cell can be enhanced by hindering the recombination of electrons and holes at the ZnO-perovskite interface with the addition of Mg ions as a dopant in ZnO lattice.

The dopant plays on an important role in improving interface interaction among ZnO and perovskite material along with enhancing their performance and characteristics [101]. This improvement in interface interaction plays an important role in the separation, recombination and generation of charge carriers at these interfaces as a result of which barrier potential has been reduced and improves the overall efficiency of the solar cell. Zheng et al. [102] suggested that doping of iodine in ZnO lattice has reduced the ionization potential; required for the removal of electrons at ZnO- perovskite interface which is because of inhibition of charge carrier recombination there as the decay lifetime of photoluminescence has been reduced with I-doping. These iodine dopant ions enhance the transportation of facile charge carriers at the interface by restricting the accumulation of charge carriers at such interfaces by providing better electrical contact between perovskite-ZnO.

2.2   Inorganic hetero-junction solar cell

There is a variety of inorganic hetero-junction solar cells which includes excitonic solar cells, quantum dot solar cells as well as thin film solar cells. The inorganic semiconducting materials could be used for the fabrication of such kinds of solar cells among which ZnO was found to be most suitable because of its small effective mass and high electron mobility. Also, ZnO is polymorph (that has different morphology) depending on the synthesis procedure, easy to fabricate, and widely available material [103]. These exclusive attributes of ZnO attracted lots of researchers to use them in the fabrication of solar cells. Varying morphologies of ZnO help in changing the surface area and charge carrier transport properties. The greater surface area helps in increasing the light scattering and hence harvesting of light which improves the overall efficiency of solar cells. All these characteristics motivate the researcher to use ZnO as an n-type semiconductor component for the formation of bulk hetero-junction layer during the synthesis of perovskite solar cells.

A mixture of ZnO and lead chalcogenides (PbZ; where Z could be S, Te, Se) was found to be the most commonly used inorganic hetero-junction solar cell. PbZ materials have a very high dielectric constant and hence possess large Bohr radii, which give rise to considerable quantum confinement in these materials making them suitable for photovoltaic applications. Also, the energy levels of PbZ can be easily tuned with those of ZnO to enhance its conductivity by reducing the band gap. On comparing the hetero-junction solar cell made up of a similar type of PbSe nanocrystals, it was observed cells with a ZnO layer have higher efficiency of power conversion by 1.6% [104]. It was observed by Leschkies et al. [105] that thermal annealing of ZnO up to a temperature range of 450 °C will improve the conductivity and mobility of charge carriers by diminishing the ratio of defects present. It was suggested that nanocrystals are more efficient than thin films as they have a larger surface area for dissociation of excitons which improves the ability of such devices. In further research supportive results were obtained.

Copper-based semiconductors are also used in combination with ZnO because of their high potential, wider availability, and non-toxic nature. Bhaumik et al. [106] synthesized highly efficient ZnO/CuO-based thin films solar cells by hydrothermal process. The integration of CuO onto its thin film improves the efficiency of such solar cells by 2.88%. This increase in efficiency is attributed to surface plasmon resonance as well as a scattering of multi-phonon in the nanostructures. Fuzimoto et al. [107] synthesized hetero-junction solar cells with n-type ZnO and p-type copper oxide via the electro-deposition method to obtain an efficiency of 0.25%. The thickness of the active ZnO layer and electro-deposition conditions could be varied to improve the efficiency of these cells. Though nanomaterial provides a large surface an uneven distribution of carriers blocks the path of electron transfer and gives rise to a high recombination rate. So there is a requirement for some novel methods for synthesis of such materials to control their morphology in a defined way.

The nanoparticulate structures gather so much attention due to their high light scattering ability along with large surface area. These Nano particulates are usually poly-dispersed which further increases this scattering effect. But these structures have some disadvantages too such as the presence of defects at grain boundaries which give rise to recombination on interfaces [108]. This agglomeration of charge carriers can be somehow reduced by using electro-deposition techniques as they allow the particles to make bonding without agglomerating which enhances the porosity of such films and reduce the recombination rate at interfaces [109].

The 3-dimensional nanostructures were explored nowadays for such applications as they tend to separate the electrical and optical characteristics of these materials and hence result in the light trapping feature of such solar cells. Zhu et al. [110] reported piezo-phototronic effect for n-ZnO/p-SnS nanowire flexible solar cells. It was observed in the report that the flexibility of these solar cells enhanced their efficiency. Also, this piezo-phototronic effect not only improves the performance efficiency by 37.3% but also has a great potential in the fabrication of such light weighing, flexible solar cells on large scale. Synthesis of superstrate liquid junction solar cell via sequential growth of vertically aligned Al; ZnO nanorods on ZnO blocking layer has been achieved. Such solar cell exhibits high efficiency in comparison to thin film structured solar cells due to higher interfacial area, more amount of light harvesting, and high rate of electric charge transport [111]. It was demonstrated by Lai et al. that with the help of nanorods, about a 14% increase in efficiency of Copper-zinc-tin selenide solar cell can be achieved by reducing the surface reflection to 2.97% from 7.76% [112]. This array of nanorods acts as a shield layer that protects the device from humidity and provides effective resistance against reflection at its surface. Among varying dimensions; 900nm heightened ZnO nanorod provided the best anti-reflection effect due to the formation of a hydrophobic layer on the surface. This suggested that ZnO can not only be used for the preparation of an active layer but also for making such kind of buffering jackets or coatings to enhance the efficiency and life of solar cells.

3.   Sensitized solar cells

O'Regan and Gratzel proposed an alternative to traditional silicon-based solar cells by introducing dye-sensitized solar cells (DSSC) in 1991 for the first time and since then it has attracted great attention. DSSCs are found to have better stability and energy conversion efficiency greater than 10% and are easy to fabricate and economical [113]. In comparison to Ru-based dyes, the use of semiconductor quantum dots provides a better alternative owing to the expense related to typical dyes (N3, N719). Additionally, QD-sensitized solar cells (QDSCs) are proficient in creating multiple electron–hole pairs through impact ionization [114-115]. In earlier reports, the efficiency of QD-sensitized solar cells was quite low in two electrode configurations however Diguna et al. reported an enhanced efficiency of 2.7 % which triggered the interest of researchers in QDSCs [116-117]. The light-conversion efficiency has also been found to be improved depending on different morphologies of photoanode. The use of various ZnO-based nanostructures and their role in DSSCs and QDSCs is elaborated on in further sections [118].

3.1 Dye-sensitized solar cells

The light conversion efficiency of 2.5 % was reported using rose bengal sensitized ZnO porous disks for monochromatic light of wavelength 562 nm [48]. The use of ruthenium-complex dye under 56 mWcm^{-2} illuminations has increased the efficiency of ZnO thin film to 2% while it was reported to be 2.5 % with the mercurochrome sensitizer under 99 mWcm^{-2} illuminations [120-121]. The bonding of mercurochrome dye with ZnO surface is similar to TiO$_2$ via carboxylate linkage. The poor performance of ZnO in comparison to TiO$_2$ is attributed to the varying absorption rate of the ruthenium-based dye. The nanopores of the ZnO electrode were filled with the agglomerates of Zn^{2+} ions and Ru-dye molecules during dye adsorption causing blockage of injected electrons from the dye molecules to the semiconductor by forming an insulating layer [122-123]. In the occurrence of an acidic dye, ZnO electrode displays poor chemical stability as ZnO has ~9 isoelectric points in comparison to TiO$_2$ which has the same parameter ~ 6 [124]. The ruthenium dye contains carboxyl groups which get deprotonated during dye adsorption making the dye solution comparatively acidic leading to the etching of ZnO surfaces. The formation of Zn^{2+}/dye complexes takes place due to the dissolution of ZnO in the presence of acidic carboxyl groups which further is responsible for preventing electron injection efficiently. However, enhanced efficiency up to 5% was reported for nanoporous ZnO electrode when sensitized with ruthenium bipyridial complex. It was attributed to suppressed formation of Zn^{2+}/dye complexes by adding a basic solution of KOH to the dye solution. Additionally, organic additives were excluded while the preparation of ZnO films for improved interfacial kinetics [125-126].

A large surface area is provided by the customary nanofilm in DSSCs for the adsorption of light-gathering dye molecules. However, the slow mechanism of trap restricted diffusion for the electron transport through the nanoparticle confines the efficiency of the device (Fig 3). Therefore, in order to escalate the electron diffusion length in the anode, a favourable solution was offered in the form of oriented 1D nanostructures, such as nanowires, nanorods, and nanotubes. A seeded growth process was reported in 2005 to synthesize ZnO nanowire photo anode having a surface area of about 1/5th of nanoparticle film and an array length between 20–25 mm. The dip coating technique was used to deposit 10-15 nm thick film of ZnO quantum dots onto the FTO substrate. Then, thermal decomposition of zinc complex was used to grow nanowires from these nuclei. It exhibited an efficiency of 1.5% with sensitization of the ruthenium dye [127].

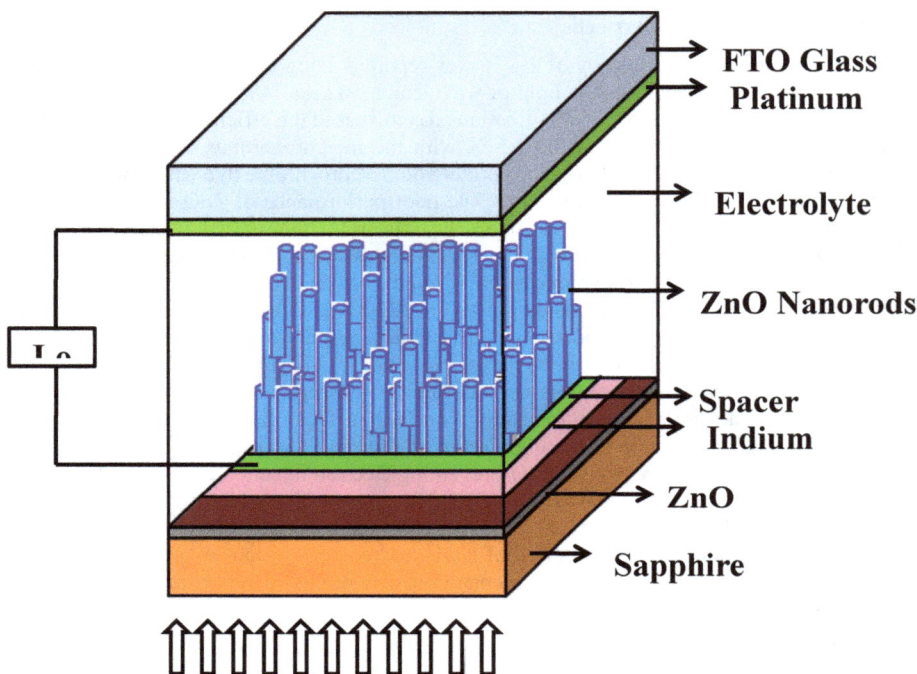

Figure 3. Schematic representation of dye-sensitized solar cell.

The orientation of nanorods along with the size, affects the photovoltaic properties of the ZnO nanorods [128]. Very low power efficiency was shown by hydrothermally grown (0.22%) and vapour deposited (0.09%) vertically aligned ZnO nanorods with N179 sensitization. Increased efficiency up to 1.2% was reported in the case of ZnO nanorods surface modified with gold nanoparticles forming a Schottky barrier and thus blocking the electron transfer back from ZnO to the N-719 dye and electrolyte [129]. With Al doping in ZnO nanorods the efficiency has been increased from 0.05% to 1.34% for the N719-sensitized ZnO nanorods [130]. A trivalent ion of Aluminium (Al^{3+}) replacing the divalent ion of Zinc (Zn^{2+}) results in more electrons which increases their electrical conductivity and hence allows electrons to shift easily into the conduction band of Al-doped ZnO lattice.

ZnO and Their Hybrid Nano-Structures   Materials Research Forum LLC
Materials Research Foundations 146 (2023) 132-172   https://doi.org/10.21741/9781644902394-5

Table 5: Effect of ZnO structure, synthesis method, and type of dye on the performance of ZnO DSSCs.

Type of Nanostructure	Method of prep.	Dye	V_{OC} [V]	J_{SC} [mAcm^{-2}]	Intensity [mWcm^{-2}]	Fill Factor [FF]	η [%]	Ref.
Nanoparticles	CBD, DBT	CYC-B1	0.57	16.09	100	0.59	5.40	[131]
Nanorods	DCTP, CBD	N-719	—	—	100	—	—	[132]
Nanoplates	HTM, DBT	N-719	0.554	8.4	100	0.41	1.90	[133]
Nanorings	PVD	—	—	—	—	—	—	[134]
Nanoflakes	HTM, DBT	N-719	0.63	11.6	100	0.50	3.64	[135]
Nanotubes	EDM	Z-907	0.69	11.51	100	0.22	1.60	[136]
Nanoflowers	HTM, SCM	N-719	0.65	5.50	100	0.53	1.90	[137]
Nanosheets	CBD	D-149	0.58	19.53	100	0.63	7.07	[138]
Nanospikes decorated sheets	EDM, HTM	N-719	0.68	6.07	100	0.60	2.51	[139]
Nanofibres	S	D-149	0.49	8.45	100	0.40	1.66	[140]
Nanofiber mats	BRP, ES	N-719	0.60	3.58	100	0.62	1.34	[141]
Nanocones	CBD, DBT	N-719	0.64	15.00	100	0.45	4.36	[142]
Nanoburgers	CBD, DBT	N-719	0.70	8.71	100	0.66	4.03	[143]
Ribbons	PVD	N-3	—	—	150	—	—	[144]
Hedgehog-like needle-clusters	HTM, DBT	N-719	0.58	7.62	100	0.50	2.22	[145]
Spindle-shaped	HTM, DBT	N-719	0.79	5.10	100	0.45	1.82	[146]
Tripods	PVD, DBT	N-719	0.55	2.80	97	0.54	0.88	[147]
Garland	SGM, DCM	Hibiscusros asinensis dye	0.27	4.20	50	0.30	0.67	[148]
	CBD	N-719	0.73	8.91	100	0.51	3.30	[149]

Rectangular prisms								
Tetra pods	DCTP	D-149	0.63	11.60	100	0.64	4.68	[15
Hexa branched nanowires	ALD, E, CBD	—	—	—	—	—	—	[15
Hexagonal clubs	HTM, DBT	N-719	0.58	11.50	100	0.65	4.28	[15
See-urchins	HTM	N-3	—	—	100	—	—	[15
Bush-like	SCM, HTM	N-719	0.69	3.46	100	0.35	0.82	[15
Bottlebrush	DCM, CBD	N-3	0.60	4.02	100	—	1.00	[15
Caterpillar-like	SCM, HTM	N-719	0.69	15.20	100	0.50	5.20	[15
Disk-like	HTM	N-719	0.69	6.92	100	52.50	2.49	[15
Comb-like	CVD	N-719	0.67	3.14	100	0.34	0.68	[15
Blanket-like	CBD	—	—	—	—	—	—	[15
Cauliflower	HTM	N-719	0.66	6.08	100	0.55	2.18	[16
Panel	CBD, DBT	N-719	0.70	11.09	100	0.72	5.59	[16
Pomegranate	DBT	N-719	0.69	8.8	100	0.72	4.35	[16
Paddlewheel-like nanorods	SGM, HTM	N-719	0.76	6.00	100	0.70	3.25	[16
Microspheres	HTM, DBT	N-719	0.57	14.73	100	0.61	5.16	[16
Hollow spheres	CBD	—	—	—	—	—	—	[16
Sponge	S	N-719	—	—	100	—	4.83	[16
Coral-shaped	S	N-719	0.60	13.44	100	0.54	4.58	[16

4.   Perovskite-sensitized Solar Cells

The idea for perovskite-sensitized solar cells was coined by Kojima et al. to lay down a new milestone in the advancement of solar cells in 2009. Presently, it is familiar as a perovskite-solar cell (PSC) [168-171]. Basically, a perovskite solar cell is a compound of the absorber material having the general formula ABX_3, which is referred to as perovskite structure, where A and B are cations with varying radii and X is an anion. Among, different halide perovskite absorber, with the general formula $CH_3NH_3PbX_3$ (methylammonium lead trihalide), where X stands for different halogen. Two organo-lead halide perovskite nanocrystals with X=Br and I, have shown superior photochemical cells for the visible light conversion due to effective sensitization with TiO_2 [172]. But, the comparative analysis of $CH_3NH_3PbBr_3$ and $CH_3NH_3PbI_3$ absorber material for the solar energy conversion efficiency was found to be higher for Iodide halide with a value 3.8% than Bromide halide with a value 3.1%. These results, opened a new zone of research to escalate the efficiency of the methyl ammonium lead iodide ($CH_3NH_3PbI_3$) centered perovskite-solar cell.

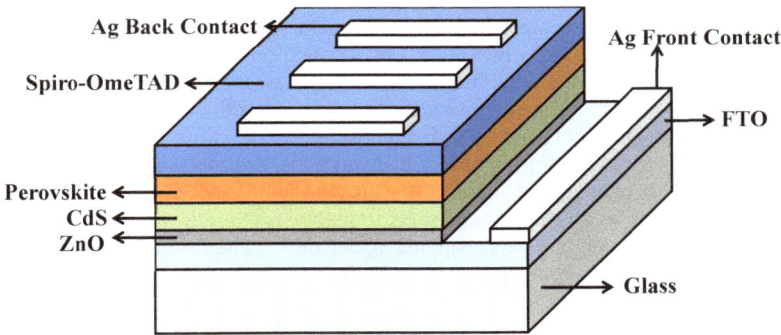

Figure 4: Schematic representation of ZnO-based perovskite solar cell.

Literature reported that the efficiency of methyl ammonium lead iodide was escalated up to 20.7% in less than ten years of research [173]. To reduce the charge recombination problem that occurred in perovskite bulk film due to photogenerated holes are removed by the electron transporting materials like TiO_2, ZnO (n-Type semiconductor oxides) to extract and transport the photogenerated electrons [174-175]. Hence, in perovskite solar cells n-Type semiconductor oxides played a vital role to refine the photovoltaic performance. Thus, requisite conditions are energy level alignment, morphology, interfacial properties, and trap states required to control the characteristics of ZnO for making electron-transporting materials layer [176]. Since ZnO nanostructure has shown major influence on the three phases of perovskite films. The results are listed as the eminence of the ZnO/perovskite interface (Figure 4), loading and on the perovskite layer morphology, and the superiority of the perovskite itself [175]. Different nanostructures of ZnO showed different kinds of effects on the performance of perovskite solar cells by interacting differently with perovskite. Results showed that 1-dimensional ZnO nano rods structure in comparison with ZnO nanoparticles gave better performance, because ZnO nano rods deliver direct pathways for the electronic transportation in perovskite solar cell, though in the case of nanoparticles electron transfer rate slow down due to many trapping and de-trapping in the nanostructure, particularly at the grain boundaries [174, 177-179]. Further, in the literature was found that there are many other factors also responsible for the power conversion efficiency along with the interacting properties of perovskite and their electron transport materials in perovskite solar cells. These are found due to the different deposition and processing methods of ZnO, which might affect the interaction between the perovskite and electron transport materials, resulting in the power conversion efficiency of the perovskite solar cells. Zheng et al. reported the reason of the low power conversion efficiency based on ZnO nanoparticles deposition by using the spin coating method. Results showed the loss of carriers mobility reducing the efficiency, because of the defective interface produced due to the formation of a pinhole surface [180]. Duan et al. reported the enhanced power conversion efficiency of the perovskite solar cells to 13.1% by a post–treatment method in which there is the addition of the in situ-thermal decomposition on the spin coated ZnO. Improved efficiency is caused due to change in the morphologies of the ZnO from nanoparticles to interconnecting the net-like structure by this method [181]. Though, ZnO nanorod in 1-dimensional mode showed improved power conversion efficiency by providing the direct pathways for electron mobility. However, another way adopted to improve their efficiency, Mahmood et al. modified the conventional low aspect –ratio of ZnO nanorods to high aspect –ratio ZnO nanorods by introducing directly PEI polymer as a capping agent during the hydrothermal growth process. These changes enhance the power conversion efficiency from 10.3% to 11.5%. Because this modification hampering the perovskite infiltration in the electron transport materials due to the large diameter of the nanorods. Further, more power conversion efficiency can be achieved with the post-treatment of solvent-annealing in ethanol vapour and by passivating the ZnO nanorods layers with Al_2O_3. By, this method efficiency increased to 17.3% due to an increase in the carrier diffusion length and the recombination resistance in perovskite solar cells [182, 183]. Doping in electron transport materials plays a significant role to make a better output perovskite solar cells device by improving the electron transport properties of the ZnO. Because, doping caused to alter the morphology of ZnO and electronic properties of the metal oxide semiconductor to increase the free charge mobility and result in the conductivity of the solar cells [184, 175]. The doping of ZnO can be done by either replacing the Zn^{2+} cation, where cationic dopants are typically metals or by the O^{2-} anion, where anionic dopants are the non-metals respectively. The replacement of Zn^{2+} caused to affect the conduction band structure. Further, valance band energy is affected due to the replacement of the O^{2-} 2p band belonging to the upside of the valance band with a different anion. Hence, the overall effect of the dopant in ZnO caused to shift of the Fermi level towards the conduction band leading to the conductivity factor and helping the working functionality [175,185]. The successful doping in ZnO for the perovskite applications with various n-type dopants like Mg, Ga, Al, N, I was reported [182, 186-188]. Mahmood et al. studied the effect of altering the aspect –ratio of the ZnO nano rods and the effect of doping of ZnO with electron rich nitrogen results shown to enhanced power conversion efficiency to 16.1%. This modification caused to increase in the conductivity of the oxide layer and reduced the internal resistance resulting to the enhancement the electron density of the electron transport materials [186]. The modification with iodine with hydrothermal process results in the wide- hexagonal structure of ZnO-I nanopillars, which is used to make the even planer ZnO-I thin film surface with good compactness having few voids obtained in comparison with ZnO nanorods arrays. Further, results showed that doping of ZnO with Iodine leads to better functionality of the electron transport materials by providing electron extraction from the perovskite layer, which caused to improvement in the power conversion efficiency of the device as high as 18.24%. Literature also suggests the importance of the doped ZnO-Al and ZnO-Ga films as the best candidates for the transparent conducting oxide materials due to their numerous features like suitable ionic radii, low cost, and outstanding optical transmission performance. Mahmood et al. studied the effect of Al doping with virgin ZnO and the results showed increased carrier concentration and electron mobility which featured excellent conductivity [187]. Further, Dong et al. reported the Al-doped ZnO doping effect, the study shown higher electron mobility, higher conduction band, and improved electron density than pure ZnO.

Conclusion and Future scope

In the present chapter, we discussed the role of ZnO as an active participant in current and upcoming solar cell applications including hybrid organic/inorganic hetero-junction solar cells, sensitized solar cells and perovskite sensitized solar cells. In comparison to polymer-based solar cells, inorganic solar cells depicted enhanced performance and efficiency and become the new generation of solar cells. Several combinations of metal oxides, alloys, and metals with specific fabrication techniques, elemental substitutions, unique nano structures, and interfacial changes have been investigated to enhance the stability and performance of the solar cell. Besides there exist a few issues related to toxicity and abundance of the material used in solar cell fabrication because most of the highest performing solar cell consists of toxic elements such as lead (Pb) and Cadmium (Cd). Presently research work is being carried out on safer alternatives that can compete with the efficiency, stability, and performance of the next generation solar cells. In this regard, a new generation solar cell architecture has been designed involving a hybrid inorganic/organic solar cell, perovskite solar cell, and complete inorganic solar cell, and their cell performance was optimized by suitable modification with dopants and experimental conditions pre-post treatment processes.

In spite of many indications of progress conquered so far in terms of the introduction of ZnO for solar cell applications but still there exists plenty of room for modification, improvement, and development. Few major challenges such as large surface area, induction of effective interfaces, improved light harvesting abilities, and increased power convergence efficiency need to be overcome. The performance of ZnO-based solar cells also depends upon environmental conditions such as temperature, humidity, and mechanical wear and tear. For longer use of solar cells the performance of the device should not be affected by these factors. to boost the positive impact on socio-economic parameters an approach can be developed by considering a low-cost methodology to synthesize ZnO by making use of a green route which can provide a platform for environmental friendliness and sustainability.

References

[1] Satpute Anand Vijay and E. Vijay Kumar. Modern development and potential uses of solar energy utilization in India: A review. WEENTECH Proc. Energy 6 (2020): 1-13. https://doi.org/10.32438/WPE.060203

[2] Kafafi Zakya H. Barry P. Rand Kwanghee Lee and René Janssen. Introduction to the issue on next-generation organic and hybrid solar cells. IEEE Journal of Selected Topics in Quantum Electronics 16 no. 6 (2010): 1512-1513. https://doi.org/10.1109/JSTQE.2010.2077690

[3] Irfan Ahmad Ruifa Jin Abdullah G. Al-Sehemi and Abdullah M. Asiri. Quantum chemical study of the donor-bridge-acceptor triphenylamine based sensitizers. Spectrochimica Acta Part A: Molecular and Biomolecular Spectroscopy 110 (2013): 60-66. https://doi.org/10.1016/j.saa.2013.02.045

[4] McEvoy Augustin Luis Castaner and Tom Markvart. Solar cells: materials manufacture and operation. Academic Press 2012.

[5] Chander A. Hema M. Krishna and Y. Srikanth. Comparision of different types of Solar cells-a review. IOSR Journal of Electrical Engineering 10 no. 6 (2015): 151-154.

[6] O'regan B. & Grätzel M. (1991). A low-cost high-efficiency solar cell based on dye-sensitized colloidal TiO2 films. nature 353(6346) 737-740. https://doi.org/10.1038/353737a0

[7] Zhang Xinyu Jiaqian Qin Yanan Xue Pengfei Yu Bing Zhang Limin Wang and Riping Liu. Effect of aspect ratio and surface defects on the photocatalytic activity of ZnO nanorods. Scientific reports 4 no. 1 (2014): 1-8. https://doi.org/10.1038/srep04596

[8] Look David C. Donald C. Reynolds J. R. Sizelove R. L. Jones Cole W. Litton G. Cantwell and W. C. Harsch. Electrical properties of bulk ZnO. Solid state communications 105 no. 6 (1998): 399-401. https://doi.org/10.1016/S0038-1098(97)10145-4

[9] Albrecht J. D. P. P. Ruden Sukit Limpijumnong W. R. L. Lambrecht and K. F. Brennan. High field electron transport properties of bulk ZnO. Journal of Applied Physics 86 no. 12 (1999): 6864-6867. [9] https://doi.org/10.1063/1.371764

[10] Noack Volker Horst Weller and Alexander Eychmüller. Electron transport in particulate ZnO electrodes: a simple approach. The Journal of Physical Chemistry B 106 no. 34 (2002): 8514-8523. https://doi.org/10.1021/jp0200270

[11] Sobczyk-Guzenda Anna Bożena Pietrzyk Hieronim Szymanowski Maciej Gazicki-Lipman and Witold Jakubowski. Photocatalytic activity of thin TiO2 films deposited using sol-gel and plasma enhanced chemical vapor deposition methods. Ceramics International 39 no. 3 (2013): 2787-2794. https://doi.org/10.1016/j.ceramint.2012.09.046

[12] Anand Vikky and Vimal Chandra Srivastava. Zinc oxide nanoparticles synthesis by electrochemical method: Optimization of parameters for maximization of productivity and characterization. Journal of Alloys and Compounds 636 (2015): 288-292. https://doi.org/10.1016/j.jallcom.2015.02.189

[13] Ghosh Saptarshi Deblina Majumder Amarnath Sen and Somenath Roy. Facile sonochemical synthesis of zinc oxide nanoflakes at room temperature. Materials Letters 130 (2014): 215-217. https://doi.org/10.1016/j.matlet.2014.05.112

[14] Kumar V. R. P. R. S. Wariar V. S. Prasad and J. Koshy. A novel approach for the synthesis of nanocrystalline zinc oxide powders by room temperature co-precipitation method. Materials Letters 65 no. 13 (2011): 2059-2061. https://doi.org/10.1016/j.matlet.2011.04.015

[15] Lim Sang Kyoo Sung-Ho Hwang Soonhyun Kim and Hyunwoong Park. Preparation of ZnO nanorods by microemulsion synthesis and their application as a CO gas sensor. Sensors and Actuators B: Chemical 160 no. 1 (2011): 94-98. https://doi.org/10.1016/j.snb.2011.07.018

[16] Ba-Abbad Muneer M. Abdul Amir H. Kadhum Abu Bakar Mohamad Mohd S. Takriff and Kamaruzzaman Sopian. Optimization of process parameters using D-optimal design for synthesis of ZnO nanoparticles via sol-gel technique. Journal of Industrial and Engineering Chemistry 19 no. 1 (2013): 99-105. https://doi.org/10.1016/j.jiec.2012.07.010

[17] Hasanpoor Meisam M. Aliofkhazraei and H. Delavari. Microwave-assisted synthesis of zinc oxide nanoparticles. Procedia Materials Science 11 (2015): 320-325. https://doi.org/10.1016/j.mspro.2015.11.101

[18] Šarić Ankica Goran Štefanić Goran Dražić and Marijan Gotić. Solvothermal synthesis of zinc oxide microspheres. Journal of Alloys and Compounds 652 (2015): 91-99. https://doi.org/10.1016/j.jallcom.2015.08.200

[19] Laurenti M. N. Garino S. Porro M. Fontana and C. Gerbaldi. Zinc oxide nanostructures by chemical vapour deposition as anodes for Li-ion batteries. Journal of Alloys and Compounds 640 (2015): 321-326. https://doi.org/10.1016/j.jallcom.2015.03.222

[20] Samanta Pijus Kanti and Santanu Mishra. Wet chemical growth and optical property of ZnO nanodiscs. Optik-International Journal for Light and Electron Optics 124 no. 17 (2013): 2871-2873. https://doi.org/10.1016/j.ijleo.2012.08.066

[21] Ivetić T. B. M. R. Dimitrievska N. L. Finčur Lj R. Đačanin I. O. Gúth B. F. Abramović and S. R. Lukić-Petrović. Effect of annealing temperature on structural and optical properties of Mg-doped ZnO nanoparticles and their photocatalytic efficiency in alprazolam degradation. Ceramics International 40 no. 1 (2014): 1545-1552. https://doi.org/10.1016/j.ceramint.2013.07.041

[22] Wang Yajun Xiaoru Zhao Libing Duan Fenggui Wang Hongru Niu Wenrui Guo and Amjed Ali. Structure luminescence and photocatalytic activity of Mg-doped ZnO nanoparticles prepared by auto combustion method. Materials Science in Semiconductor Processing 29 (2015): 372-379. https://doi.org/10.1016/j.mssp.2014.07.034

[23] Selvam N. Clament Sagaya S. Narayanan L. John Kennedy and J. Judith Vijaya. Pure and Mg-doped self-assembled ZnO nano-particles for the enhanced photocatalytic degradation of 4-chlorophenol. Journal of environmental sciences 25 no. 10 (2013): 2157-2167. https://doi.org/10.1016/S1001-0742(12)60277-0

[24] Patil Ashokrao B. Kashinath R. Patil and Satish K. Pardeshi. Ecofriendly synthesis and solar photocatalytic activity of S-doped ZnO. Journal of Hazardous Materials 183 no. 1-3 (2010): 315-323. https://doi.org/10.1016/j.jhazmat.2010.07.026

[25] Bechambi Olfa Sami Sayadi and Wahiba Najjar. Photocatalytic degradation of bisphenol A in the presence of C-doped ZnO: effect of operational parameters and photodegradation mechanism. Journal of Industrial and Engineering Chemistry 32 (2015): 201-210. https://doi.org/10.1016/j.jiec.2015.08.017

[26] Haibo Ouyang Huang Jian Feng Li Cuiyan Cao Liyun and Fei Jie. Synthesis of carbon doped ZnO with a porous structure and its solar-light photocatalytic properties. Materials Letters 111 (2013): 217-220. https://doi.org/10.1016/j.matlet.2013.08.081

[27] Wu Changle. Facile one-step synthesis of N-doped ZnO micropolyhedrons for efficient photocatalytic degradation of formaldehyde under visible-light irradiation. Applied surface science 319 (2014): 237-243. https://doi.org/10.1016/j.apsusc.2014.04.217

[28] Rajbongshi Biju Mani Anjalu Ramchiary and S. K. Samdarshi. Influence of N-doping on photocatalytic activity of ZnO nanoparticles under visible light irradiation. Materials Letters 134 (2014): 111-114. https://doi.org/10.1016/j.matlet.2014.07.073

[29] Gu Pengfei Xudong Wang Tao Li and Huimin Meng. Investigation of defects in N-doped ZnO powders prepared by a facile solvothermal method and their UV photocatalytic properties. Materials Research Bulletin 48 no. 11 (2013): 4699-4703. https://doi.org/10.1016/j.materresbull.2013.08.034

[30] Yu Zongbao Li-Chang Yin Yingpeng Xie Gang Liu Xiuliang Ma and Hui-Ming Cheng. Crystallinity-dependent substitutional nitrogen doping in ZnO and its improved visible light photocatalytic activity. Journal of colloid and interface science 400 (2013): 18-23. https://doi.org/10.1016/j.jcis.2013.02.046

[31] Kaneva Nina V. Dimitre T. Dimitrov and Ceco D. Dushkin. Effect of nickel doping on the photocatalytic activity of ZnO thin films under UV and visible light. Applied Surface Science 257 no. 18 (2011): 8113-8120. https://doi.org/10.1016/j.apsusc.2011.04.119

[32] Jia Xiaohua Huiqing Fan Mohammad Afzaal Xiangyang Wu and Paul O'Brien. Solid state synthesis of tin-doped ZnO at room temperature: characterization and its enhanced gas sensing and photocatalytic properties. Journal of hazardous materials 193 (2011): 194-199. https://doi.org/10.1016/j.jhazmat.2011.07.049

[33] Su C. Y. C. T. Lu W. T. Hsiao W. H. Liu and F. S. Shieu. Evaluation of the microstructural and photocatalytic properties of aluminum-doped zinc oxide coatings deposited by plasma spraying. Thin Solid Films 544 (2013): 170-174. https://doi.org/10.1016/j.tsf.2013.03.129

[34] bo Zhong Jun Jian zhang Li Yan Lu Xi yang He Jun Zeng Wei Hu and Yue cheng Shen. Fabrication of Bi3+-doped ZnO with enhanced photocatalytic performance. Applied Surface Science 258 no. 11 (2012): 4929-4933. https://doi.org/10.1016/j.apsusc.2012.01.121

[35] Ahmad M. E. Ahmed Z. L. Hong Z. Iqbal N. R. Khalid T. Abbas Imran Ahmad A. M. Elhissi and W. Ahmed. Structural optical and photocatalytic properties of hafnium doped zinc oxide nanophotocatalyst. Ceramics International 39 no. 8 (2013): 8693-8700. https://doi.org/10.1016/j.ceramint.2013.04.051

[36] Sanoop P. K. S. Anas S. Ananthakumar V. Gunasekar R. Saravanan and V. Ponnusami. Synthesis of yttrium doped nanocrystalline ZnO and its photocatalytic activity in methylene blue degradation. Arabian Journal of chemistry 9 (2016): S1618-S1626. https://doi.org/10.1016/j.arabjc.2012.04.023

[37] Zhong Jun Jian zhang Li Xi yang He Jun Zeng Yan Lu Wei Hu and Kun Lin. Improved photocatalytic performance of Pd-doped ZnO. Current Applied Physics 12 no. 3 (2012): 998-1001. https://doi.org/10.1016/j.cap.2012.01.003

[38] Khalil A. M. A. Gondal and M. A. Dastageer. Augmented photocatalytic activity of palladium incorporated ZnO nanoparticles in the disinfection of Escherichia coli microorganism from water. Applied Catalysis A: General 402 no. 1-2 (2011): 162-167. https://doi.org/10.1016/j.apcata.2011.05.041

[39] Divband B. M. Khatamian GR Kazemi Eslamian and M. Darbandi. Synthesis of Ag/ZnO nanostructures by different methods and investigation of their photocatalytic efficiency for 4-nitrophenol degradation. Applied surface science 284 (2013): 80-86. https://doi.org/10.1016/j.apsusc.2013.07.015

[40] Shinde S. S. C. H. Bhosale and K. Y. Rajpure. Oxidative degradation of acid orange 7 using Ag-doped zinc oxide thin films. Journal of Photochemistry and Photobiology B: Biology 117 (2012): 262-268. https://doi.org/10.1016/j.jphotobiol.2012.10.011

[41] Whang Thou-Jen Mu-Tao Hsieh and Huang-Han Chen. Visible-light photocatalytic degradation of methylene blue with laser-induced Ag/ZnO nanoparticles. Applied Surface Science 258 no. 7 (2012): 2796-2801. https://doi.org/10.1016/j.apsusc.2011.10.134

[42] Ba-Abbad Muneer M. Abdul Amir H. Kadhum Abu Bakar Mohamad Mohd S. Takriff and Kamaruzzaman Sopian. Visible light photocatalytic activity of Fe3+-doped ZnO nanoparticle prepared via sol-gel technique. Chemosphere 91 no. 11 (2013): 1604-1611. https://doi.org/10.1016/j.chemosphere.2012.12.055

[43] Rekha K. M. Nirmala Manjula G. Nair and A. Anukaliani. Structural optical photocatalytic and antibacterial activity of zinc oxide and manganese doped zinc oxide nanoparticles. Physica B: Condensed Matter 405 no. 15 (2010): 3180-3185. https://doi.org/10.1016/j.physb.2010.04.042

[44] Ullah Ruh and Joydeep Dutta. Photocatalytic degradation of organic dyes with manganese-doped ZnO nanoparticles. Journal of Hazardous materials 156 no. 1-3 (2008): 194-200. https://doi.org/10.1016/j.jhazmat.2007.12.033

[45] Ullah Ruh and Joydeep Dutta. Photocatalytic degradation of organic dyes with manganese-doped ZnO nanoparticles. Journal of Hazardous materials 156 no. 1-3 (2008): 194-200. https://doi.org/10.1016/j.jhazmat.2007.12.033

[46] Mahmood Mohammad Abbas Sunandan Baruah and Joydeep Dutta. Enhanced visible light photocatalysis by manganese doping or rapid crystallization with ZnO nanoparticles. Materials Chemistry and Physics 130 no. 1-2 (2011): 531-535. https://doi.org/10.1016/j.matchemphys.2011.07.018

[47] Thennarasu G. and A. Sivasamy. Enhanced visible photocatalytic activity of cotton ball like nano structured Cu doped ZnO for the degradation of organic pollutant. Ecotoxicology and Environmental Safety 134 (2016): 412-420. https://doi.org/10.1016/j.ecoenv.2015.10.030

[48] Mittal Manish Manoj Sharma and O. P. Pandey. UV-Visible light induced photocatalytic studies of Cu doped ZnO nanoparticles prepared by co-precipitation method. Solar Energy 110 (2014): 386-397. https://doi.org/10.1016/j.solener.2014.09.026

[49] Mohan Rajneesh Karthikeyan Krishnamoorthy and Sang-Jae Kim. Enhanced photocatalytic activity of Cu-doped ZnO nanorods. Solid State Communications 152 no. 5 (2012): 375-380. https://doi.org/10.1016/j.ssc.2011.12.008

[50] Jongnavakit P. P. Amornpitoksuk S. Suwanboon and N. J. A. S. S. Ndiege. Preparation and photocatalytic activity of Cu-doped ZnO thin films prepared by the

sol-gel method. Applied Surface Science 258 no. 20 (2012): 8192-8198.
https://doi.org/10.1016/j.apsusc.2012.05.021

[51] Fu Min Yalin Li Peng Lu Jing Liu and Fan Dong. Sol-gel preparation and enhanced photocatalytic performance of Cu-doped ZnO nanoparticles. Applied Surface Science 258 no. 4 (2011): 1587-1591. https://doi.org/10.1016/j.apsusc.2011.10.003

[52] He Rongliang Rosalie K. Hocking and Takuya Tsuzuki. Co-doped ZnO nanopowders: location of cobalt and reduction in photocatalytic activity. Materials Chemistry and Physics 132 no. 2-3 (2012): 1035-1040. https://doi.org/10.1016/j.matchemphys.2011.12.061

[53] Yayapao Oranuch Somchai Thongtem Anukorn Phuruangrat and Titipun Thongtem. Sonochemical synthesis photocatalysis and photonic properties of 3% Ce-doped ZnO nanoneedles. Ceramics International 39 (2013): S563-S568. https://doi.org/10.1016/j.ceramint.2012.10.136

[54] Rezaei M. and A. Habibi-Yangjeh. Simple and large scale refluxing method for preparation of Ce-doped ZnO nanostructures as highly efficient photocatalyst. Applied surface science 265 (2013): 591-596. https://doi.org/10.1016/j.apsusc.2012.11.053

[55] Faisal M. Adel A. Ismail Ahmed A. Ibrahim Houcine Bouzid and Saleh A. Al-Sayari. Highly efficient photocatalyst based on Ce doped ZnO nanorods: Controllable synthesis and enhanced photocatalytic activity. Chemical engineering journal 229 (2013): 225-233. https://doi.org/10.1016/j.cej.2013.06.004

[56] Karunakaran Chockalingam Paramasivan Gomathisankar and Govindasamy Manikandan. Preparation and characterization of antimicrobial Ce-doped ZnO nanoparticles for photocatalytic detoxification of cyanide. Materials Chemistry and Physics 123 no. 2-3 (2010): 585-594. https://doi.org/10.1016/j.matchemphys.2010.05.019

[57] Khataee Alireza Reza Darvishi Cheshmeh Soltani Atefeh Karimi and Sang Woo Joo. Sonocatalytic degradation of a textile dye over Gd-doped ZnO nanoparticles synthesized through sonochemical process. Ultrasonics Sonochemistry 23 (2015): 219-230. https://doi.org/10.1016/j.ultsonch.2014.08.023

[58] Zong Yanqing Zhe Li Xingmin Wang Jiantao Ma and Yi Men. Synthesis and high photocatalytic activity of Eu-doped ZnO nanoparticles. Ceramics international 40 no. 7 (2014): 10375-10382. https://doi.org/10.1016/j.ceramint.2014.02.123

[59] Yayapao Oranuch Titipun Thongtem Anukorn Phuruangrat and Somchai Thongtem. Ultrasonic-assisted synthesis of Nd-doped ZnO for photocatalysis. Materials Letters 90 (2013): 83-86. https://doi.org/10.1016/j.matlet.2012.09.027

[60] Kumar Surender and P. D. Sahare. Nd-doped ZnO as a multifunctional nanomaterial. Journal of rare earths 30 no. 8 (2012): 761-768. https://doi.org/10.1016/S1002-0721(12)60126-4

[61] Zhen Z. H. A. O. Ji-ling Song Jia-hong Zheng and Jian-she Lian. Optical properties and photocatalytic activity of Nd-doped ZnO powders. Transactions of Nonferrous Metals Society of China 24 no. 5 (2014): 1434-1439. https://doi.org/10.1016/S1003-6326(14)63209-X

[62] Sin Jin-Chung Sze-Mun Lam Keat-Teong Lee and Abdul Rahman Mohamed. Photocatalytic performance of novel samarium-doped spherical-like ZnO hierarchical nanostructures under visible light irradiation for 2 4-dichlorophenol degradation. Journal of colloid and interface science 401 (2013): 40-49. https://doi.org/10.1016/j.jcis.2013.03.043

[63] Sin Jin-Chung Sze-Mun Lam Keat-Teong Lee and Abdul Rahman Mohamed. Preparation and photocatalytic properties of visible light-driven samarium-doped ZnO nanorods. Ceramics International 39 no. 5 (2013): 5833-5843. https://doi.org/10.1016/j.ceramint.2013.01.004

[64] Anandan S. A. Vinu KLP Sheeja Lovely N. Gokulakrishnan P. Srinivasu T. Mori V. Murugesan V. Sivamurugan and K. Ariga. Photocatalytic activity of La-doped ZnO for the degradation of monocrotophos in aqueous suspension. Journal of Molecular Catalysis A: Chemical 266 no. 1-2 (2007): 149-157. https://doi.org/10.1016/j.molcata.2006.11.008

[65] Khataee Alireza Reza Darvishi Cheshmeh Soltani Atefeh Karimi and Sang Woo Joo. Sonocatalytic degradation of a textile dye over Gd-doped ZnO nanoparticles synthesized through sonochemical process. Ultrasonics Sonochemistry 23 (2015): 219-230. https://doi.org/10.1016/j.ultsonch.2014.08.023

[66] Mascolo G. R. Comparelli M. L. Curri G. Lovecchio A. Lopez and A. Agostiano. Photocatalytic degradation of methyl red by TiO2: Comparison of the efficiency of immobilized nanoparticles versus conventional suspended catalyst. Journal of Hazardous Materials 142 no. 1-2 (2007): 130-137. https://doi.org/10.1016/j.jhazmat.2006.07.068

[67] Hu Bing Jin Zhou and Xiu-Min Wu. Decoloring methyl orange under sunlight by a photocatalytic membrane reactor based on ZnO nanoparticles and polypropylene macroporous membrane. International Journal of Polymer Science 2013 (2013). https://doi.org/10.1155/2013/451398

[68] Brezova V. M. Jankovičová M. Soldan A. Blažková M. Rehakova I. Šurina M. Čeppan and B. Havlinova. Photocatalytic degradation of p-toluenesulphonic acid in aqueous systems containing powdered and immobilized titanium dioxide. Journal of Photochemistry and Photobiology A: Chemistry 83 no. 1 (1994): 69-75. https://doi.org/10.1016/1010-6030(94)03804-X

[69] Dijkstra M. F. J. A. Michorius H. Buwalda H. J. Panneman J. G. M. Winkelman and A. A. C. M. Beenackers. Comparison of the efficiency of immobilized and suspended systems in photocatalytic degradation. Catalysis Today 66 no. 2-4 (2001): 487-494. https://doi.org/10.1016/S0920-5861(01)00257-7

[70] Mozia Sylwia. Photocatalytic membrane reactors (PMRs) in water and wastewater treatment. A review. Separation and purification technology 73 no. 2 (2010): 71-91. https://doi.org/10.1016/j.seppur.2010.03.021

[71] Mansilla H. D. C. Bravo R. Ferreyra M. I. Litter W. F. Jardim C. Lizama J. Freer and J. Fernandez. Photocatalytic EDTA degradation on suspended and immobilized TiO2. Journal of Photochemistry and Photobiology A: Chemistry 181 no. 2-3 (2006): 188-194. https://doi.org/10.1016/j.jphotochem.2005.11.023

[72] González Verónica Israel López Raul Martín Palma Yolanda Peña and Idalia Gómez. Organic-inorganic hybrid solar cells based on 1D ZnO/P3HT active layers and 0D Au as cathode. Materials Research Express 7 no. 7 (2020): 075005. https://doi.org/10.1088/2053-1591/ab9cec

[73] Zhou Yunfei Michael Eck and Michael Krüger. Bulk-heterojunction hybrid solar cells based on colloidal nanocrystals and conjugated polymers. Energy & Environmental Science 3 no. 12 (2010): 1851-1864. https://doi.org/10.1039/c0ee00143k

[74] Halim Mohammad A. Harnessing sun's energy with quantum dots based next generation solar cell. Nanomaterials 3 no. 1 (2012): 22-47. https://doi.org/10.3390/nano3010022

[75] Saboor Abdus Syed Mujtaba Shah and Hazrat Hussain. Band gap tuning and applications of ZnO nanorods in hybrid solar cell: Ag-doped verses Nd-doped ZnO nanorods. Materials Science in Semiconductor Processing 93 (2019): 215-225. https://doi.org/10.1016/j.mssp.2019.01.009

[76] Maragliano C. S. Lilliu M. S. Dahlem M. Chiesa T. Souier and M. Stefancich. Quantifying charge carrier concentration in ZnO thin films by Scanning Kelvin Probe Microscopy. Scientific reports 4 no. 1 (2014): 1-7. https://doi.org/10.1038/srep04203

[77] Consonni Vincent Joe Briscoe Erki Kärber Xuan Li and Thomas Cossuet. ZnO nanowires for solar cells: a comprehensive review. Nanotechnology 30 no. 36 (2019): 362001. https://doi.org/10.1088/1361-6528/ab1f2e

[78] Gonzalez-Valls Irene and Monica Lira-Cantu. Vertically-aligned nanostructures of ZnO for excitonic solar cells: a review. Energy & Environmental Science 2 no. 1 (2009): 19-34. https://doi.org/10.1039/B811536B

[79] Hames Yakup Zühal Alpaslan Arif Kösemen Sait Eren San and Yusuf Yerli. Electrochemically grown ZnO nanorods for hybrid solar cell applications. Solar Energy 84 no. 3 (2010): 426-431. https://doi.org/10.1016/j.solener.2009.12.013

[80] Qi Juanjuan Junwei Chen Weili Meng Xiaoyan Wu Changwen Liu Wenjin Yue and Mingtai Wang. Recent advances in hybrid solar cells based on metal oxide nanostructures. Synthetic Metals 222 (2016): 42-65. https://doi.org/10.1016/j.synthmet.2016.04.027

[81] Beek Waldo JE Martijn M. Wienk and Rene AJ Janssen. Efficient hybrid solar cells from zinc oxide nanoparticles and a conjugated polymer. Advanced Materials 16 no. 12 (2004): 1009-1013. https://doi.org/10.1002/adma.200306659

[82] Wang Huan Guobin Yi Xihong Zu Pei Qin Miao Tan and Hongsheng Luo. Photoelectric characteristics of the p-n junction between ZnO nanorods and polyaniline nanowires and their application as a UV photodetector. Materials Letters 162 (2016): 83-86. https://doi.org/10.1016/j.matlet.2015.09.128

[83] Greene Lori E. Matt Law Benjamin D. Yuhas and Peidong Yang. ZnO– TiO2 core–shell nanorod/P3HT solar cells. The Journal of Physical Chemistry C 111 no. 50 (2007): 18451-18456. https://doi.org/10.1021/jp0775931

[84] Oosterhout Stefan D. Martijn M. Wienk Svetlana S. Van Bavel Ralf Thiedmann L. Jan Anton Koster Jan Gilot Joachim Loos Volker Schmidt and René AJ Janssen. The effect of three-dimensional morphology on the efficiency of hybrid polymer solar cells. Nature materials 8 no. 10 (2009): 818-824. https://doi.org/10.1038/nmat2533

[85] Li Shao-Sian and Chun-Wei Chen. Polymer-metal-oxide hybrid solar cells. Journal of Materials Chemistry A 1 no. 36 (2013): 10574-10591. https://doi.org/10.1039/c3ta11998j

[86] Helgesen Martin Roar Søndergaard and Frederik C. Krebs. Advanced materials and processes for polymer solar cell devices. Journal of Materials Chemistry 20 no. 1 (2010): 36-60. https://doi.org/10.1039/B913168J

[87] Beek Waldo JE Martijn M. Wienk Martijn Kemerink Xiaoniu Yang and René AJ Janssen. Hybrid zinc oxide conjugated polymer bulk heterojunction solar cells. The Journal of Physical Chemistry B 109 no. 19 (2005): 9505-9516. https://doi.org/10.1021/jp050745x

[88] Zhou Yunfei Michael Eck and Michael Krüger. Bulk-heterojunction hybrid solar cells based on colloidal nanocrystals and conjugated polymers. Energy & Environmental Science 3 no. 12 (2010): 1851-1864. https://doi.org/10.1039/c0ee00143k

[89] Ravirajan Punniamoorthy Ana M. Peiró Mohammad K. Nazeeruddin Michael Graetzel Donal DC Bradley James R. Durrant and Jenny Nelson. Hybrid polymer/zinc oxide photovoltaic devices with vertically oriented ZnO nanorods and an amphiphilic molecular interface layer. The Journal of Physical Chemistry B 110 no. 15 (2006): 7635-7639. https://doi.org/10.1021/jp0571372

[90] Gonzalez-Valls Irene and Monica Lira-Cantu. Vertically-aligned nanostructures of ZnO for excitonic solar cells: a review. Energy & Environmental Science 2 no. 1 (2009): 19-34. https://doi.org/10.1039/B811536B

[91] Jung Seungon Junghyun Lee Jihyung Seo Ungsoo Kim Yunseong Choi and Hyesung Park. Development of annealing-free solution-processable inverted organic solar cells

with N-doped graphene electrodes using zinc oxide nanoparticles. Nano letters 18 no. 2 (2018): 1337-1343. https://doi.org/10.1021/acs.nanolett.7b05026

[92] Olson Dana C. Jorge Piris Reuben T. Collins Sean E. Shaheen and David S. Ginley. Hybrid photovoltaic devices of polymer and ZnO nanofiber composites. Thin solid films 496 no. 1 (2006): 26-29. https://doi.org/10.1016/j.tsf.2005.08.179

[93] Olson Dana C. Yun-Ju Lee Matthew S. White Nikos Kopidakis Sean E. Shaheen David S. Ginley James A. Voigt and Julia WP Hsu. Effect of ZnO processing on the photovoltage of ZnO/poly (3-hexylthiophene) solar cells. The Journal of Physical Chemistry C 112 no. 26 (2008): 9544-9547. https://doi.org/10.1021/jp802626u

[94] Baeten Linny Bert Conings Hans-Gerd Boyen Jan D'Haen An Hardy Marc D'Olieslaeger Jean V. Manca and Marlies K. Van Bael. Towards efficient hybrid solar cells based on fully polymer infiltrated ZnO nanorod arrays. Advanced Materials 23 no. 25 (2011): 2802-2805. https://doi.org/10.1002/adma.201100414

[95] Han Jianhua Zhifeng Liu Xuerong Zheng Keying Guo Xueqi Zhang Tiantian Hong Bo Wang and Junqi Liu. Trilaminar ZnO/ZnS/Sb 2 S 3 nanotube arrays for efficient inorganic-organic hybrid solar cells. Rsc Advances 4 no. 45 (2014): 23807-23814. https://doi.org/10.1039/c4ra02554g

[96] Alshanableh Abdelelah Sin Tee Tan Chi Chin Yap Hock Beng Lee Hind Fadhil Oleiwi Kai Jeat Hong Mohd Hafizuddin Hj Jumali and Muhammad Yahaya. Surface engineering of ZnO nanorod for inverted organic solar cell. Materials Science and Engineering: B 238 (2018): 136-141. https://doi.org/10.1016/j.mseb.2018.12.024

[97] Huang Jia Zhigang Yin and Qingdong Zheng. Applications of ZnO in organic and hybrid solar cells. Energy & Environmental Science 4 no. 10 (2011): 3861-3877. https://doi.org/10.1039/c1ee01873f

[98] Zhang Rong Chengbin Fei Bo Li Haoyu Fu Jianjun Tian and Guozhong Cao. Continuous size tuning of monodispersed ZnO nanoparticles and its size effect on the performance of perovskite solar cells. ACS Applied Materials & Interfaces 9 no. 11 (2017): 9785-9794. https://doi.org/10.1021/acsami.7b00726

[99] Zhang Huiyin Jiangjian Shi Xin Xu Lifeng Zhu Yanhong Luo Dongmei Li and Qingbo Meng. Mg-doped TiO 2 boosts the efficiency of planar perovskite solar cells to exceed 19%. Journal of Materials Chemistry A 4 no. 40 (2016): 15383-15389. https://doi.org/10.1039/C6TA06879K

[100] Fan Sheng-Qiang Baizeng Fang Jung Ho Kim Jeum-Jong Kim Jong-Sung Yu and Jaejung Ko. Hierarchical nanostructured spherical carbon with hollow core/mesoporous shell as a highly efficient counter electrode in CdSe quantum-dot-sensitized solar cells. Applied Physics Letters 96 no. 6 (2010): 063501. https://doi.org/10.1063/1.3313948

[101] Zheng Yan-Zhen Er-Fei Zhao Fan-Li Meng Xue-Sen Lai Xue-Mei Dong Jiao-Jiao Wu and Xia Tao. Iodine-doped ZnO nanopillar arrays for perovskite solar cells with

high efficiency up to 18.24%. Journal of Materials Chemistry A 5 no. 24 (2017): 12416-12425. https://doi.org/10.1039/C7TA03150E

[102] Loh Leonard and Steve Dunn. Recent progress in ZnO-based nanostructured ceramics in solar cell applications. Journal of nanoscience and nanotechnology 12 no. 11 (2012): 8215-8230. https://doi.org/10.1166/jnn.2012.6680

[103] Miskin Caleb K. Swapnil D. Deshmukh Venkata Vasiraju Kevin Bock Gaurav Mittal Angela Dubois-Camacho Sreeram Vaddiraju and Rakesh Agrawal. Lead chalcogenide nanoparticles and their size-controlled self-assemblies for thermoelectric and photovoltaic applications. ACS Applied Nano Materials 2 no. 3 (2019): 1242-1252. https://doi.org/10.1021/acsanm.8b02125

[104] Leschkies Kurtis S. Timothy J. Beatty Moon Sung Kang David J. Norris and Eray S. Aydil. Solar cells based on junctions between colloidal PbSe nanocrystals and thin ZnO films. ACS nano 3 no. 11 (2009): 3638-3648. https://doi.org/10.1021/nn901139d

[105] Bhaumik Anagh A. Haque P. Karnati M. F. N. Taufique R. Patel and Kartik Ghosh. Copper oxide based nanostructures for improved solar cell efficiency. Thin Solid Films 572 (2014): 126-133. https://doi.org/10.1016/j.tsf.2014.09.056

[106] Fujimoto Kazuya Takeo Oku Tsuyoshi Akiyama and Atsushi Suzuki. Fabrication and characterization of copper oxide-zinc oxide solar cells prepared by electrodeposition. In Journal of Physics: Conference Series vol. 433 no. 1 p. 012024. IOP Publishing 2013. https://doi.org/10.1088/1742-6596/433/1/012024

[107] Guillén Elena Eneko Azaceta Laurence M. Peter Arnost Zukal Ramón Tena-Zaera and Juan A. Anta. ZnO solar cells with an indoline sensitizer: a comparison between nanoparticulate films and electrodeposited nanowire arrays. Energy & Environmental Science 4 no. 9 (2011): 3400-3407. https://doi.org/10.1039/c0ee00500b

[108] Kumari J. M. K. W. N. Sanjeevadharshini M. A. K. L. Dissanayake G. K. R. Senadeera and C. A. Thotawatthage. The effect of TiO2 photo anode film thickness on photovoltaic properties of dye-sensitized solar cells. Ceylon Journal of Science 45 no. 1 (2016). https://doi.org/10.4038/cjs.v45i1.7362

[109] Wang Lijing Hongju Zhai Gan Jin Xiaoying Li Chunwei Dong Hao Zhang Bai Yang Haiming Xie and Haizhu Sun. 3D porous ZnO-SnS p-n heterojunction for visible light driven photocatalysis. Physical Chemistry Chemical Physics 19 no. 25 (2017): 16576-16585. https://doi.org/10.1039/C7CP01687E

[110] Peksu Elif and Hakan Karaagac. Synthesis of ZnO nanowires and their photovoltaic application: znO nanowires/AgGaSe2 thin film core-shell solar cell. Journal of Nanomaterials 2015 (2015). https://doi.org/10.1155/2015/516012

[111] Irfan Ahmad. First principle investigations to enhance the charge transfer properties by bridge elongation. Journal of Theoretical and Computational Chemistry 13 no. 02 (2014): 1450013. https://doi.org/10.1142/S0219633614500138

[112] Nozik Arthur J. Quantum dot solar cells. Physica E: Low-dimensional Systems and Nanostructures 14 no. 1-2 (2002): 115-120. https://doi.org/10.1016/S1386-9477(02)00374-0

[113] Schaller Richard D. and Victor I. Klimov. High efficiency carrier multiplication in PbSe nanocrystals: implications for solar energy conversion. Physical review letters 92 no. 18 (2004): 186601. https://doi.org/10.1103/PhysRevLett.92.186601

[114] Robel István Vaidyanathan Subramanian Masaru Kuno and Prashant V. Kamat. Quantum dot solar cells. Harvesting light energy with CdSe nanocrystals molecularly linked to mesoscopic TiO2 films. Journal of the American Chemical Society 128 no. 7 (2006): 2385-2393. https://doi.org/10.1021/ja056494n

[115] Fan Sheng-Qiang Baizeng Fang Jung Ho Kim Jeum-Jong Kim Jong-Sung Yu and Jaejung Ko. Hierarchical nanostructured spherical carbon with hollow core/mesoporous shell as a highly efficient counter electrode in CdSe quantum-dot-sensitized solar cells. Applied Physics Letters 96 no. 6 (2010): 063501. https://doi.org/10.1063/1.3313948

[116] Diguna Lina J. Motonobu Murakami Akira Sato Yuki Kumagai Taishi Ishihara Naoki Kobayashi Qing Shen and Taro Toyoda. Photoacoustic and photoelectrochemical characterization of inverse opal TiO2 sensitized with CdSe quantum dots. Japanese journal of applied physics 45 no. 6S (2006): 5563. https://doi.org/10.1143/JJAP.45.5563

[117] Matsumura Michio Shigeyuki Matsudaira Hiroshi Tsubomura Masasuke Takata and Hiroaki Yanagida. Dye sensitization and surface structures of semiconductor electrodes. Industrial & Engineering Chemistry Product Research and Development 19 no. 3 (1980): 415-421. https://doi.org/10.1021/i360075a025

[118] Xie Xueping Jinfeng Liao Xiaoru Shao Qianshun Li and Yunfeng Lin. The effect of shape on cellular uptake of gold nanoparticles in the forms of stars rods and triangles. Scientific reports 7 no. 1 (2017): 1-9. https://doi.org/10.1038/s41598-017-04229-z

[119] Rensmo Håkan Karin Keis Henrik Lindström Sven Södergren Anita Solbrand Anders Hagfeldt S-E. Lindquist L. N. Wang and M. Muhammed. High light-to-energy conversion efficiencies for solar cells based on nanostructured ZnO electrodes. The Journal of Physical Chemistry B 101 no. 14 (1997): 2598-2601. https://doi.org/10.1021/jp962918b

[120] Horiuchi Hiroaki Ryuzi Katoh Kohjiro Hara Masatoshi Yanagida Shigeo Murata Hironori Arakawa and M. Tachiya. Electron injection efficiency from excited N3 into nanocrystalline ZnO films: effect of (N3− Zn2+) aggregate formation. The Journal of Physical Chemistry B 107 no. 11 (2003): 2570-2574. https://doi.org/10.1021/jp0220027

[121] Hara Kohjiro Takaro Horiguchi Tohru Kinoshita Kazuhiro Sayama Hideki Sugihara and Hironori Arakawa. Highly efficient photon-to-electron conversion with mercurochrome-sensitized nanoporous oxide semiconductor solar cells. Solar energy materials and solar cells 64 no. 2 (2000): 115-134. https://doi.org/10.1016/S0927-0248(00)00065-9

[122] Keis Karin Jan Lindgren Sten-Eric Lindquist and Anders Hagfeldt. Studies of the adsorption process of Ru complexes in nanoporous ZnO electrodes. Langmuir 16 no. 10 (2000): 4688-4694. https://doi.org/10.1021/la9912702

[123] Parks George A. The isoelectric points of solid oxides solid hydroxides and aqueous hydroxo complex systems. Chemical Reviews 65 no. 2 (1965): 177-198. https://doi.org/10.1021/cr60234a002

[124] Keis Karin C. Bauer Gerrit Boschloo Anders Hagfeldt K. Westermark Håkan Rensmo and Hans Siegbahn. Nanostructured ZnO electrodes for dye-sensitized solar cell applications. Journal of Photochemistry and photobiology A: Chemistry 148 no. 1-3 (2002): 57-64. https://doi.org/10.1016/S1010-6030(02)00039-4

[125] Keis K. C. Bauer G. Boschloo and A. Hagfeldt. K. westermark H. Rensmo H. Siegbahn. J. Photochem. Photobiol. A 148 (2002): 57. https://doi.org/10.1016/S1010-6030(02)00039-4

[126] Law M. L. E. Greene and J. C. Johnson. R. saykally PD Yang. Nat. Mater 4 (2005): 455. https://doi.org/10.1038/nmat1387

[127] Zhao Qidong Tengfeng Xie Linlin Peng Yanhong Lin Ping Wang Liang Peng and Dejun Wang. Size-and orientation-dependent photovoltaic properties of ZnO nanorods. The Journal of Physical Chemistry C 111 no. 45 (2007): 17136-17145. https://doi.org/10.1021/jp075368y

[128] Chen Z. H. Y. B. Tang C. P. Liu Y. H. Leung G. D. Yuan L. M. Chen Y. Q. Wang et al. Vertically aligned ZnO nanorod arrays sentisized with gold nanoparticles for Schottky barrier photovoltaic cells. The Journal of Physical Chemistry C 113 no. 30 (2009): 13433-13437. https://doi.org/10.1021/jp903153w

[129] Yun Sining Juneyoung Lee Jooyoung Chung and Sangwoo Lim. Improvement of ZnO nanorod-based dye-sensitized solar cell efficiency by Al-doping. Journal of Physics and Chemistry of Solids 71 no. 12 (2010): 1724-1731 https://doi.org/10.1016/j.jpcs.2010.08.020

[130] Lee Chuan-Pei Chen-Yu Chou Chia-Yuan Chen Min-Hsin Yeh Lu-Yin Lin R. Vittal Chun-Guey Wu and Kuo-Chuan Ho. Zinc oxide-based dye-sensitized solar cells with a ruthenium dye containing an alkyl bithiophene group. Journal of Power Sources 246 (2014): 1-9. https://doi.org/10.1016/j.jpowsour.2013.05.101

[131] Martinson Alex BF James E. McGarrah Mohammed OK Parpia and Joseph T. Hupp. Dynamics of charge transport and recombination in ZnO nanorod array dye-

sensitized solar cells. Physical Chemistry Chemical Physics 8 no. 40 (2006): 4655-4659. https://doi.org/10.1039/b610566a

[132] Akhtar M. Shaheer M. Alam Khan Myung Seok Jeon and O-Bong Yang. Controlled synthesis of various ZnO nanostructured materials by capping agents-assisted hydrothermal method for dye-sensitized solar cells. Electrochimica Acta 53 no. 27 (2008): 7869-7874. https://doi.org/10.1016/j.electacta.2008.05.055

[133] Kong Xiang Yang Yong Ding Rusen Yang and Zhong Lin Wang. Single-crystal nanorings formed by epitaxial self-coiling of polar nanobelts. Science 303 no. 5662 (2004): 1348-1351. https://doi.org/10.1126/science.1092356

[134] Mou Jixia Weiguang Zhang Jun Fan Hong Deng and Wei Chen. Facile synthesis of ZnO nanobullets/nanoflakes and their applications to dye-sensitized solar cells. Journal of alloys and compounds 509 no. 3 (2011): 961-965. https://doi.org/10.1016/j.jallcom.2010.09.148

[135] Abd-Ellah Marwa Nafiseh Moghimi Lei Zhang Nina F. Heinig Liyan Zhao Joseph P. Thomas and K. T. Leung. Effect of electrolyte conductivity on controlled electrochemical synthesis of zinc oxide nanotubes and nanorods. The Journal of Physical Chemistry C 117 no. 13 (2013): 6794-6799. https://doi.org/10.1021/jp312321t

[136] Jiang C. Y. X. W. Sun G. Q. Lo D. L. Kwong and J. X. Wang. Improved dye-sensitized solar cells with a ZnO-nanoflower photoanode. Applied Physics Letters 90 no. 26 (2007): 263501. https://doi.org/10.1063/1.2751588

[137] Lin Chia-Yu Yi-Hsuan Lai Hsin-Wei Chen Jian-Ging Chen Chung-Wei Kung L. R. Vittal and Kuo-Chuan Ho. Highly efficient dye-sensitized solar cell with a ZnO nanosheet-based photoanode. Energy & Environmental Science 4 no. 9 (2011): 3448-3455. https://doi.org/10.1039/c0ee00587h

[138] Ameen Sadia M. Shaheer Akhtar and Hyung Shik Shin. Growth and characterization of nanospikes decorated ZnO sheets and their solar cell application. Chemical engineering journal 195 (2012): 307-313. https://doi.org/10.1016/j.cej.2012.04.081

[139] Lupan Oleg Victoire Marie Guerin Lidia Ghimpu I. M. Tiginyanu and Thierry Pauporté. Nanofibrous-like ZnO layers deposited by magnetron sputtering and their integration in dye-sensitized solar cells. Chemical Physics Letters 550 (2012): 125-129. https://doi.org/10.1016/j.cplett.2012.08.071

[140] Kim Il-Doo Jae-Min Hong Byong Hong Lee Dong Young Kim Eun-Kyung Jeon Duck-Kyun Choi and Dae-Jin Yang. Dye-sensitized solar cells using network structure of electrospun ZnO nanofiber mats. Applied Physics Letters 91 no. 16 (2007): 163109. https://doi.org/10.1063/1.2799581

[141] Chang Jin Rasin Ahmed Hongxia Wang Hongwei Liu Renzhi Li Peng Wang and Eric R. Waclawik. ZnO nanocones with high-index $\{10\bar{1}1\}$ facets for enhanced energy

conversion efficiency of dye-sensitized solar cells. The Journal of Physical Chemistry C 117 no. 27 (2013): 13836-13844. https://doi.org/10.1021/jp402742n

[142] Chang Wei-Chen Lu-Yin Lin and Wan-Chin Yu. Bifunctional zinc oxide nanoburger aggregates as the dye-adsorption and light-scattering layer for dye-sensitized solar cells. Electrochimica Acta 169 (2015): 456-461. https://doi.org/10.1016/j.electacta.2015.04.056

[143] Wang Feifei Ruibin Liu Anlian Pan Sishen Xie and Bingsuo Zou. A simple and cheap way to produce porous ZnO ribbons and their photovoltaic response. Materials Letters 61 no. 23-24 (2007): 4459-4462. https://doi.org/10.1016/j.matlet.2007.02.021

[144] Qu Jie Yongan Yang Qingduan Wu Paul R. Coxon Yingjun Liu Xiong He Kai Xi Ningyi Yuan and Jianning Ding. Hedgehog-like hierarchical ZnO needle-clusters with superior electron transfer kinetics for dye-sensitized solar cells. RSC Advances 4 no. 22 (2014): 11430-11437. https://doi.org/10.1039/C3RA45929B

[145] Ameen Sadia M. Shaheer Akhtar Hyung-Kee Seo Young Soon Kim and Hyung Shik Shin. Influence of Sn doping on ZnO nanostructures from nanoparticles to spindle shape and their photoelectrochemical properties for dye sensitized solar cells. Chemical Engineering Journal 187 (2012): 351-356. https://doi.org/10.1016/j.cej.2012.01.097

[146] Bahadur Lal and Suman Kushwaha. Structural and optical properties of tripod-like ZnO thin film and its application in dye-sensitized solar cell. Journal of Solid State Electrochemistry 17 no. 7 (2013): 2001-2008. https://doi.org/10.1007/s10008-013-2053-z

[147] Thambidurai M. N. Muthukumarasamy Dhayalan Velauthapillai and Changhee Lee. Synthesis and characterization of flower like ZnO nanorods for dye-sensitized solar cells. Journal of Materials Science: Materials in Electronics 24 no. 7 (2013): 2367-2371. https://doi.org/10.1007/s10854-013-1103-8

[148] Al-Agel F. A. M. Shaheer Akhtar H. Alshammari A. Alshammari and Shamshad A. Khan. Solution processed ZnO rectangular prism as an effective photoanode material for dye sensitized solar cells. Materials Letters 147 (2015): 119-122. https://doi.org/10.1016/j.matlet.2015.02.025

[149] Lee Kun-Mu Wei-Hao Chiu Chih-Yu Hsu Hsin-Ming Cheng Chia-Hua Lee and Chun-Guey Wu. Ionic liquid diffusion properties in tetrapod-like ZnO photoanode for dye-sensitized solar cells. Journal of Power Sources 216 (2012): 330-336. https://doi.org/10.1016/j.jpowsour.2012.05.079

[150] Kozhummal Rajeevan Yang Yang Firat Güder Andreas Hartel Xiaoli Lu Umut M. Küçükbayrak Aurelio Mateo-Alonso Miko Elwenspoek and Margit Zacharias. Homoepitaxial branching: an unusual polymorph of zinc oxide derived from seeded solution growth. ACS nano 6 no. 8 (2012): 7133-7141. https://doi.org/10.1021/nn302188q

[151] Lee Chuan-Pei Jen-Chieh Lin Yi-Chun Wang Chen-Yu Chou Min-Hsin Yeh R. Vittal and Kuo-Chuan Ho. Synthesis of hexagonal ZnO clubs with opposite faces of unequal dimensions for the photoanode of dye-sensitized solar cells. Physical Chemistry Chemical Physics 13 no. 47 (2011): 20999-21008. https://doi.org/10.1039/c1cp21762c

[152] Zhou Yi Ce Liu Mengyao Li Hongyan Wu Xian Zhong Dang Li and Difa Xu. Fabrication and optical properties of ordered sea urchin-like ZnO nanostructures by a simple hydrothermal process. Materials Letters 106 (2013): 94-96. https://doi.org/10.1016/j.matlet.2013.04.102

[153] Qin Zi Yunhua Huang Junjie Qi Huifeng Li Jia Su and Yue Zhang. Facile synthesis and photoelectrochemical performance of the bush-like ZnO nanosheets film. Solid state sciences 14 no. 1 (2012): 155-158 https://doi.org/10.1016/j.solidstatesciences.2011.11.014

[154] Pawar R. C. J. S. Shaikh P. S. Shinde and P. S. Patil. Dye sensitized solar cells based on zinc oxide bottle brush. Materials Letters 65 no. 14 (2011): 2235-2237. https://doi.org/10.1016/j.matlet.2011.04.045

[155] McCune Mallarie Wei Zhang and Yulin Deng. High efficiency dye-sensitized solar cells based on three-dimensional multilayered ZnO nanowire arrays with caterpillar-like structure. Nano letters 12 no. 7 (2012): 3656-3662. https://doi.org/10.1021/nl301407b

[156] Wang J. X. Chi Man Lawrence Wu Wing Sze Cheung L. B. Luo Z. B. He G. D. Yuan W. J. Zhang Chun Sing Lee and Shuit Tong Lee. Synthesis of hierarchical porous ZnO disklike nanostructures for improved photovoltaic properties of dye-sensitized solar cells. The Journal of Physical Chemistry C 114 no. 31 (2010): 13157-13161. https://doi.org/10.1021/jp100637c

[157] Umar Ahmad. Growth of comb-like ZnO nanostructures for dye-sensitized solar cells applications. Nanoscale research letters 4 no. 9 (2009): 1004-1008. https://doi.org/10.1007/s11671-009-9353-3

[158] Hu Xiulan Yoshitake Masuda Tatsuki Ohji and Kazumi Kato. Fabrication of Blanket-Like Assembled ZnO Nanowhiskers Using an Aqueous Solution. Journal of the American Ceramic Society 92 no. 4 (2009): 922-926. https://doi.org/10.1111/j.1551-2916.2009.03024.x

[159] Wang Yuqiao Xia Cui Yuan Zhang Xiaorui Gao and Yueming Sun. Preparation of cauliflower-like ZnO films by chemical bath deposition: photovoltaic performance and equivalent circuit of dye-sensitized solar cells. Journal of Materials Science & Technology 29 no. 2 (2013): 123-127. https://doi.org/10.1016/j.jmst.2012.12.019

[160] Shi Yantao Chao Zhu Lin Wang Wei Li Kwok Kwong Fung and Ning Wang. Asymmetric ZnO Panel-Like Hierarchical Architectures with Highly Interconnected

Pathways for Free-Electron Transport and Photovoltaic Improvements. Chemistry-A European Journal 19 no. 1 (2013): 282-287. https://doi.org/10.1002/chem.201202527

[161] Chauhan Ratna Manish Shinde Abhinav Kumar Suresh Gosavi and Dinesh P. Amalnerkar. Hierarchical zinc oxide pomegranate and hollow sphere structures as efficient photoanodes for dye-sensitized solar cells. Microporous and Mesoporous Materials 226 (2016): 201-208. https://doi.org/10.1016/j.micromeso.2015.11.054

[162] Senthil T. S. N. Muthukumarasamy and Misook Kang. Applications of highly ordered paddle wheel like structured ZnO nanorods in dye sensitized solar cells. Materials Letters 102 (2013): 26-29. https://doi.org/10.1016/j.matlet.2013.03.097

[163] Li Z Zhou Y Xue G Yu T Liu J Zou Z. Fabrication of hierarchically assembled microspheres consisting of nanoporousZnOnanosheets for high-efficiency dye-sensitized solar cells. J Mater Chem 2012;22(29):14341-5. https://doi.org/10.1039/c2jm32823b

[164] He Chun-Xiu Bing-Xin Lei Yu-Fen Wang Cheng-Yong Su Yue-Ping Fang and Dai-Bin Kuang. Sonochemical preparation of hierarchical ZnO hollow spheres for efficient dye-sensitized solar cells. Chemistry-A European Journal 16 no. 29 (2010): 8757-8761. https://doi.org/10.1002/chem.201000264

[165] Lamberti A. A. Sacco M. Laurenti M. Fontana C. F. Pirri and S. Bianco. Sponge-like ZnO nanostructures by low temperature water vapor-oxidation method as dye-sensitized solar cell photoanodes. Journal of alloys and compounds 615 (2014): S487-S490 https://doi.org/10.1016/j.jallcom.2013.12.091

[166] Lamberti Andrea Rossana Gazia Adriano Sacco Stefano Bianco Marzia Quaglio Angelica Chiodoni Elena Tresso and Candido Fabrizio Pirri. Coral-shaped ZnO nanostructures for dye-sensitized solar cell photoanodes. Progress in Photovoltaics: Research and Applications 22 no. 2 (2014): 189-197. https://doi.org/10.1002/pip.2251

[167] Yun, Sining, Yong Qin, Alexander R. Uhl, Nick Vlachopoulos, Min Yin, Dongdong Li, Xiaogang Han, and Anders Hagfeldt. "New-generation integrated devices based on dye-sensitized and perovskite solar cells." Energy & Environmental Science 11, no. 3 (2018): 476-526. https://doi.org/10.1039/C7EE03165C

[168] Park, Nam-Gyu. "Research direction toward scalable, stable, and high efficiency perovskite solar cells." Advanced Energy Materials 10, no. 13 (2020): 1903106. https://doi.org/10.1002/aenm.201903106

[169] Ouedraogo, Nabonswende Aida Nadege, Yichuan Chen, Yue Yue Xiao, Qi Meng, Chang Bao Han, Hui Yan, and Yongzhe Zhang. "Stability of all-inorganic perovskite solar cells." Nano Energy 67 (2020): 104249. https://doi.org/10.1016/j.nanoen.2019.104249

[170] Schwenzer, Jonas A., Lucija Rakocevic, Tobias Abzieher, Diana Rueda-Delgado, Somayeh Moghadamzadeh, Saba Gharibzadeh, Ihteaz M. Hossain et al. "Toward stable perovskite solar cell architectures: robustness against temperature variations of

real-world conditions." IEEE Journal of Photovoltaics 10, no. 3 (2020): 777-784. https://doi.org/10.1109/JPHOTOV.2020.2969785

[171] Kojima, Akihiro, Kenjiro Teshima, Yasuo Shirai, and Tsutomu Miyasaka. "Organometal halide perovskites as visible-light sensitizers for photovoltaic cells." Journal of the american chemical society 131, no. 17 (2009): 6050-6051. https://doi.org/10.1021/ja809598r

[172] Zhao, Jingjing, Xiaopeng Zheng, Yehao Deng, Tao Li, Yuchuan Shao, Alexei Gruverman, Jeffrey Shield, and Jinsong Huang. "Is Cu a stable electrode material in hybrid perovskite solar cells for a 30-year lifetime?." Energy & Environmental Science 9, no. 12 (2016): 3650-3656. https://doi.org/10.1039/C6EE02980A

[173] Pauportè, Thierry. "Synthesis of ZnO nanostructures for solar cells-a focus on dye-sensitized and perovskite solar cells." In The Future of Semiconductor Oxides in Next-Generation Solar Cells, pp. 3-43. Elsevier, 2018. https://doi.org/10.1016/B978-0-12-811165-9.00001-6

[174] Luo, Jun, Yanxiang Wang, and Qifeng Zhang. "Progress in perovskite solar cells based on ZnO nanostructures." Solar Energy 163 (2018): 289-306. https://doi.org/10.1016/j.solener.2018.01.035

[175] Yang, Guang, Hong Tao, Pingli Qin, Weijun Ke, and Guojia Fang. "Recent progress in electron transport layers for efficient perovskite solar cells." Journal of Materials Chemistry A 4, no. 11 (2016): 3970-3990. https://doi.org/10.1039/C5TA09011C

[176] Liu, Dianyi, and Timothy L. Kelly. "Perovskite solar cells with a planar heterojunction structure prepared using room-temperature solution processing techniques." Nature photonics 8, no. 2 (2014): 133-138. https://doi.org/10.1038/nphoton.2013.342

[177] Yang, Xiaohui, Ruixue Wang, Changjun Fan, Guoqing Li, Zuhong Xiong, and Ghassan E. Jabbour. "Ethoxylated polyethylenimine as an efficient electron injection layer for conventional and inverted polymer light emitting diodes." Organic Electronics 15, no. 10 (2014): 2387-2394. https://doi.org/10.1016/j.orgel.2014.07.009

[178] Zuo, Lijian, Zhuowei Gu, Tao Ye, Weifei Fu, Gang Wu, Hanying Li, and Hongzheng Chen. "Enhanced photovoltaic performance of CH3NH3PbI3 perovskite solar cells through interfacial engineering using self-assembling monolayer." Journal of the American Chemical Society 137, no. 7 (2015): 2674-2679. https://doi.org/10.1021/ja512518r

[179] Zheng, Enqiang, Yaqin Wang, Jiaxing Song, Xiao-Feng Wang, Wenjing Tian, Gang Chen, and Tsutomu Miyasaka. "ZnO/ZnS core-shell composites for low-temperature-processed perovskite solar cells." Journal of energy chemistry 27, no. 5 (2018): 1461-1467. https://doi.org/10.1016/j.jechem.2017.09.026

[180] Duan, Jinxia, Qiu Xiong, Hao Wang, Jun Zhang, and Jinghua Hu. "ZnO nanostructures for efficient perovskite solar cells." Journal of Materials Science: Materials in Electronics 28, no. 1 (2017): 60-66. https://doi.org/10.1007/s10854-016-5492-3

[181] Mahmood, Khalid, Bhabani Sankar Swain, and Aram Amassian. "16.1% Efficient hysteresis-free mesostructured perovskite solar cells based on synergistically improved ZnO nanorod arrays." Advanced Energy Materials 5, no. 17 (2015): 1500568. https://doi.org/10.1002/aenm.201500568

[182] Li, Shibin, Peng Zhang, Hao Chen, Yafei Wang, Detao Liu, Jiang Wu, Hojjatollah Sarvari, and Zhi David Chen. "Mesoporous PbI2 assisted growth of large perovskite grains for efficient perovskite solar cells based on ZnO nanorods." Journal of Power Sources 342 (2017): 990-997. https://doi.org/10.1016/j.jpowsour.2017.01.024

[183] Zhuiykov, Serge. Nanostructured semiconductor oxides for the next generation of electronics and functional devices: properties and applications. Woodhead Publishing, 2014.

[184] Norton, D. P., Y. W. Heo, and M. P. Lvill. "K. lp, SJ Pearton, MF Chisholm, T. Steiner." Mater. Today 34 (2004). https://doi.org/10.1016/S1369-7021(04)00287-1

[185] Zheng, Yan-Zhen, Er-Fei Zhao, Fan-Li Meng, Xue-Sen Lai, Xue-Mei Dong, Jiao-Jiao Wu, and Xia Tao. "Iodine-doped ZnO nanopillar arrays for perovskite solar cells with high efficiency up to 18.24%." Journal of Materials Chemistry A 5, no. 24 (2017): 12416-12425. https://doi.org/10.1039/C7TA03150E

[186] Mahmood, Khalid, Bhabani Sankar Swain, and Hyun Suk Jung. "Controlling the surface nanostructure of ZnO and Al-doped ZnO thin films using electrostatic spraying for their application in 12% efficient perovskite solar cells." Nanoscale 6, no. 15 (2014): 9127-9138. https://doi.org/10.1039/C4NR02065K

[187] Zhang, Rong, Chengbin Fei, Bo Li, Haoyu Fu, Jianjun Tian, and Guozhong Cao. "Continuous size tuning of monodispersed ZnO nanoparticles and its size effect on the performance of perovskite solar cells." ACS Applied Materials & Interfaces 9, no. 11 (2017): 9785-9794. https://doi.org/10.1021/acsami.7b00726

[188] Zhang, Huiyin, Jiangjian Shi, Xin Xu, Lifeng Zhu, Yanhong Luo, Dongmei Li, and Qingbo Meng. "Mg-doped TiO 2 boosts the efficiency of planar perovskite solar cells to exceed 19%." Journal of Materials Chemistry A 4, no. 40 (2016): 15383-15389. https://doi.org/10.1039/C6TA06879K

Materials Research Forum LLC
https://doi.org/10.21741/9781644902394-6

Chapter 6

Advances on ZnO Hetro-Structure as Nanoadsorbant for Heavy Metal Removals

Garima Rana[1*], Pooja Dhiman[1], Anand Sharma[2]

[1]International Research Centre of Nanotechnology for Himalayan Sustainability (IRCNHS), Shoolini University, India

[2]School of Physics & Materials Science, Shoolini University of Biotechnology and Management Sciences, Bajhol, Solan (H.P.) 173229

* mrs.garimarana@gmail.com

Abstract

Industrialization is going at an incredibly fast rate, which is putting more and more heavy metals in the water we drink. Almost all heavy metals are very toxic, and even a small amount of these metals in water can be very bad for humans and for the aquatic ecosystems that live in the water. As a result, the removal of heavy metals from industrial effluents is a big deal. Due to their high surface area to volume ratio, nanoadsorbents have received substantial attention in the past decade for their ability to remove heavy metals from water. Due to its good biocompatibility, low toxicity, negative zeta potential, surface changes during development, and redox reactions resulting from the production of efficient photoinduced electron-hole pairs in ZnO nanoparticles, ZnO is a suitable material for heavy metal remediation. In this chapter, we have discussed the synthesis method of ZnO NPs and their nanocomposites. Also in this chapter, we'll go over how ZnO nanostructures can be used to remove heavy metal ions from water. Various ZnO-based nanostorbents, including virgin ZnO NPs, doped ZnO nanostructures, ZnO nanocomposites, and surface-modified ZnO NPs, are fully examined, with statistical analyses of their maximum adsorption capacity for various heavy metal ions (Cd^{2+}, Hg^{2+}, Pb^{2+}, Cr^{6+}, and Cu^{2+}).

Keywords

ZnO, Heavy Metals Removal, Adsorption, Pollutants

Contents

1.   Introduction

There are two basic requirements for the survival of human civilization that are energy and a clean environment. There was one prime factor in the rapid development of human society and it was the industrial revolution. The industrial revolution led to a number of problems such as global warming, and air and water pollution [1]. These problems have been caused due to the increased use of fossil fuels like petroleum and coal. The most important problem related to urban living is due to water pollution and in turn, it results in irreparable damage to the earth and living organism [2]. A recent WHO report shows that about 3.7 million people globally die each year due to hazards of air and water pollution and 92% of the world's population still does not have pure water access [3].

One of the goals of decontamination is to detect and remove substances that are toxic from the contaminated water in an affordable way [4] because there are widely distributed substances; for example, heavy metals, which are known to cause harm to the environment and the humans [5-7]. Heavy metals are any metallic chemical element that has a relatively high density and is toxic at low concentrations [8]. Heavy metal is a very dangerous contaminant that is found in water as a result of manufacturing, mining [2], smelting, fertilizers, and municipal waste.

There are numerous methods proposed for heavy metal removal from the water such as chemical precipitation, ion exchange, adsorption, membrane filtration, and electrochemical technologies [9, 10]. One of the best methods is adsorption for water treatment because of its low cost and regeneration ability of adsorbents [11, 12].

The adsorption capability of nanomaterials is very high due to their active surface area and their accessible pores [11, 13-17]. For the removal of heavy metals, the adsorption process is one of the major processes. For the removal of heavy metal ions from water, the adsorbent should have the following properties [18] (I) rich active sites; (II) low cost; (III) good mechanical properties, and (IV) environmentally friendly.

Recently, metal oxide nanoparticles are used for the removal of metal ions because they are simple to synthesize and they have high porosity [19].

ZnO nanoparticles are the best materials amongst the different metal oxide nanoparticles, because the reason that it can be crystallized into a variety of nanostructure configurations that have a high surface-to-volume ratio, which includes nanorods, nanowires, nanobelts, nanorings [20-22]. The use of ZnO on the nanoscale can improve the electrical, optical, and magnetic properties of ZnO [23, 24]. The ZnO is environment friendly and has no such harmful effects on living organisms, which is why the ZnO has a broad range of applications that are no threat to living organisms and to the environment [25, 26].

The synthesis of ZnO heterostructures with other unique materials has been found to improve the properties of ZnO [27]. The use of nanoparticles in toxic heavy metal ion removal from wastewater has become very interesting [28]. ZnO-based nanocomposites and heterojunctions can be fabricated using the hydrothermal method, sol-gel method, ultrasonic and microwave-assisted methods, thermal decomposition method, and chemical co-precipitation technique.

The best candidate for making heterostructures with ZnO nanoparticles are PAni and Graphene Oxide (GO) due to their unique conductivity, which can be controlled, and also their structure is very unique for example long chain of PAni and two-dimensional Graphene Oxide (GO) [27, 29-32].

The most promising adsorbent is GO, an organic polymer that can adsorb various heavy metal ions [33-36], because of the reason that it has a large surface area, high negative charge density, surface hydrophobic π-π interaction, and it can be easily synthesized from the abundant natural graphite in large scale [37-39].

PAni was discovered more than 160 years ago but was paid attention to recently. PAni is one of the best conducting organic polymers. PAni has high electrical conductance, environmental stability, and ease of synthesis [40].

CuO- ZnO nanocomposites (defined as CZ) is a good pair because of their high catalytic efficiency [41].

ZnO@Chitosan is an important nanocomposite helpful in the removal of heavy metals from the water. Chitosan (CS) is a polymer that has low cost and has several properties: - biocompatibility, biodegradability, non-toxicity, and activity for the adsorbent of toxic metals [42]. The Chitosan has the ability to adsorb heavy metal ions due to the presence of amino acid and hydroxyl group in the Chitosan structure which work as many active sites [43]. In this chapter, various ZnO-based nanostorbents are used for the removal of heavy metals.

2.   Basic Parameters of ZnO

Wurtzite (space group $P63mc$, a = 3.25 nm, and 5.20nm) is the most frequent crystal structure for ZnO. $P63mc$ is the space group in which the wurtzite structure has a hexagonal unit cell with lattice parameters a and c in the ratio c/a = 1.633. This is because ZnO likes to crystallize in hexagonal columns, which makes it possible to make stimulated emissions or "lasers" with relatively low threshold power because of the six-sided Fabry–Pérot cavities [44, 45] that are found naturally in ZnO. As a result of this anisotropic development, 1D nanowires or nanorods (NRs) were generated far more easily than NPs [46, 47]. ZnO is used to degrade industrial and agricultural wastewater, just like other photocatalysts. Real wastewater, on the other hand, is still a very complicated system comprising a variety of harmful compounds, the amounts of which are difficult to analyze or quantify. Phenol, formaldehyde, heavy metal ions (e.g., Pb^{2+}, Cd^{2+}, Cr^{6+}), and other noxious chemicals are among these [48-51]. The first two contaminants (phenol and formaldehyde) are organic, whereas the heavy metal ions are inorganic. Photocatalysts could effectively eliminate all of them. However, because these compounds are frequently colourless, their concentrations must be determined using other methods, such as liquid or gaseous chromatograms.

3.   Methods to synthesize ZnO nanomaterials

Different parameters, such as shape, particle size, surface chemistry, and other characteristics influence the properties of nanoparticles (NPs). As a result, for chosen applications, the creation of NPs with regulated properties such as homogeneity in morphology, size, and functionality is required. These characteristics are largely reliant on the synthesis procedure in which they are calculated. To manufacture well-defined ZnO nanostructures, a variety of physical and chemical approaches have been developed in recent years [52], including chemical precipitation, sol-gel, vapor-deposition, hydrothermal deposition, green synthesis, mechano-chemical methods, and others. Detailed explanations of these procedures are provided below in Table 1-4. This section also includes a summary of some of the well-established ways for creating stable ZnO nanostructures, as well as the primary benefits and problems associated with each of these synthesis processes, as well as some of the most recent research on these topics.

3.1   Sol-gel method

Example 1:- Benhebal et al. [53] used the sol-gel technique to manufacture zinc oxide powder from zinc acetate dihydrate and oxalic acid, with ethanol serving as the solvent. A variety of methods, including nitrogen adsorption isotherms, X-ray diffraction (XRD), scanning electron microscopy (SEM), and ultraviolet-visible spectroscopy (UV-Vis) were used to analyze the final result (Figure 1). The zinc oxide that has been synthesized has a hexagonal wurtzite structure, with the particles having a spherically shaped form. Using the BET technique, the surface area of the calcined ZnO powder was measured and found to be equivalent to 10 m^2/g, which is indicative of a material at low porosity or crystalline material.

Example 2:- Yue et al. [54] used the sol-gel approach to generate zinc oxide as well. The fabrication of high-filling, uniform, ordered ZnO nanotubes onto an ultrathin AAO membrane using the sol-gel process has been completed. Integrating ultrathin AAO membranes with the sol-gel process may aid in the fabrication of high-quality 1D nanomaterials and the extension of the technology's utility as a template for nanostructures development, among other things.

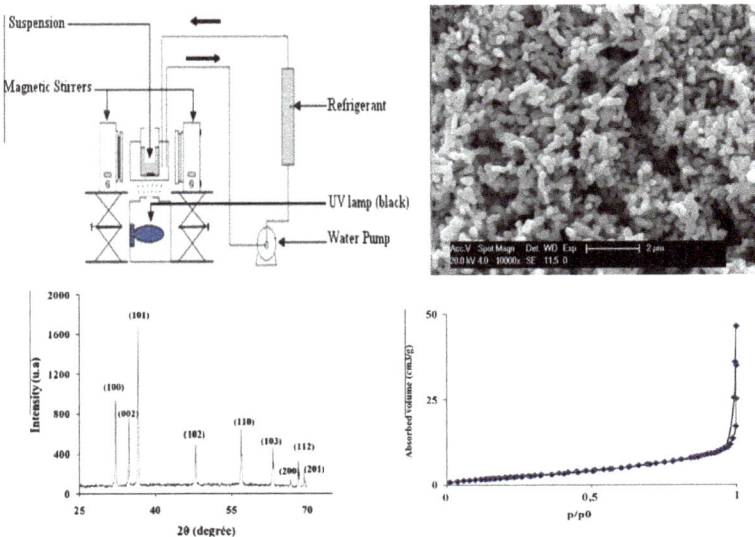

Figure 1: Reactor used in the photocatalytic experiment, SEM image of ZnO, XRD and nitrogen adsorption desorption isotherm of ZnO [53].

Table 1: Summary of sol-gel method of obtaining zinc oxide nanomaterials

Method	Precursors	Synthesis conditions	Properties and applications	Ref.
Sol-Gel Method	$Zn(CH_3COO)_2$, oxalic acid $(C_2H_2O_4)$, ethanol	Reaction: 50 °C, 60 min; Drying: 80 °C, 20 h; Calcined: under flowing air for 4 h at 650 °C	Hexagonal wurtize structure; uniform, spherically shaped particles	[53]
	$Zn(CH_3COO)_2$, diethanolamine, ethanol	reaction: room temperature; annealed of sol: 2 h, 500 °C	hexagonal wurtize structure; particles: nanotubes of 70 nm	[54]

There are many advantages of using a sol-gel approach, including:

(1) low cost

(2) simplicity

(3) relatively moderate synthesis conditions

(4) repeatability

(5) the use of organic chemicals that result in an effective surface modification in ZnO NPs [55, 56].

Challenges in sol-gel method:

(1) Sol-gel matrix elements may be present in the samples, necessitating the necessity for extra purification [56].

3.2   Precipitation method

Example 1:- Zinc oxide also has been precipitated from zinc chloride & zinc acetate aqueous solutions. The reagent concentration, substrate addition rate, and reaction temperature were all under strict control throughout this procedure. Figure 2 present the SEM micrograph of the ZnO. The ZnO sample that formed when KOH was added to $Zn(CH_3COO)_2$ at a rate of 11 cm^3 / min has small, almost spherical particles (Figure 2c) that don't tend to clump together. The ZnO samples that were made at higher or lower rates of dosing have a lot of tendency to clump together (Figure 2a and d). ZnO has a large surface area and a monomodal particle size distribution [57].

Example 2:- Lanje et al. [58] employed a cost-competitive and easy precipitation approach to synthesize zinc oxide, which was shown to be effective. For the cost-effective creation of ZnO nanoparticles, it is preferable to use a one-step approach that allows for large-scale manufacturing without the introduction of undesirable contaminants. As a result, low-cost ingredients like zinc nitrate and sodium hydroxide were employed to produce the ZnO nanoparticles (about 40 nm) in order to reduce costs. During the first nucleation stage, the starch molecule was utilized to minimize the agglomeration of the smaller particles since it includes several O-H functional groups and has the ability to bind to the surface of the nanoparticles during the early nucleation stage.

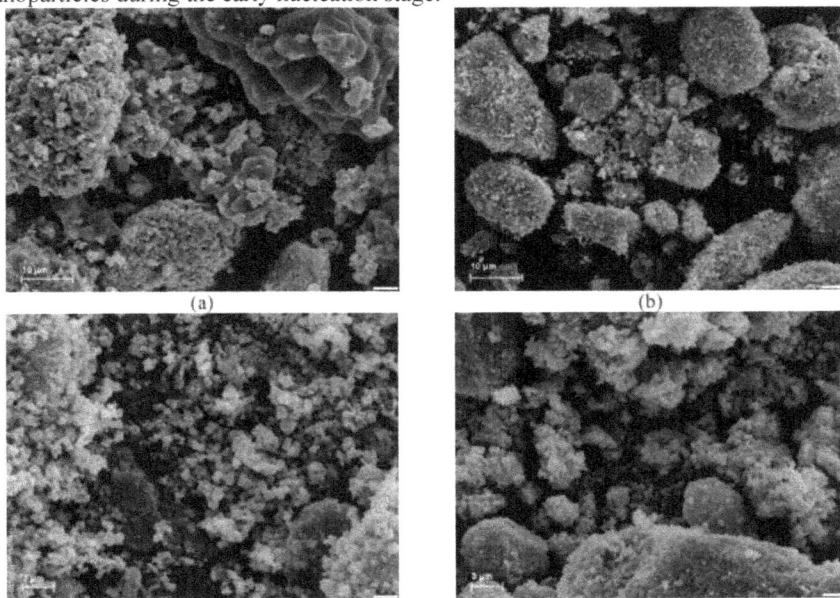

Figure 2: SEM pictures of ZnO samples precipitated at varying rates of loading of KOH solution to Zn(CH₃COO)₂ solution of (a) 1.1 cm³ /min, (b) 3.0 cm³ /min, (c) 11 cm³ /min, and (d) 15 cm³ /min [57]

Table 2: Summary of Precipitation method of obtaining zinc oxide nanomaterials.

Method	Precursors	Synthesis conditions	Properties and applications	Ref.
Precipitation method	$Zn(CH_3COO)_2$, and KOH as a water solutions	The temperature of the process: 20–80 °C; Drying: 120 °C	Particles diameter: 160–500 nm, BET: 4–16 m2/g	[57]
	$Zn(NO_3)_2$, NaOH	Synthesis: 2 h; Drying: 2 h, 100 °C	Particles of the spherical size of around 40 nm	[58]

The precipitation approach has the following advantages:

1. it is simple.

2. it is fast.

3. it is inexpensive [55].

Difficulties in the technique of precipitation:

1. Because the process is very quick, nucleation and development occur simultaneously, making it difficult to regulate the growth process.

2. High-temperature thermal treatment is often necessary[55].

3.3   Hydrothermal synthesis method

Example 1:- By using hydrothermal methods, Dem'Yanets and colleagues [59] produced zinc oxide nanoparticles with varying morphologies. For example, a hydroxide (LiOH, NH_4OH) and zinc acetate (or nitrate) reaction resulted in $Zn(OH)_2nH_2O$. It was done in an autoclave, where the temperature ranged from 120 to 250 degrees Fahrenheit. Crystallites of ZnO with a hexagonal arrangement and diameters ranging from 100 nm to 20 m were formed by dehydrating the precursor and then recrystallizing it. The diameter of the ZnO particles increased when the hydrothermal process was extended for longer periods of time. The experiment was able to be completed in four times less time when the temperature was increased by 50–70 °C, which is a great benefit.

Example 2: ZnO particles were studied by Musi et al. [60], who investigated the influence of chemical products on the size and characteristics of ZnO particles. Underwent hydrothermal processing in an autoclave at 160 degrees Celsius after being prepared from a solution of $Zn(CH_3COO)_2H_2O$ and neutralized with varying amounts of a solution of $NH4OH$ in varying proportions. A significant effect of pH on the size and morphology of ZnO particles was discovered. The formation of aggregates comprised of ZnO particles with diameters ranging from 20 to 60 nm after 7 months of fermentation (at a pH of 10 as well as at room temperature) was seen after the initial aqueous solution was fermented. A sol-gel approach, developed by Musi and colleagues, was used to produce zinc oxide, which included the fast hydrolysis of zinc 2-ethylhexanoate dissolved in propan-2-ol. The nanoparticles formed as a consequence of this process produce unique alterations in the usual Raman spectra of zinc oxide (Figure3(a)). ZnO particles of micron dimensions are shown in Figure 3 c, whereas ZnO particles of tape-like dimensions appear in Figure 3 b. At pH 10, only prismatic ZnO rods can be generated by increasing the pH of the suspension (Figure 3b). 72 hours of autoclaving at 160 -C improved the form of the prismatic ZnO rods (sample C9), however some of them are noticeably longer along the c-axis (Figure 3c).

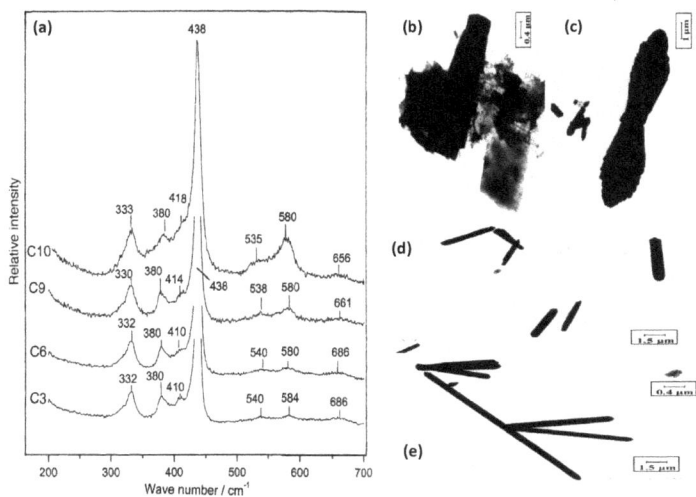

Figure :3 (a) Raman spectra (b-e) TEM images of prepared sample [60].

Table 3: Summary of hydrothermal method of obtaining zinc oxide nanomaterials.

Method	Precursors	Synthesis conditions	Properties and applications	Ref.
Hydrothermal Synthesis Method	$Zn(CH_3COO)_2$, $Zn(NO_3)_2$, LiOH, KOH, NH4OH	Reaction: 10–48 h, 120–250 °C	Hexagonal (wurtize) structure, size of microcrystallites: 100 nm– 20 μm	[59]
	$Zn(CH_3COO)_2$, NH3, zinc 2-ethylhexanoate, TMAH, ethanol, 2-propano	Time of autoclaving: 15 min, 2–72 h; final pH: 7–10	Particles with irregular ends and holes; aggregates consist of particles of 20–60 nm, BET: 0.49–6.02 m2/g	[60]

The following are the advantages of the hydrothermal method:

1. it requires only a small amount of equipment (an autoclave).

2. it results in the growth of high-quality crystals with a tightly controlled chemical composition (high purity).

3. it does not necessitate any additional processing such as grinding and calcination of the product[55].

The hydrothermal process has a number of difficulties.

Among the drawbacks are the need for expensive autoclaves, the inability to monitor the crystal during its formation, and a lengthy procedure[55].

3.4   Mechanochemical synthesis method

Example 1: Stankovi et al. [61] investigated mechanical-thermal synthesis (MTS), which involves mechanical activation followed by heat activation of ZnO from $ZnCl_2$ & oxalic acid as reactants in order to get pure ZnO nanopowder. ZnO nanopowders' crystal structure, size of the particles, and shape were all examined, as were the impacts of oxalic acid, an organic PCA, and various billing periods. The original reactant combination was milled for 30 minutes to 4 hours and then annealed for 1 hour at 450 °C. Analyzing the powders' quality was done using XRD and Raman spectroscopy after they had been manufactured.

The XRD study of the produced ZnO powders demonstrated flawless long-range order as well as the pure wurtzite structure, regardless of the milling time. Raman spectroscopy, on the other hand, shows that ZnO particles have a distinct middle-range order. Particle shape greatly relies on milling duration independent of subsequent heat treatment as shown by SEM pictures obtained (Figure 4a&b). The ZnO particles have nearly perfect spherical forms, are well dispersed, and are non-agglomerated, with diameters ranging from 15 to 50 nm (Figure 4a–d). At the bottom of Figure 4, the PSD over number and volume, with d10, d50 and d90 inserted, is shown. It took a longer period of time for the milling process to produce a smaller particle diameter.

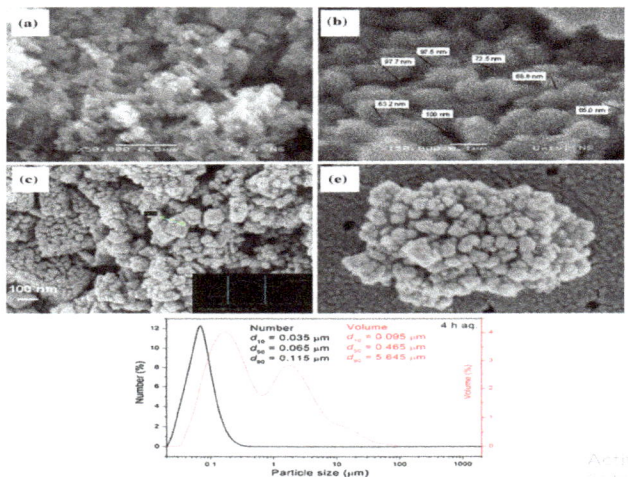

Figure 4: SEM (a, b) and FE SEM (c, d) pictures of *ZnO-240, as well as the associated particle size distributions (bottom)[61].

Table 4: Summary of mechanochemical method of obtaining zinc oxide nanomaterials.

Method	Precursors	Synthesis conditions	Properties and applications	Ref.
Mechanochemical process	$ZnCl_2$, Na_2CO_3, NaCl	calcination: 2 h, 600 °C	hexagonal structure; particles diameter: 18–35 nm	[61]

Mechano-chemical methods provide many advantages, including:

(1) low production costs.

(2) tiny particles with no aggregation.

(3) homogeneous crystal size and morphology among others [56].

Challenges in mechano-chemical method:

Increasing the milling time and energy also increases the number of contaminants in a sample [56].

3.6 Biosynthesis

Advantages of the biosynthesis process are:

1. Eco-friendly nature,

2. Nontoxicity

3. Use of inexpensive organic solvents rather than dangerous chemicals

Challenges in the biosynthesis process:

1. Uncertainty of mechanism.

2. stability of nanoparticles [62-64].

4. Removal of heavy metals

HMs are naturally occurring elements that have a specific density greater than 5 g/cm^3 and a higher atomic weight [65]. It refers to a group of different types of heavy metals, such as vanadium, chromium, manganese, iron, cobalt, nickel, copper, zinc, gallium, silver, cadmium, antimony, tellurium, cerium, platinum, gold, silver, mercury, thallium, lead, uranium, and so on. Trace elements are essential to all living things and play an important role in metabolic processes. Examples of these elements include iron, zinc, cobalt, copper, manganese, magnesium, chrome, nickel, and many more. There is a wide range of health issues that can result from a lack or excessive intake of these metals, which are normally far lower than 100 mg/day.

There are four more types of HMs that are very harmful to living things. These are Hg, As, Cd, and Pb. Even a small amount of these HMs can be very harmful to living things. They don't play a role in any metabolic activity. Instead, a very small amount of these can have a negative effect on a number of metabolic processes [66]. There are two ways to stop metabolic functions. In the first case, the HMs build up in important parts of the human body and stop them from working normally. In the second case, the HMs move the important nutrients from where they were supposed to be, which has a big effect on a lot of different biological processes.

According to the World Health Organization, water remediation could prevent a significant amount of diseases around the world. Every year, approximately 485,000 deaths are reported as a result of diarrhoea caused by the consumption of contaminated water [67]. As a result, finding safe drinking water that is free of heavy metals and other toxic pollutants is one of the most pressing issues facing the world today. Figure 5 depicts some of the industrial sources of HMs that severely damage water supplies [68].

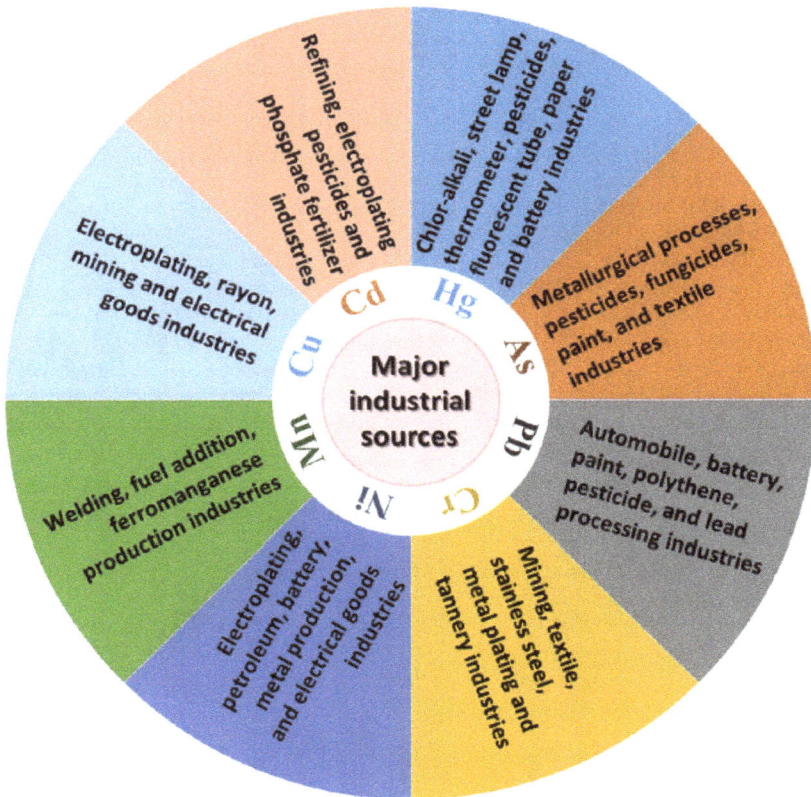

Figure 5: Major industrial sources of the toxic HM ions [68].

4.1   Mercury (Hg) removal

Mercury is a persistent bioaccumulative toxic metal that comes from a variety of anthropogenic sources, including chloralkali industries, batteries, combustion of fossil fuels, the disposal of thermometers, street lamps, and fluorescent tubes, gold mining, cement and metals production, pesticides, and the use of dental amalgam [69]. The Agency for Toxic Substances and Disease Registry (ATSDR) has listed it third on its "Priority List of hazardous substances." Mercury has also been identified by the World Health Organization as one of the 10 toxins of serious public health concern [70]. It is possible that the discharge of Hg-containing industrial effluents into various water bodies may provide a health danger to the public because of the accumulation and biological magnification of mercury in the seafood chain, which will eventually result in mercury entering the human body [71]. As a result, severe restrictions have been enacted addressing the quantity of Hg that is bearable in drinking water as well as the limit for Hg in waterborne industrial waste. Because of this, it is imperative to find ways to reduce Hg concentrations in wastewater that are both effective and affordable. ZnO nanocrystallites generated by the mix of small ZnO nanocrystallites were employed to concurrently remove heavy metal ions, such as Hg^{2+}, Pb^2, As^{3+}, and Cd^2, from waste-water that included 47.16 mg/L Hg^2, 42.0 mg/L Pb^2, 19.6 mg/L As^{3+}, 47.8 mg/L Cd^{2+}, 28.83 mg/L Cu2, and 15.75 mg/L Co^{2+}, respectively[72]. The nano adsorbents were found to eliminate all of the heavy metal ions concurrently. The RE for Hg^{2+} was 63.5 percent when 200 mg of ZnO nano assemblies were mixed with 40 mL of water that had been made. It was also found that the magnetic semiconductor nanocomposites made with Fe_3O_4 and ZnO could be used to remove metal ions from wastewater that had been mixed with 45.06 mg/L Hg^{2+}, 36.22 mg/L Pb^{2+}, 40.11 mg/L Cd^{2+}, 22.55 mg/L As^{3+}, 23.83 mg/L Cu^{2+}, 15.81 mg/L Ni^{2+}, and 15.75 mg/L Co^{2+}. 50 mg Fe_3O_4-ZnO nanocomposites were mixed with wastewater and shook for 24 hours at pH 6 and room temperature to get 100% RE for Hg^{2+}ions.

ZnO-NiO nanocomposites prepared by precipitation technique with excess NaOH and a capping agent were studied by Kumar et al. [73] for their ability to remove Hg^{2+}ions (triton X-100). It was found that the Langmuir model best fit the adsorption isotherm data, and the sorption capacity was predicted to be 1474.9 mg/g using the Langmuir model. According to the PSO kinetic model, Hg^{2+}ion adsorption on ZnO-NiO nanocomposites is endothermic and spontaneous. The RE of ZnO-NiO nanocomposites was 80.36 percent when used for the first time, but it dropped to 53.17 percent and 21.63 percent after the first and second regenerations, respectively. Using regent (0.1 M HCl solution) to regenerate ZnO-NiO nanocomposites results in poor regeneration because of the strong interaction between the nanocomposites and Hg^{2+} ions.

4.2   Cadmium (Cd) removal

Cd is one of the most dangerous HM. If you drink water, you can't have more than 0.03 parts per million (ppm) of it. This is the lowest limit set by the World Health Organization (WHO). People who drink water that has more cadmium than this can have problems with their blood pressure, stomach problems, chronic renal failure, bone defects, bone marrow cancer, gastrointestinal problems, kidney damage, risks of stillbirth, DNA and membrane damage, and sometimes even death [71]. However, the use of cadmium in a wide range of industrial processes, such as the production of Ni-Cd batteries, paints, phosphate fertilizers, pesticides, metal refineries, electroplating, and alloys, has led to an increase in the

Materials Research Forum LLC
https://doi.org/10.21741/9781644902394-6

concentration of cadmium in water resources. This means that new techniques must be found to remove cadmium from industrial waste. Many scientists have looked into how ZnO nano adsorbents can remove Cd^{2+} ions from the water.

When Mahdavi et al. conducted their first ZnO nano adsorbents investigation, they compared the adsorption of single and multiple ZnO nanoparticles in aqueous solutions for the elimination of Cd^{2+}, Cu^{2+}, Ni^{2+}, and Pb^{2+} from water samples. In comparison to Fe_3O_4 (114.5 mg/g) and CuO (73 mg/g) nanoparticles, ZnO (360.6 mg/g) had a substantially higher adsorption capacity in a multi-component solution (including Cd^{2+}, Cu^{2+}, Ni^{2+}, and Pb^{2+} ions). According to ZnO nanoparticles' adsorption capacity, they are cited as one of the most potential sorbents. Additionally, the elimination capacity of single metal ions solutions was determined to be in the following order: $Cd^{2+} > Pb^{2+} > Cu^{2+} > Ni^{2+}$, whereas the elimination capacity of multiple metal ions solutions was shown to be in the following order: $Pb^{2+} > Cu^{2+} > Cd^{2+} > Ni^+$. ZnO NPs had the highest Cd^{2+} adsorption capacity of 119.1 mg/g in a single metal solution. According to Khezami et al. [74], the rate and amount of adsorption vary in accordance with temperature, and Cd^{2+} adsorption on ZnO NPs is endothermic; however, Somu et al. [75] revealed that ZnO NPs manufactured utilizing casein as a biogenic reducing and capping agent result in exothermic Cd^{2+} adsorption, which could be related to surface changes of ZnO NPs. After the first, second, and third cycles of regeneration, Khezami et al. [74] found that 90%, 85%, and 79% of Cd^{2+} ions were removed, respectively. Somu et al. [75] confirmed this excellent removal ability following regeneration, reporting just a 13% decline in RE after five rounds of adsorption and desorption. Based on comparisons between the RE recorded under dark, visible, and UV illumination, Le et al. [76] concluded that physical adsorption was the primary mechanism responsible for Cd^{2+} ion elimination.

NPs containing various metal ions, notably indium (In), calcium (Ca), and gallium (Ga), have also been studied. Ghiloufi et al. [77] looked into the effect of indium concentration on the Cd^{2+} adsorption capacity of In doped ZnO NPs. They found that the adsorption capacity of In doped ZnO NPs went down when In was added. Afterwards, they studied the impacts of calcium doping with distinct [Ca]/[Zn] atomic ratios (0.00, 0.01, 0.03 and 0.05) [78]. Doping with Ca led to an increase in adsorption capacity, which peaked in comparison to another concentration analyzed at [Ca]/[Zn] of 0.03. pH 7 and 298 K are reported to have Cd^2 sorption +'s capacity of 56.53 mg/g at pH 7. Ca doped ZnO NPs' adsorption capacity was found to increase with pyrolysis, and an adsorption capacity of 92.21 mg/g was confirmed for pyrolysis in temperatures of 1000°C. To find out how well these Ga doped ZnO NPs adsorb Cd^{2+}, the atomic ratios of Ga to the Zn in the particles were varied (0, 0, 0, 0, 03, and 0.05)[79].

4.3 Lead (Pb) removal

Lead is an extremely poisonous heavy metal that has no metabolic activity. Lead exposure, on the other hand, is a public health concern since it can produce a variety of negative health impacts, including neurological, hematological, gastrointestinal, cardiovascular, and renal diseases [80]. It has been discovered that even a quite small amount of lead can harm a child's growing neurological system. The Food and Agriculture Organization (FAO) created effluent discharge rules that limit the highest amount of lead in industrial wastewater to 0.05 mg/L. Furthermore, lead has been added to the WHO's list of substances of the highest public health concern, with a maximum limit of 0.01 mg/L for lead in drinking water [70]. As a result, in order to maintain environmental quality, the amount of Pd existing in industrial effluent must be reduced to an appropriate standard. Krasovska et al. [81] examined the Pd^{2+} removal effectiveness of ZnO nanotubes and nanorods grown at different temperatures while keeping other parameters constant. Due to their large specific surface area, hollow nanostructures outperform solid nanostructures given the same morphology. The adsorption capacity of ZnO nanotubes (611 mg/g) was found to be twice that of ZnO nanorods (256 mg/g). This is because the micro/nanostructured porous nanosheet-built ZnO hollow microspheres had a larger Pd^{2+} adsorption capacity (158 mg/g) than commercial ZnO nanopowders (33 mg/g) [82]. Hierarchically constructed mesoporous ZnO nanorods with 15.75 $m^2\,g^{-1}$ high surface area and 0.038 $cm^3\,g^{-1}$ porosity have an adsorption capacity of 160.7 mg/g [83]. The Langmuir and Freundlich models were the right fit for the equilibrium isotherm, and the adsorption kinetics were determined to be PSO. In addition, the IPD model showed that the adsorption rate was firstly influenced by film diffusion but later by pore diffusion. Pb removal has also been investigated using ZnO NPs doped with other metal ions such as Al, Ca, Ga, and La. Jawed et al. [84] discovered that Al doping increased the surface area (28.16 m^2 /g) of Al-doped ZnO NPs fourfold above ZnO NPs (8.69 m^2 /g), which maximized their adsorption capability. The adsorption technique was developed to be two-step, with rapid adsorption occurring during the first 60 minutes of contact time, followed by delayed adsorption until 300 minutes. The Al-doped ZnO NPs demonstrated an adsorption capacity of 56 mg/g as well as a good regeneration performance with a RE of 86 percent after the third regeneration cycle. They also employed the Al-doped ZnO NPs to remediate genuine industrial wastewater from the IIT Guwahati Sewage Treatment Plant. The elimination of Pb from wastewater with 0.15 0.02 mg/L Pb concentration by Al-doped ZnO NPs resulted in a percentage of 0.08 0.01 mg/L, which is substantially lower than the allowed limit of Pb in potable water (0.1 mg/L). The Pb^{2+} adsorption capability of nanocomposites of ZnO with other oxide materials such as Co_3O_4, SnO_2, and TiO_2 has also been investigated. Sheikhshoaie et al. [85] investigated the RE of Co_3O_4 and ZnO nanocomposites (5:5, 8:2, 7:3, and 9:1). They discovered that pure Co_3O_4 did not eliminate lead from the solution, but that adding ZnO to Co_3O_4 increased adsorption performance. The 5:5 Co_3O_4/ZnO nanocomposites had the best Pb^{2+}adsorption capacity of 185.2 mg/g. This was achieved in 3 hours with 0.125g/15mL of initial metal concentration. More than 100% adsorption efficiency was achieved in 1 hour using SnO_2 and ZnO nanocomposites with varied [Sn]/[Zn] mole ratios (0.00,0.01,0.02,0.04,0.05) [86]. When compared to other samples such as ZnO (97.64 percent), 0.01 M SnO_2/ZnO (97.94 percent), 0.02 M SnO_2/ZnO (99.33 percent), and 0.05 M SnO_2/ZnO, 0.04 M SnO_2/ZnO had the highest adsorption efficiency (99.92 percent).

4.4   Arsenic (As) removal

As a possible carcinogen, arsenic has a number of negative health impacts on humans [87]. Toxic Substances and Disease Registry (ATSDR) has it at the top of their "Priority List of Hazardous Substances". Arsenic poisoning is a major public health issue that is expected to affect hundreds of thousands of people worldwide. The first signs of arsenic poisoning are primarily skin-related, such as colour abnormalities and hyperkeratosis on the soles and palms. Even though As exposure is both acute and long-term, it also causes a person to not be able to eat or drink, as well as respiratory problems and neurological disorders. It can also cause cancers in the skin, bladder, prostate, lung, kidney, and liver. Rehman et al. [88] used precipitation to create a flower-like structure of ZnO. They reported that the best arsenic elimination from drinking water is accomplished by employing an adsorbent dosage of 4 g/L, a solution pH of 5, a contact period of 90 minutes, and a temperature of 50 degrees Celsius. Furthermore, they stated that the synthesis of oxyanions by arsenic oxide in aqueous media at high pH values resulted in the highest elimination of As ions in an acidic medium and that the accessibility of hydroxyl ions is abundant on the surface of produced NPs. They achieved a RE of 46 percent with a starting metal concentration of 1 ppm, and adsorbent dosage of 4 g/L, a contact time of 90 minutes, a pH of < 5, and a temperature of 50 degrees Celsius. To remove Hg^{2+}, Pb^{2+}, As^+, Cd^{2+}, Co^{2+}, Cu^{2+}, Ni^{2+} and Co^{2+} from wastewater, Singh et al. [72] employed mesoporous ZnO nano adsorbents produced by combining tiny ZnO nanocrystallites. The ZnO nano-assemblies (200 mg) removed 100% As^{3+} from 40 mL of prepared effluent. The simultaneous removal of metal ions from wastewater using Fe_3O_4 contained ZnO magnetic semiconductor nanocomposites was also explored [50]. Adding 50 mg Fe_3O_4-ZnO nanocomposites in 40 mL wastewater and mixing for 24 hours at pH 6 and ambient temperature gave 100% RE for As^{3+} ions. After half a day of treatment with 0.01 M HCl solution (10 mL) and then washing with water, the Fe_3O_4-ZnO nanocomposites (50 mg) removed 95% of As^{3+}. For the removal of dissolved Ar from water, Nath et al. [89] explored a bio nanocomposite composed of $ZnO:CeO_2$:nanocellulose:polyaniline. By changing the pH from 2 to 12, the bionanocomposite demonstrated an average adsorption performance of 97.5%, with a maximum adsorption performance of 99.5% at pH 8. In this study, it was found that the ZnO-CeO_2-nanocellulose-polyaniline nanoadsorbent has a high adsorption performance that is independent of the pH of the solution, unlike other ZnO nanoadsorbents.

4.5   Chromium (Cr) removal

Chromite is one of the twenty-first most common elements in the Earth's crust [90]. The most common oxidation states are trivalent Cr(III) and hexavalent Cr(II), however, there are many others (VI). The chemical properties and toxicity of Cr(III) and Cr(VI) are very different. Cr(III) is an important nutrient that helps the body make lipids and glucose. Cr(VI) is a toxicant that can have a number of negative effects on cells, including epigenetic changes, DNA damage, microsatellite instability, and chromosomal disorder [91]. Further, it has been found that Cr(VI) is very rare in nature and is mostly cleared by industrial processes [92]. Electroplating, pigments, welding of chromium-containing materials, tanning agents, alloys such as stainless steels and high chromium steels, insecticides, chromic acid, paints, and catalysts are some of the common industrial processes that produce Cr(VI). Chromium must be removed from industrial effluent before it can be dumped into the environment. ZnO NPs with other metal ions such as In, Ca, and Ga have also been used to remove Cr^{6+}. Ghiloufi et al. [77] investigated the effect of indium concentration (1% - 5% of [In]/[Zn]) on Cr^{6+} adsorption capacity of In doped ZnO NPs and found that doping ZnO NPs with In increased adsorption capacity. In ZnO, the adsorption capacity was determined to be the highest with a value of 24.78 mg/g at a concentration of 3 at.% In. Following that, they studied the effect of calcium doping with various [Ca]/[Zn] atomic ratios (0.00, 0.01, 0.03 and 0.05) [78]. Here, too, the adsorption capacity went up when Ca was added. It was at its highest for [Ca]/[Zn] of 0.03 when compared to other concentrations. It can adsorb 123.22 mg/g of Cr^{6+} at a pH of 3 and a temperature of 298 K. In the instance of Ga doping, however, ZnO NPs were found to be a more effective Cr^{6+} remover at 1 wt% Ga than the other samples (0, 3, and 5 wt% Ga) [79]. Further, the kinetic process was found to be PSO, and Langmuir's model was the best fit for the isotherms. When the atomic ratio of Ga to Zn was 0.01, the adsorption capacity of ZnO NPs was 220.7 mg/g at a pH of 3 and 298 K. Table 5 lists the reported work on ZnO as nanoadsorbents for removal of different heavy metals.

Table 5: Various ZnO nanostructures reported for various heavy metals removal from water.

Catalysts	Synthesis Method	Heavy metal	Time (min.)	Removal Efficiency(%)	Best isotherm & Kinetic model fitted	References
ZnO	Semi green synthesis	Cadmium	120	85.63	Langmuir & PSO	[75]
Self-aggregated spherical ZnO NPs	Soft chemical route at low temperature	Mercury	1440	63.5	-	[79]
ZnO NPs	Gel combustion method	Lead	180	>95	Langmuir and Freundlich	[93]
Self-aggregated spherical ZnO NPs	Soft chemical route at low temperature	Cobalt	1440	15.5	-	[72]
ZnO	Commercial	Cadmium	150	95.75	PSO	[94]
ZnO-NiO nanocomposites	Co-precipitation method	Mercury	90	99	Langmuir & PSO	[73]

ZnO NPs	Semi-green method	Lead	120	95.35	Langmuir & PSO	[75]
Hierarchically assembled mesoporous ZnO nanorods	Hydrothermal method	Cadmium	90	91	Langmuir & PSO	[83]
ZnO:CeO$_2$:nanocellulose: polyaniline bionanocomposite	Modified polymerization process	Mercury	720	99.5	Freundlich and Dubinin-Radushkevich & PSO	[11]
ZnO Nanorod-Reduced Graphene Oxide (GO) Hybrid Nanocomposites	Hydrothermal method	Cobalt	120	90.1	Langmuir & PSO	[95]
ZnO	Precipitation method	Copper	20	98.71	Freundlich & PSO	[96]
ZnO@activated carbon (AC) nanocomposites	ZnO NPs + AC (magnetic stirring & ultrasonication)	Cadmium	1.5 (Sonications)	80	Langmuir, Freundlich	[97]
ZnO NPs and CC derived activated carbon	-	Mercury	300	98.24	-	[98]
ZnO NPs	Semi-green method	Cobalt	120	71.23	Langmuir & PSO	[84]
Polyacrylonitrile (PAN) nanofibers loaded with ZnO NPs	Electrospinning	Cadmium	60	-	Langmuir & PSO	[99]
Hierarchically assembled mesoporous ZnO nanorods	Hydrothermal method	Lead	90	96	Langmuir & PSO	[83]
ZnO	Precipitation method	Copper	120	97.6	Freundlich & PSO	[100]
ZnO and montmorillonite nanocomposites	Heating method	Copper	90	89.5	Langmuir& PSO	[101]
Co$_3$O$_4$ @ ZnO NPs (Co$_3$O$_4$: ZnO =5:5)	Gel combustion method	Lead	180	77.17	-	[85]
Sewage sludge-based carbon (SSC)/TiO$_2$/ZnO nanocomposite	-	Copper	120	68.3	Redlich–Peterson& PSO	[102]
Al doped ZnO NPs	Co-precipitation method	Lead	300	94	Langmuir & PSO	[84]
ZnO	Precipitation method		60	92	-	[103]

Conclusion

In contaminated water, ZnO-based nanoadsorbents have demonstrated good adsorption capability to remove heavy metal ions from the solution. ZnO nanoadsorbents are useful for heavy metal cleanup because of their abundance, economic feasibility, non-toxic behavior, and great biocompatibility. Small-scale laboratory trials with ZnO nanoadsorbents are promising, but large-scale wastewater treatment with ZnO nanoparticles has not been studied scientifically. Rather than relying on hazardous chemicals to make ZnO nanoparticles, early research suggests that a green process that uses extracts from plants or microbes as biological reducing and capping agents is a better option for the environment.

References

[1] H. Ali, E. Khan, Environmental chemistry in the twenty-first century, Environmental Chemistry Letters, 15 (2017) 329-346. https://doi.org/10.1007/s10311-016-0601-3

[2] B. Draszawka-Bołzan, Effect of heavy metals on living organisms, World Scientific News, (2014) 26-34.

[3] A.C. Rai, P. Kumar, F. Pilla, A.N. Skouloudis, S. Di Sabatino, C. Ratti, A. Yasar, D. Rickerby, End-user perspective of low-cost sensors for outdoor air pollution monitoring, Science of The Total Environment, 607 (2017) 691-705. https://doi.org/10.1016/j.scitotenv.2017.06.266

[4] M.A. Shannon, P.W. Bohn, M. Elimelech, J.G. Georgiadis, B.J. Mariñas, A.M. Mayes, Science and technology for water purification in the coming decades, Nature, 452 (2008) 301-310. https://doi.org/10.1038/nature06599

[5] B. Jia, W. Jia, F. Qu, X. Wu, General strategy for self assembly of mesoporous SnO 2 nanospheres and their applications in water purification, RSC advances, 3 (2013) 12140-12148. https://doi.org/10.1039/c3ra41638k

[6] J. Wang, F. Qu, X. Wu, Photocatalytic Degradation of Organic Dyes with Hierarchical Ag2 O/ZnO Heterostructures, Science of Advanced Materials, 5 (2013) 1364-1371. https://doi.org/10.1166/sam.2013.1597

[7] P. Dhiman, A. Kumar, M. Shekh, G. Sharma, G. Rana, D.-V.N. Vo, N. AlMasoud, M. Naushad, Z.A. ALOthman, Robust magnetic ZnO-Fe2O3 Z-scheme hetereojunctions with in-built metal-redox for high performance photo-degradation of sulfamethoxazole and electrochemical dopamine detection, Environmental Research, 197 (2021) 111074. https://doi.org/10.1016/j.envres.2021.111074

[8] P. Dhiman, S. Sharma, A. Kumar, M. Shekh, G. Sharma, M. Naushad, Rapid visible and solar photocatalytic Cr(VI) reduction and electrochemical sensing of dopamine using solution combustion synthesized ZnO-Fe2O3 nano heterojunctions: Mechanism Elucidation, Ceramics International, 46 (2020) 12255-12268. https://doi.org/10.1016/j.ceramint.2020.01.275

[9] M.C.F. Magalhães, Arsenic. An environmental problem limited by solubility, Pure and Applied Chemistry, 74 (2002) 1843-1850. https://doi.org/10.1351/pac200274101843

[10] P. Dhiman, G. Rana, A. Kumar, G. Sharma, D.-V.N. Vo, M. Naushad, ZnO-based heterostructures as photocatalysts for hydrogen generation and depollution: a review, Environmental Chemistry Letters, 20 (2022) 1047-1081. https://doi.org/10.1007/s10311-021-01361-1

[11] N.J. Vickers, Animal communication: when i'm calling you, will you answer too?, Current biology, 27 (2017) R713-R715. https://doi.org/10.1016/j.cub.2017.05.064

[12] A. Kumar, A. Kumar, G. Sharma, M. Naushad, F.J. Stadler, A.A. Ghfar, P. Dhiman, R.V. Saini, Sustainable nano-hybrids of magnetic biochar supported g-C3N4/FeVO4 for solar powered degradation of noxious pollutants- Synergism of adsorption, photocatalysis & photo-ozonation, Journal of Cleaner Production, 165 (2017) 431-451. https://doi.org/10.1016/j.jclepro.2017.07.117

[13] Y. Wang, G. Wang, H. Wang, C. Liang, W. Cai, L. Zhang, Chemical-template synthesis of micro/nanoscale magnesium silicate hollow spheres for waste-water treatment, Chemistry-A European Journal, 16 (2010) 3497-3503. https://doi.org/10.1002/chem.200902799

[14] F. Salehi-Babarsad, E. Derikvand, M. Razaz, R. Yousefi, A. Shirmardi, Heavy metal removal by using ZnO/organic and ZnO/inorganic nanocomposite heterostructures, International Journal of Environmental Analytical Chemistry, 100 (2020) 702-719. https://doi.org/10.1080/03067319.2019.1639685

[15] M.-R. Huang, S. Li, X.-G. Li, Longan shell as novel biomacromolecular sorbent for highly selective removal of lead and mercury ions, The Journal of Physical Chemistry B, 114 (2010) 3534-3542. https://doi.org/10.1021/jp910697s

[16] A. Bhatnagar, M. Sillanpää, Applications of chitin-and chitosan-derivatives for the detoxification of water and wastewater-a short review, Advances in colloid and Interface science, 152 (2009) 26-38. https://doi.org/10.1016/j.cis.2009.09.003

[17] S. Sharma, G. Sharma, A. Kumar, P. Dhiman, T.S. AlGarni, M. Naushad, Z.A. Alothman, F.J. Stadler, Controlled synthesis of porous Zn/Fe based layered double hydroxides: Synthesis mechanism, and ciprofloxacin adsorption, Separation and Purification Technology, 278 (2021) 119481. https://doi.org/10.1016/j.seppur.2021.119481

[18] L. Zhang, M. Fang, Nanomaterials in pollution trace detection and environmental improvement, Nano Today, 5 (2010) 128-142. https://doi.org/10.1016/j.nantod.2010.03.002

[19] Z. Ganjiani, F. Jamali-Sheini, R. Yousefi, Electrochemical synthesis and physical properties of Sn-doped CdO nanostructures, Superlattices and Microstructures, 100 (2016) 988-996. https://doi.org/10.1016/j.spmi.2016.10.064

[20] X. Wang, J. Song, J. Liu, Z.L. Wang, Direct-current nanogenerator driven by ultrasonic waves, Science, 316 (2007) 102-105. https://doi.org/10.1126/science.1139366

[21] Z.L. Wang, Zinc oxide nanostructures: growth, properties and applications, Journal of physics: condensed matter, 16 (2004) R829. https://doi.org/10.1088/0953-8984/16/25/R01

[22] Z. Wang, Mater Today 7: 26, doi: 10.1016, S1369-7021 (04), (2004). https://doi.org/10.1016/S1369-7021(04)00286-X

[23] L.-E. Shi, Z.-H. Li, W. Zheng, Y.-F. Zhao, Y.-F. Jin, Z.-X. Tang, Synthesis, antibacterial activity, antibacterial mechanism and food applications of ZnO nanoparticles: a review, Food Additives & Contaminants: Part A, 31 (2014) 173-186. https://doi.org/10.1080/19440049.2013.865147

[24] P. Dhiman, J. Chand, A. Kumar, R.K. Kotnala, K.M. Batoo, M. Singh, Synthesis and characterization of novel Fe@ZnO nanosystem, Journal of Alloys and Compounds, 578 (2013) 235-241. https://doi.org/10.1016/j.jallcom.2013.05.015

[25] L. Schmidt-Mende, J.L. MacManus-Driscoll, ZnO-nanostructures, defects, and devices, Materials today, 10 (2007) 40-48. https://doi.org/10.1016/S1369-7021(07)70078-0

[26] P. Dhiman, M. Naushad, K.M. Batoo, A. Kumar, G. Sharma, A.A. Ghfar, G. Kumar, M. Singh, Nano FexZn1−xO as a tuneable and efficient photocatalyst for solar powered degradation of bisphenol A from aqueous environment, Journal of Cleaner Production, 165 (2017) 1542-1556. https://doi.org/10.1016/j.jclepro.2017.07.245

[27] W. Peng, H. Li, Y. Liu, S. Song, A review on heavy metal ions adsorption from water by graphene oxide and its composites, Journal of Molecular Liquids, 230 (2017) 496-504. https://doi.org/10.1016/j.molliq.2017.01.064

[28] Y. Feng, J.-L. Gong, G.-M. Zeng, Q.-Y. Niu, H.-Y. Zhang, C.-G. Niu, J.-H. Deng, M. Yan, Adsorption of Cd (II) and Zn (II) from aqueous solutions using magnetic hydroxyapatite nanoparticles as adsorbents, Chemical engineering journal, 162 (2010) 487-494. https://doi.org/10.1016/j.cej.2010.05.049

[29] R. Pandimurugan, S. Thambidurai, Synthesis of seaweed-ZnO-PANI hybrid composite for adsorption of methylene blue dye, Journal of Environmental Chemical Engineering, 4 (2016) 1332-1347. https://doi.org/10.1016/j.jece.2016.01.030

[30] S. Daikh, F. Zeggai, A. Bellil, A. Benyoucef, Chemical polymerization, characterization and electrochemical studies of PANI/ZnO doped with hydrochloric acid and/or zinc chloride: differences between the synthesized nanocomposites, Journal of Physics and Chemistry of Solids, 121 (2018) 78-84. https://doi.org/10.1016/j.jpcs.2018.02.003

[31] N.S. Singh, L. Kumar, A. Kumar, S. Vaisakh, S.D. Singh, K. Sisodiya, S. Srivastava, M. Kansal, S. Rawat, T.A. Singh, Fabrication of zinc oxide/polyaniline (ZnO/PANI) heterojunction and its characterisation at room temperature, Materials Science in Semiconductor Processing, 60 (2017) 29-33. https://doi.org/10.1016/j.mssp.2016.12.021

[32] R. Saravanan, E. Sacari, F. Gracia, M.M. Khan, E. Mosquera, V.K. Gupta, Conducting PANI stimulated ZnO system for visible light photocatalytic degradation of coloured dyes, Journal of molecular liquids, 221 (2016) 1029-1033. https://doi.org/10.1016/j.molliq.2016.06.074

[33] A.K. Mishra, S. Ramaprabhu, Functionalized graphene sheets for arsenic removal and desalination of sea water, Desalination, 282 (2011) 39-45. https://doi.org/10.1016/j.desal.2011.01.038

[34] H. Wang, X. Yuan, Y. Wu, H. Huang, G. Zeng, Y. Liu, X. Wang, N. Lin, Y. Qi, Adsorption characteristics and behaviors of graphene oxide for Zn (II) removal from aqueous solution, Applied Surface Science, 279 (2013) 432-440. https://doi.org/10.1016/j.apsusc.2013.04.133

[35] J.Y. Lim, N. Mubarak, E. Abdullah, S. Nizamuddin, M. Khalid, Recent trends in the synthesis of graphene and graphene oxide based nanomaterials for removal of heavy metals-A review, Journal of Industrial and Engineering Chemistry, 66 (2018) 29-44. https://doi.org/10.1016/j.jiec.2018.05.028

[36] G. Zhao, J. Li, X. Ren, C. Chen, X. Wang, Few-layered graphene oxide nanosheets as superior sorbents for heavy metal ion pollution management, Environmental science & technology, 45 (2011) 10454-10462. https://doi.org/10.1021/es203439v

[37] X. Yang, C. Chen, J. Li, G. Zhao, X. Ren, X. Wang, Graphene oxide-iron oxide and reduced graphene oxide-iron oxide hybrid materials for the removal of organic and inorganic pollutants, RSC advances, 2 (2012) 8821-8826. https://doi.org/10.1039/c2ra20885g

[38] X.Y.Y. Hu, Hou W. Zhou J. Song L, Appl. Surf. Sci, 273 (2013) 118-121. https://doi.org/10.1016/j.apsusc.2013.01.201

[39] D. Gu, J.B. Fein, Adsorption of metals onto graphene oxide: Surface complexation modeling and linear free energy relationships, Colloids and Surfaces A: Physicochemical and Engineering Aspects, 481 (2015) 319-327. https://doi.org/10.1016/j.colsurfa.2015.05.026

[40] B. Kim, V. Koncar, C. Dufour, Polyaniline-coated pet conductive yarns: Study of electrical, mechanical, and electro-mechanical properties, Journal of Applied Polymer Science, 101 (2006) 1252-1256. https://doi.org/10.1002/app.22799

[41] A. Ananth, S. Dharaneedharan, M.-S. Heo, Y.S. Mok, Copper oxide nanomaterials: synthesis, characterization and structure-specific antibacterial performance, Chemical Engineering Journal, 262 (2015) 179-188. https://doi.org/10.1016/j.cej.2014.09.083

[42] K.-S. Huang, C.-H. Yang, S.-L. Huang, C.-Y. Chen, Y.-Y. Lu, Y.-S. Lin, Recent advances in antimicrobial polymers: a mini-review, International journal of molecular sciences, 17 (2016) 1578. https://doi.org/10.3390/ijms17091578

[43] M.H. Farzana, S. Meenakshi, Photocatalytic aptitude of titanium dioxide impregnated chitosan beads for the reduction of Cr (VI), International journal of biological macromolecules, 72 (2015) 1265-1271. https://doi.org/10.1016/j.ijbiomac.2014.09.029

[44] Z. Tang, G.K. Wong, P. Yu, M. Kawasaki, A. Ohtomo, H. Koinuma, Y. Segawa, Room-temperature ultraviolet laser emission from self-assembled ZnO microcrystallite thin films, Applied physics letters, 72 (1998) 3270-3272. https://doi.org/10.1063/1.121620

[45] M.H. Huang, S. Mao, H. Feick, H. Yan, Y. Wu, H. Kind, E. Weber, R. Russo, P. Yang, Room-temperature ultraviolet nanowire nanolasers, science, 292 (2001) 1897-1899. https://doi.org/10.1126/science.1060367

[46] Z.W. Pan, Z.R. Dai, Z.L. Wang, Nanobelts of semiconducting oxides, Science, 291 (2001) 1947-1949. https://doi.org/10.1126/science.1058120

[47] W.-Z. Wang, B.-Q. Zeng, J. Yang, B. Poudel, J.Y. Huang, M.J. Naughton, Z. Ren, Aligned ultralong ZnO nanobelts and their enhanced field emission, Advanced Materials, 18 (2006) 3275-3278. https://doi.org/10.1002/adma.200601274

[48] C. Adán, A. Bahamonde, I. Oller, S. Malato, A. Martínez-Arias, Influence of iron leaching and oxidizing agent employed on solar photodegradation of phenol over nanostructured iron-doped titania catalysts, Applied Catalysis B: Environmental, 144 (2014) 269-276. https://doi.org/10.1016/j.apcatb.2013.07.027

[49] Y. Tang, H. Zhou, K. Zhang, J. Ding, T. Fan, D. Zhang, Visible-light-active ZnO via oxygen vacancy manipulation for efficient formaldehyde photodegradation, Chemical Engineering Journal, 262 (2015) 260-267. https://doi.org/10.1016/j.cej.2014.09.095

[50] S. Singh, K. Barick, D. Bahadur, Fe 3 O 4 embedded ZnO nanocomposites for the removal of toxic metal ions, organic dyes and bacterial pathogens, Journal of Materials Chemistry A, 1 (2013) 3325-3333. https://doi.org/10.1039/c2ta01045c

[51] Y. Yang, G. Wang, Q. Deng, H. Wang, Y. Zhang, D.H. Ng, H. Zhao, Enhanced photocatalytic activity of hierarchical structure TiO_2 hollow spheres with reactive (001) facets for the removal of toxic heavy metal Cr (vi), (2014). https://doi.org/10.1039/C4RA04787G

[52] J. Theerthagiri, S. Salla, R. Senthil, P. Nithyadharseni, A. Madankumar, P. Arunachalam, T. Maiyalagan, H.-S. Kim, A review on ZnO nanostructured materials: energy, environmental and biological applications, Nanotechnology, 30 (2019) 392001. https://doi.org/10.1088/1361-6528/ab268a

[53] H. Benhebal, M. Chaib, T. Salmon, J. Geens, A. Leonard, S.D. Lambert, M. Crine, B. Heinrichs, Photocatalytic degradation of phenol and benzoic acid using zinc oxide powders prepared by the sol-gel process, Alexandria Engineering Journal, 52 (2013) 517-523. https://doi.org/10.1016/j.aej.2013.04.005

[54] S. Yue, Z. Yan, Y. Shi, G. Ran, Synthesis of zinc oxide nanotubes within ultrathin anodic aluminum oxide membrane by sol-gel method, Materials Letters, 98 (2013) 246-249. https://doi.org/10.1016/j.matlet.2013.02.037

[55] X. Wang, M. Ahmad, H. Sun, Three-dimensional ZnO hierarchical nanostructures: Solution phase synthesis and applications, Materials, 10 (2017) 1304. https://doi.org/10.3390/ma10111304

[56] A. Kołodziejczak-Radzimska, T. Jesionowski, Zinc oxide-from synthesis to application: a review, Materials, 7 (2014) 2833-2881. https://doi.org/10.3390/ma7042833

[57] A. Kołodziejczak-Radzimska, T. Jesionowski, A. Krysztafkiewicz, Obtaining zinc oxide from aqueous solutions of KOH and Zn (CH3COO) 2, Physicochemical Problems of Mineral Processing, 44 (2010) 93-102.

[58] A.S. Lanje, S.J. Sharma, R.S. Ningthoujam, J.-S. Ahn, R.B. Pode, Low temperature dielectric studies of zinc oxide (ZnO) nanoparticles prepared by precipitation method, Advanced Powder Technology, 24 (2013) 331-335. https://doi.org/10.1016/j.apt.2012.08.005

[59] L. Dem'Yanets, L. Li, T. Uvarova, Zinc oxide: hydrothermal growth of nano-and bulk crystals and their luminescent properties, Journal of materials science, 41 (2006) 1439-1444. https://doi.org/10.1007/s10853-006-7457-z

[60] S. Musić, Đ. Dragčević, S. Popović, M. Ivanda, Precipitation of ZnO particles and their properties, Materials letters, 59 (2005) 2388-2393. https://doi.org/10.1016/j.matlet.2005.02.084

[61] A. Stanković, L. Veselinović, S. Škapin, S. Marković, D. Uskoković, Controlled mechanochemically assisted synthesis of ZnO nanopowders in the presence of oxalic acid, Journal of materials science, 46 (2011) 3716-3724. https://doi.org/10.1007/s10853-011-5273-6

[62] S.-E. Jin, H.-E. Jin, Synthesis, characterization, and three-dimensional structure generation of zinc oxide-based nanomedicine for biomedical applications, Pharmaceutics, 11 (2019) 575. https://doi.org/10.3390/pharmaceutics11110575

[63] N. Roy, A. Gaur, A. Jain, S. Bhattacharya, V. Rani, Green synthesis of silver nanoparticles: an approach to overcome toxicity, Environmental toxicology and pharmacology, 36 (2013) 807-812. https://doi.org/10.1016/j.etap.2013.07.005

[64] A. Gour, N.K. Jain, Advances in green synthesis of nanoparticles, Artificial cells, nanomedicine, and biotechnology, 47 (2019) 844-851. https://doi.org/10.1080/21691401.2019.1577878

[65] J.H. Duffus, " Heavy metals" a meaningless term?(IUPAC Technical Report), Pure and applied chemistry, 74 (2002) 793-807. https://doi.org/10.1351/pac200274050793

[66] K. Chand Verma, V. Pratap Singh, M. Ram, J. Shah, R.K. Kotnala, Structural, microstructural and magnetic properties of NiFe2O4, CoFe2O4 and MnFe2O4 nanoferrite thin films, Journal of Magnetism and Magnetic Materials, 323 (2011) 3271-3275. https://doi.org/10.1016/j.jmmm.2011.07.029

[67] W.H. Organization, UN-Water global analysis and assessment of sanitation and drinking-water (GLAAS) 2017 report: financing universal water, sanitation and hygiene under the sustainable development goals, (2017).

[68] V. Dhiman, N. Kondal, ZnO Nanoadsorbents: A potent material for removal of heavy metal ions from wastewater, Colloid and Interface Science Communications, 41 (2021) 100380. https://doi.org/10.1016/j.colcom.2021.100380

[69] C. Tunsu, B. Wickman, Effective removal of mercury from aqueous streams via electrochemical alloy formation on platinum, Nature communications, 9 (2018) 1-9. https://doi.org/10.1038/s41467-018-07300-z

[70] WHO, Ten chemicals of major public health concern, Geneva, (2010) 1-4.

[71] H. Ali, E. Khan, Trophic transfer, bioaccumulation, and biomagnification of non-essential hazardous heavy metals and metalloids in food chains/webs-Concepts and implications for wildlife and human health, Human and Ecological Risk Assessment: An International Journal, 25 (2019) 1353-1376. https://doi.org/10.1080/10807039.2018.1469398

[72] S. Singh, K. Barick, D. Bahadur, Novel and efficient three dimensional mesoporous ZnO nanoassemblies for envirnomental remediation, International Journal of Nanoscience, 10 (2011) 1001-1005. https://doi.org/10.1142/S0219581X11008654

[73] K.Y. Kumar, H. Muralidhara, Y.A. Nayaka, Facile synthesis of ZnO-NiO nanocomposites for the removal of Hg (II) ions: complete adsorption studies by using differential pulse anodic stripping voltammetry, Journal of Chemical and Pharmaceutical Research, 4 (2012) 5005-5019.

[74] L. Khezamia, K.K. Tahaa, E. Amamic, I. Ghiloufid, L. El Mird, Removal of cadmium (II) from aqueous solution by zinc oxide nanoparticles: kinetic and thermodynamic studies, Desalin Water Treat, 62 (2017) 346-354. https://doi.org/10.5004/dwt.2017.0196

[75] P. Somu, S. Paul, Casein based biogenic-synthesized zinc oxide nanoparticles simultaneously decontaminate heavy metals, dyes, and pathogenic microbes: a rational

strategy for wastewater treatment, Journal of Chemical Technology & Biotechnology, 93 (2018) 2962-2976. https://doi.org/10.1002/jctb.5655

[76] A.T. Le, S.-Y. Pung, S. Sreekantan, A. Matsuda, Mechanisms of removal of heavy metal ions by ZnO particles, Heliyon, 5 (2019) e01440. https://doi.org/10.1016/j.heliyon.2019.e01440

[77] I. Ghiloufi, A. Imam, M. Saud, Effect of indium concentration in zinc oxide nanoparticles on heavy metals adsorption from aqueous solution, Recent Advances in Circuits, Communications and Signal Processing, (2013).

[78] I. Ghiloufi, J. El Ghoul, A. Modwi, L. El Mir, Preparation and characterization of Ca-doped zinc oxide nanoparticles for heavy metal removal from aqueous solution, MRS Advances, 1 (2016) 3607-3612. https://doi.org/10.1557/adv.2016.511

[79] I. Ghiloufi, J. El Ghoul, A. Modwi, L. El Mir, Ga-doped ZnO for adsorption of heavy metals from aqueous solution, Materials Science in Semiconductor Processing, 42 (2016) 102-106. https://doi.org/10.1016/j.mssp.2015.08.047

[80] A. Ara, J.A. Usmani, Lead toxicity: a review, Interdisciplinary toxicology, 8 (2015) 55. https://doi.org/10.1515/intox-2015-0009

[81] M. Krasovska, V. Gerbreders, E. Tamanis, S. Gerbreders, A. Bulanovs, The study of adsorption process of Pb ions using well-aligned arrays of ZnO nanotubes as a sorbent, Latvian Journal of Physics and Technical Sciences, 54 (2017) 41. https://doi.org/10.1515/lpts-2017-0005

[82] X. Wang, W. Cai, S. Liu, G. Wang, Z. Wu, H. Zhao, ZnO hollow microspheres with exposed porous nanosheets surface: structurally enhanced adsorption towards heavy metal ions, Colloids and Surfaces A: Physicochemical and Engineering Aspects, 422 (2013) 199-205. https://doi.org/10.1016/j.colsurfa.2013.01.031

[83] K.Y. Kumar, H. Muralidhara, Y.A. Nayaka, J. Balasubramanyam, H. Hanumanthappa, Hierarchically assembled mesoporous ZnO nanorods for the removal of lead and cadmium by using differential pulse anodic stripping voltammetric method, Powder technology, 239 (2013) 208-216. https://doi.org/10.1016/j.powtec.2013.02.009

[84] A. Jawed, L.M. Pandey, Application of bimetallic Al-doped ZnO nano-assembly for heavy metal removal and decontamination of wastewater, Water Science and Technology, 80 (2019) 2067-2078. https://doi.org/10.2166/wst.2019.393

[85] I. Sheikhshoaie, A. Rezazadeh, S. Ramezanpour, Removal of Pb (II) from aqueous solution by gel combustion of a new nano sized Co3O4/ZnO composite, Asian Journal of Nanosciences and Materials, 1 (2018) 271-281.

[86] S. El-Dafrawy, S. Fawzy, S. Hassan, Preparation of modified nanoparticles of zinc oxide for removal of organic and inorganic pollutant, Trends Appl. Sci. Res, 12 (2016) 1-9. https://doi.org/10.3923/tasr.2017.1.9

[87] V.D. Martinez, E.A. Vucic, D.D. Becker-Santos, L. Gil, W.L. Lam, Arsenic exposure and the induction of human cancers, Journal of toxicology, 2011 (2011). https://doi.org/10.1155/2011/431287

[88] H. Rehman, Z. Ali, M. Hussain, S. Gilani, T. Shahzady, A. Zahra, S. Hussain, H. Hussain, I. Hussain, M. Farooq, Synthesis and characterization of ZnO nanoparticles and their use as an adsorbent for the arsenic removal from drinking water, Digest Journal of Nanomaterials and Biostructures, 14 (2019) 1033-1040.

[89] B. Nath, C. Chaliha, E. Kalita, M. Kalita, Synthesis and characterization of ZnO: CeO2: nanocellulose: PANI bionanocomposite. A bimodal agent for arsenic adsorption and antibacterial action, Carbohydrate polymers, 148 (2016) 397-405. https://doi.org/10.1016/j.carbpol.2016.03.091

[90] S. Avudainayagam, M. Megharaj, G. Owens, R.S. Kookana, D. Chittleborough, R. Naidu, Chemistry of chromium in soils with emphasis on tannery waste sites, Reviews of environmental contamination and toxicology, (2003) 53-91. https://doi.org/10.1007/0-387-21728-2_3

[91] A. Zhitkovich, Chromium in drinking water: sources, metabolism, and cancer risks, Chemical research in toxicology, 24 (2011) 1617-1629. https://doi.org/10.1021/tx200251t

[92] H. Sun, J. Brocato, M. Costa, Oral chromium exposure and toxicity, Current environmental health reports, 2 (2015) 295-303. https://doi.org/10.1007/s40572-015-0054-z

[93] V. Venkatesham, G. Madhu, S. Satyanarayana, H. Preetham, Adsorption of lead on gel combustion derived nano ZnO, Procedia Engineering, 51 (2013) 308-313. https://doi.org/10.1016/j.proeng.2013.01.041

[94] R. Anusa, C. Ravichandran, E. Sivakumar, Removal of heavy metal ions from industrial waste water by nano-ZnO in presence of electrogenerated Fenton's reagent, Int J ChemTech Res, 10 (2017) 501-508.

[95] K.S. Ranjith, P. Manivel, R.T. Rajendrakumar, T. Uyar, Multifunctional ZnO nanorod-reduced graphene oxide hybrids nanocomposites for effective water remediation: Effective sunlight driven degradation of organic dyes and rapid heavy metal adsorption, Chemical Engineering Journal, 325 (2017) 588-600. https://doi.org/10.1016/j.cej.2017.05.105

[96] K. Nalwa, A. Thakur, N. Sharma, Synthesis of ZnO nanoparticles and its application in adsorption, Advanced Materials Proceedings, 2 (2017) 697-703. https://doi.org/10.5185/amp/2017/696

[97] H. Green, J. Bailey, L. Schwarz, J. Vanos, K. Ebi, T. Benmarhnia, Impact of heat on mortality and morbidity in low and middle income countries: a review of the epidemiological evidence and considerations for future research, Environmental research, 171 (2019) 80-91. https://doi.org/10.1016/j.envres.2019.01.010

[98] G. Cruz, M. Gómez, J. Solis, J. Rimaycuna, R. Solis, J. Cruz, B. Rathnayake, R. Keiski, Composites of ZnO nanoparticles and biomass based activated carbon: adsorption, photocatalytic and antibacterial capacities, Water Science and Technology, 2017 (2018) 492-508. https://doi.org/10.2166/wst.2018.176

[99] M.Y. Haddad, H.F. Alharbi, Enhancement of heavy metal ion adsorption using electrospun polyacrylonitrile nanofibers loaded with ZnO nanoparticles, Journal of Applied Polymer Science, 136 (2019) 47209. https://doi.org/10.1002/app.47209

[100] Z. Rafiq, R. Nazir, M.R. Shah, S. Ali, Utilization of magnesium and zinc oxide nano-adsorbents as potential materials for treatment of copper electroplating industry wastewater, Journal of Environmental Chemical Engineering, 2 (2014) 642-651. https://doi.org/10.1016/j.jece.2013.11.004

[101] H.A. Sani, M.B. Ahmad, M.Z. Hussein, N.A. Ibrahim, A. Musa, T.A. Saleh, Nanocomposite of ZnO with montmorillonite for removal of lead and copper ions from aqueous solutions, Process Safety and Environmental Protection, 109 (2017) 97-105. https://doi.org/10.1016/j.psep.2017.03.024

[102] M. Khosravi, N. Mehrdadi, G. Nabi Bidhendi, M. Baghdadi, Synthesis of sewage sludge-based carbon/TiO2/ZnO nanocomposite adsorbent for the removal of Ni (II), Cu (II), and chemical oxygen demands from aqueous solutions and industrial wastewater, Water Environment Research, 92 (2020) 588-603. https://doi.org/10.1002/wer.1253

[103] V. Srivastava, D. Gusain, Y.C. Sharma, Synthesis, characterization and application of zinc oxide nanoparticles (n-ZnO), Ceramics International, 39 (2013) 9803-9808. https://doi.org/10.1016/j.ceramint.2013.04.110

Materials Research Forum LLC
https://doi.org/10.21741/9781644902394-7

Chapter 7

ZnO for Probes in Diagnostics

Debjita Mukherjee[1], Ehsan Amel Zendehdel[2#], Mojdeh Rahnama Ghahfarokhi[3#],
Minoo Alizadeh Pirposhte[4#], Azadeh Jafarizadeh Dehaghani[5#], Agnese Brangule[6,7*],
Dace Bandere[6,7*], and Jhaleh Amirian[6,7*]

[1]College of Medical, Veterinary and Life Sciences, University of Glasgow, University Ave,
Glasgow G12 8QQ, United Kingdom

[2]The faculty of Art and Architecture, Eshragh Institute of Higher Education, Bojnord, Iran

[3]Department of Materials Engineering, Faculty of Materials Processing and Fabrication, Isfahan
University of Technology, Isfahan, Iran

[4]Department of Materials Engineering, Faculty of advanced materials, Isfahan University of
Technology, Isfahan, Iran

[5]Department of Materials Engineering, Faculty of Materials Processing and Fabrication, Isfahan
University of Technology, Isfahan, Iran

[6]Department of Pharmaceutical Chemistry, Riga Stradiņš University, Dzirciema 16, LV-1007,
Riga, Latvia

[7]Baltic Biomaterials Centre of Excellence, Headquarters at Riga Technical University, Kalku
Street 1, LV-1658 Riga, Latvia

[#] These authors contributed same.
*Prof. Dace Bandere (dace.bandere@rsu.lv)
Dr. Agnese Brangule (agnese.brangule@rsu.lv)
Jhaleh Amirian (jalehamirian@gmail.com)

Abstract

Nanoparticles have revolutionized the field of diagnostics in recent years and ZnO
nanoparticles (ZnO-NPs) have been one of the most commonly used ones. These easily
synthesizable ZnO-NPs have a multitude of advantages over other metal-based
nanoparticles owing to their biocompatibility, easy functionalization through their
hydroxyl group-rich surface, and cost-effectiveness among several other benefits. Due to
their inherent luminescence and fluorescent-tag functionalizing properties, ZnO-NPs have
been useful as a probe in tumour and live cell bioimaging. ZnO-NPs have also been
identified as probes in biosensors for the detection of various clinically important
biochemical analytes like glucose and cholesterol, pathogens, drug molecules, and
antibody-antigen based detection systems. In this chapter, several of the different

applications of ZnO as probes in diagnostics will be dealt with in detail. Also, the characteristics of ZnO nanoparticles useful for such applications and the way these devices and techniques are developed will be explained.

Keywords

ZnO Nanoparticles (ZnO-NPs), Biosensors, Diagnostics

Contents

1.   Introduction

Since the advent of nanomaterials in the late 1900s, several novel applications for these materials have been discovered. As technology progressed, healthcare systems have improved greatly with nanomaterials replacing bulk materials in most sectors ranging from the detection of diseases to their treatment. Like some of the other metal oxide nanomaterials, Zinc Oxide (ZnO) nanomaterials have found extensive applications in several fields due to numerous beneficial characteristics. This chapter focuses on the various applications of ZnO nanomaterials in diagnostics, describing in detail some of the recent and most interesting systems that have been researched.

1.1   Nanomaterials in diagnostics

To be useful as probes in diagnostics, the nanomaterials need to possess most or all of the following characteristics [1-2]:

- Large surface-to-volume ratio with small size ranging from 1 to 100 nm to be able to reach very small areas and/or detect minuscule structures
- Ease of modifying the physical properties (composition, size, shape) of the nanomaterial based on application
- Good target binding properties with the ability to bind a wide range of biologically relevant molecules
- Robustness of the structures
- Biocompatible
- Does not interfere with the structural and/or functional integrity of the analyte molecule
- Cost-effectiveness and simple production procedure

Not all of these characteristics are found in all nanomaterials hence making them unsuitable for biological applications. However, several features of the different ZnO nanomaterials turn out to be useful for their application as diagnostic probes. The noteworthy features of ZnO nanomaterials are as follows [3-4]:

- These have dimensions like the bio-components which are used to form the bio-selective layer.
- The large surface-to-volume ratios are very useful for increasing sensitivity of the analysis
- Being a metal oxide material, ZnO is an n-type semiconductor with a wide band gap of 3.37 eV
- It has a Bohr exciton radius of around 2.34 nm and a high dielectric constant.
- ZnO nanomaterials possess unique electron transport ability and show multifunctionality based on electric conductance facilitating immobilization of various types of biomolecules
- Functionalization of these materials is facilitated by their hydroxyl group-rich surface
- These have a high isoelectric point (pH 9-9.5)
- These materials have a broad photoluminescence (PL) range and can show highly intense photoluminescence at room temperature (with ultraviolet (UV), infrared, microwaves, radio waves, and even visible light).
- When ZnO materials are irradiated with UV light, an electron/hole pair is created resulting in excitonic emission
- They can be synthesized following simple steps and easily available raw materials and are therefore cost-effective

The ZnO nanomaterials have been tested for various diagnostics-based approaches. Materials such as nanorods (NRs), nanoparticles (NPs), nanowires (NWs), and quantum dots (QDs) are some of the most used nanomaterials of ZnO for diagnostics [5]. The commonly detected analytes belong to the following broad categories:

- Proteins (such as albumin, immunoglobulins, haemoglobin, etc.)
- Low molecular weight compounds (such as acetone, urea, riboflavin, dopamine, etc.)
- nucleic acids (DNA or RNA)
- Cells (from tumour/cancer, somatic cells) and microbes (virus, bacteria, fungi, etc.)

2. Types of detection mechanisms for ZnO probe-based analysis

There are various methods currently available for diagnostic techniques based on ZnO nanomaterials. The basic working principles of the most commonly used ones have been briefly described below:

2.1 Photoluminescence detection

Based on the inherent PL properties of ZnO, many biosensors made of ZnO nanomaterials use this technique for detecting the final analyte using spectroscopy or microscopy. In this technique, light (photons) in the ultraviolet (UV), visible, or near infrared (NIR) range are directed to a sample. The photons are absorbed by the sample and excite the sample's electrons to higher excitation states (photoexcitation) [6]. The excited electron goes back to its equilibrium state after some time which results in the emission of the energy (photoemission) in the form of light. The resulting luminescence can be fluorescence, phosphorescence, or chemiluminescence depending on internal energy transitions, transition into a state with different spin multiplicity, or chemical reaction during transition (Figure 1). At a particular excitation wavelength, the maximum intensity of the emission can be observed for the sample which can help get a better resolution. Factors such as stability, quantum yield, energy transfer, and characteristics of excitation or absorption determine how suitable a nanomaterial is for PL applications based on its emission strength [7].

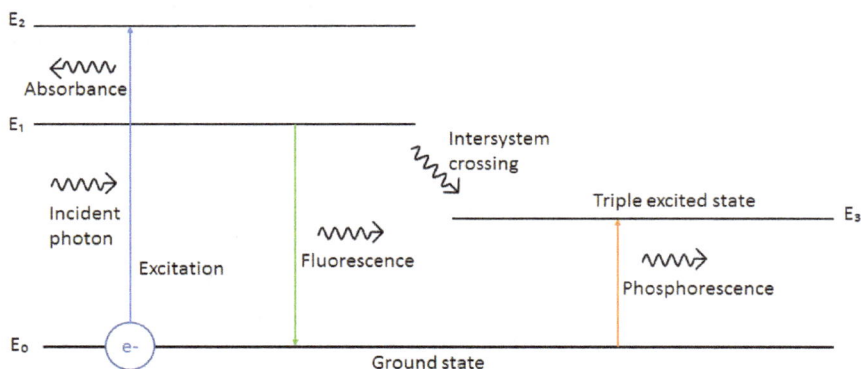

Figure 1: Principle of Photoluminescence: Fluorescence and Phosphorescence

2.2   Diagnostic imaging techniques

The ZnO-based nanocomposites for bioimaging are commonly detected by Magnetic Resonance Imaging (MRI) and tomography (Computed/Positron emission/Single Photon emission computed) techniques. The working principle of MRI is based on the nuclear spin properties in the presence of a very strong external magnet [8]. The strong external magnetic field of the magnet results in parallel (low-energy state) or perpendicular (high-energy state) alignment of the nucleus with respect to the external field. A second radio frequency (RF) magnetic field (pulses) is applied perpendicular to the external field. This RF field results in the transition of the molecules/cells from a higher to a lower energy level after energy absorption. The energy absorbed is again emitted when relaxation occurs. The absorbed/emitted energy is detected by a coiled wire as voltages. The signals are amplified, and the results are obtained as Free Induction Decay (FID). Fourier transform analysis of the averaged FID can help understand the biochemical state of the tissues and can help distinguish between cancerous and healthy tissues. A basic setup of the MRI scan system has been shown in Figure 2. (A).

Nuclear medicine or tomography-based bioimaging works with the help of radiolabelled probe-tracing with gamma radiation trackers/detectors (cameras in most cases) [9]. The Single-photon emission computed tomography (SPECT) is based on the Compton effect, where a photon beam is emitted after interacting with electrons present in the tissues (Figure 2. (C)). The photons are either deflected away from the detector's direction (attenuation) or are deflected towards the detector but with an altered direction of incidence to the detector (scattering). The photons emitted from the radiotracer are considered independent events to obtain specific information regarding the biochemical condition of

ZnO and Their Hybrid Nano-Structures   Materials Research Forum LLC
Materials Research Foundations 146 (2023) 202-233   https://doi.org/10.21741/9781644902394-7

the tissues to form an imaging matrix. Positron-emission tomography (PET) involves positively charged electrons emitted due to the radioactive disintegration of the probe nucleus. The positron keeps losing kinetic energy and gets deflected from its original direction. It finally combines with an electron, annihilates, and finally emits two photons which are detected by the detector (Figure 2. (b)). The tomography techniques are capable of showing the function of tissues in a given region even before anatomical changes occur in these tissues.

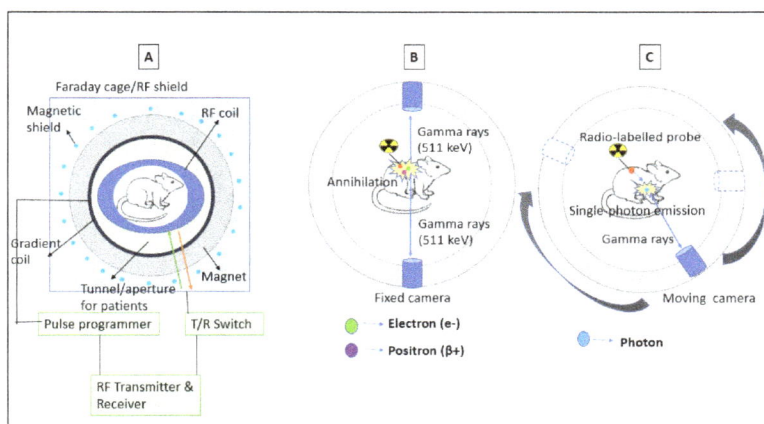

Figure 2: Basic setup of common diagnostic techniques (A) MRI, (B) PET, and (C) SPECT.

2.3   Electrochemical detection

Most of the biochemical sensors are based on electrochemical or surface plasmon resonance (SPR) techniques because of high-sensitivity detections and ease of control of parameters affecting the detection. In electrochemical systems, two or three electrodes are commonly used [10]. A standard 3-electrode system contains the following:

(i)   Working electrode: this is the main electrode which is mostly modified to improve functionality or sensitivity; it acts as a transducer for biochemical sensors

(ii) Counter electrode: this controls the flow of current to the working electrode

(iii)   Reference electrode: it provides a stable potential to the working electrode

The biological element or target is detected by the working electrode through transduction and is then converted into electrical signals for analysis. This detection method is broadly classified into current-based (amperometric), charge, or potential-based (potentiometric such as voltammetry) and based on the variation in conductance of the medium (conductometric). In most cases, the bio-recognition element (such as antibodies, antigens, or antibody fragments) is added as coatings on the working electrode for detecting analytes like specific microbes, cells, nucleic acids, or enzymes. Figure 3. (A) shows how electrochemical systems can be used in biosensing.

The other commonly used technique, SPR, is only observed at the metal-dielectric interface. In this method, a surface plasmon (electromagnetic plasma wave) propagating at the metal-dielectric interface, can be excited by an evanescent wave resulting in SPR [11]. These excited plasmons decay resulting in energy conversion to photons and a drastic decrease of the reflected wave. When this ray travels in a lower refractive index medium it undergoes total internal reflection (TIR) and the resonance angle (θ_{SPR}) changes which is an indication of the absorption-desorption or association-dissociation phenomena occurring at the interface (Figure 3. (B)). In a standard SPR system, a Kretschmann configuration is followed. A prism is used in this configuration where the incident light propagates and reaches a gold film to which a testing cell is attached containing the solutions to be analysed. TIR occurs at this interface till the point where the angle of incident light is greater than the critical angle. At a particular angle, the evanescent waves can excite the plasmons resulting in SPR at the gold film interface. The incident angle of minimum reflectivity (or the SPR angle) can change due to changes in the refractive index of the test solution. Such changes in refractive index can be induced by modifying the mass and density of the solution which happens due to foreign bodies attaching to the gold film surface. The SPR angle (θ_{SPR}) is given by the equation:

$$\theta_{SPR} = \sin^{-1}\left(\frac{1}{n_1}\sqrt{\frac{(n_2 n_g)^2}{n_2^2 + n_g^2}}\right)$$

where, n_1 is refractive index of medium 1

n_2 is refractive index of medium 2 at interface ($n_1 > n_2$)

n_g is refractive index of gold film

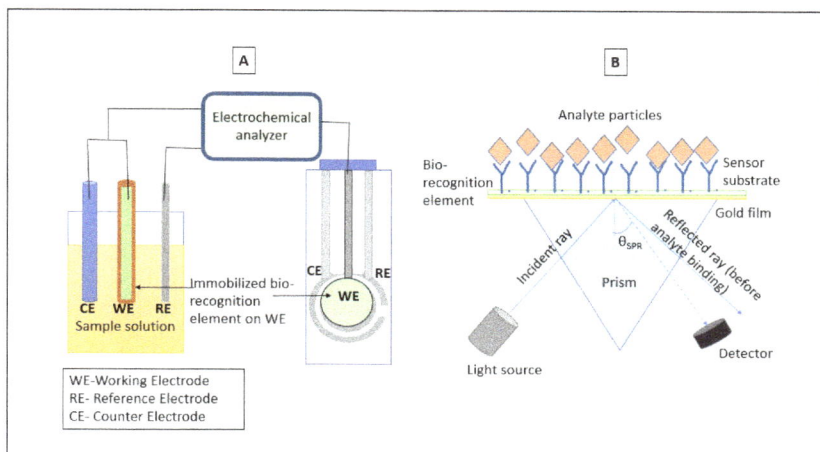

Figure 3: (a) Electrochemical detection system as electrochemical cell (left) and as screen-printed electrode (right) for biosensors ;(b) Biosensing using SPR technique.

3.   Applications

3.1   ZnO-NPs in cancer diagnostics

Cancer or uncontrolled proliferation of cells at specific sites has become a common life-threatening disease in the modern world. Incidences of cancer increase globally every year. The International Agency for Research on Cancer (IARC) published statistics about the cancer situation in 2020 where around 19.3 million people were diagnosed with cancer and around 10 million deaths attributed to cancer [12]. Most of these deaths were due to late diagnosis of the condition which leaves very little time for treatment [13]. Several improved diagnostics are being developed currently to enable timely diagnosis of cancer with high specificity and sensitivity. Cancer diagnostics mainly involve one or all of the following [14]:

(i) *In vivo* bio-imaging of tumours using techniques such as upconversion luminescence (UCL), MRI, and such other.

(ii) Detection of cancer cells based on their surface functionalization

(iii)   Detection of the moving tumour or vesicular cancerous cells

(iv)   Biomarker-based detection

In the aspect of cancer diagnosis, ZnO nanomaterials are investigated both as single-material probes or in conjunction with other nanomaterials. The ZnO nanomaterials have inherent luminescence properties and can also easily get functionalized to the surfaces of several fluorescence-emitters or other moieties useful in bioimaging *in vivo* thanks to their non-toxic properties [15]. Currently, ZnO nanocomposites are also being explored to enhance signal resolution. Nanostructures made of ZnO are also being implemented in biosensing elements for detecting cancer biomarkers due to factors such as the immobilization efficiency of antibodies to its surface, ease of fabrication, and high specificity [16].

The enzyme matrix metalloprotease 9 (MMP-9) has been identified as a biomarker for malignancy in tumours and cancer malignancy [17]. Normally, MMP-9 plays a significant role in the maintenance of the extracellular matrix (ECM) and the physiological processes of the central nervous system (CNS). The normal range for MMP-9 in blood serum ranges from 1- 100 ng/mL (up to 500 ng/mL) [18]. In 2020, Shabani and colleagues developed a label-free biosensor where ZnO NPs were spin-coated and annealed over a gold (Au) electrode substrate and the ZnO nanorods were fabricated hydrothermally. Mouse monoclonal antibodies against MMP-9 were immobilized on these nanorods. Results were analysed using cyclic voltammetry (CV) and electrochemical impedance spectroscopy (EIS) where linearity was observed for an MMP-9 concentration range of 1-1000 ng/mL which can cover the normal detection range for MMP-9. The biosensor had a detection limit of 0.15 ng/mL and showed a sensitivity of 32.5 $\mu A/(decade \times cm^2)$. Compared to the traditional ELISA method, this biosensor produced results within a much shorter time (around 35 minutes) and with only an about 8% difference in accuracy.

Several kinds of research have been conducted to find techniques for using ZnO NPs in biomarker-based diagnostics. With PVA, ZnO NPs have been used for immobilizing biomolecules for identifying the epithelial cancer biomarker, EpCAM (Epithelial cell adhesion molecule) by Zhu and colleagues [19]. In another study, ZnO NRs were used by Wang and his group to coat a quartz crystal microbalance for label-free detection of CA 15.3 (breast cancer biomarker) [20]. Murugan and co-workers fabricated a gradient triple-layered core-shell NP with ZnO as one of the core components and silica as the shell for detecting the cancer biomarker, acetylcholine [21].

Some ELISA-based detection methods involving ZnO nanomaterials have also been used for antigen-based cancer diagnostics. In 2015, Pal and Bhand immobilized monoclonal antibodies for carcinoembryonic antigen (CEA) (capture antibodies) to ZnO NPs and added them to microwell plates [22]. A polyclonal antibody was used for detection while an HRP-labelled tertiary antibody was used for colorimetric detection. They observed a 3-fold increase in chemiluminescence compared to standard ELISA along with improvements in thermal stability. The system could detect CEA levels in blood serum from 1 pg/mL to 20 ng/mL.

Due to their inherent PL properties and semiconductor-like properties, ZnO nanomaterials have found extensive applications in the bioimaging of cancer cells and tumours. The chemiluminescence (CL) properties were applied for bioimaging of HeLa cells (cervical cancer cell-line) by Liu and co-workers in 2020 [23]. The CL activity is observed on addition of the compound bis (2,4,5-trichloro-6-carbopentoxyphenyl) oxalate (CPPO) and

H_2O_2 to the ZnO NPs to produce strong blue luminescence. The intermediate dioxetanedione intermediate produced during the reaction has high energy and can react with the interstitial zinc (Zn_i) to result in CL. To enhance the CL properties, silica (SiO_2) shell was added to the ZnO NPs which increases the quantum yield of the system from 6.2 x 10^{-6} to 3.72 x 10^{-4} E mol^{-1}. The ZnO@SiO_2 NPs allow dual-mode imaging of live cells as the ZnO NPs have inherent PL properties and yellow fluorescence can be observed when illuminated with 365 nm UV light. The ZnO@SiO_2 NPs show high specificity and a very low level of cytotoxicity (more than 90% viability) till a concentration of 500 μM.

For *in vivo* bioimaging of tumours, ZnO NPs have been used as part of the probes in conjunction with other materials to enable multi-modal imaging to obtain better specificity and detailed information about the tumour. A very interesting system was developed by Hong and colleagues in 2015 for multi-modal tumour imaging in mice [24]. The ZnO NPs showed red fluorescence and allowed detection by Positron emission tomography (PET) thereby enabling dual-mode imaging. The ZnO NPs were synthesized by calcination, and they were PEGylated after thiolation. The ZnO-PEG-NH_2 (1 part) was reacted with S-2-(4-isothiocyanatobenzyl)- 1,4,7-triazacyclononane-1,4,7-triacetic acid (p-SCN-Bn-NOTA) (10 parts). The NOTA-ZnO-PEG-NH_2 (1 part) and succinimidyl carboxymethyl PEG maleimide (SCM-PEG-Mal) (30 parts) to obtain the intermediate NOTA-ZnO-PEG-Mal. Both of these reactions occur at pH 8.5. A biomarker, CD 105 is commonly detected in proliferative tumour cells which were targeted using the TRC-105 antibody which was conjugated to the NOTA-ZnO-PEG-Mal intermediate to form the NOTA-ZnO-TRC105. Finally, radio-labelled copper (^{64}Cu) was added to the antibody-conjugated system to form the final PET-detection probe which was shown to detect 4T1 tumour in mice. Addition of the TRC105 antibody to the nanocomposite improved accumulation (almost by 3-fold) of the composites specifically at the tumour site. Figure 4 shows the step-by-step synthesis of the ZnO NP-based probe and PET scan results in mice using the synthesized probe.

Most of the bioimaging techniques that have been developed for cancer diagnostics focus on serving more than the purpose of just bioimaging. These bioimaging techniques also target therapy or drug delivery applications and thereby act as theranostic agents. This topic has been covered in detail in the last section of this chapter (Section 2.4).

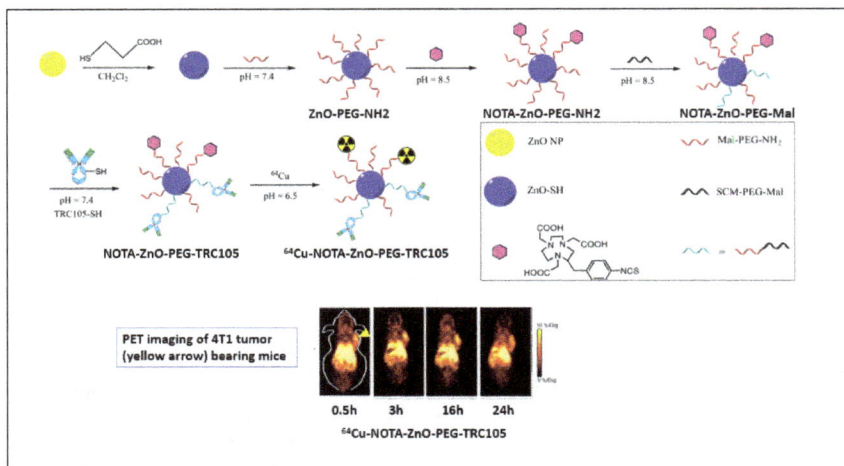

Figure 4: The stepwise synthesis of the ZnO NP-based probe and in vivo 4T1 tumour analysis in mice using PET scan with the synthesized probe.
(Reprinted (adapted) with permission from {Hong et al. (2015). Red fluorescent zinc oxide nanoparticle: a novel platform for cancer targeting. ACS Appl. Mater. Interfaces 7, 3373-3381.}. Copyright © {2015} American Chemical Society)

3.2   ZnO-NPs in microbial disease diagnosis

Microbial diseases are currently the cause of mortality in a large part of the global population. Over the past few years, several viral diseases such as Ebola, swine flu, SARS, and several bacterial diseases such as pneumonia, tuberculosis, cholera, and meningitis, have significantly affected life expectancy [25]. Detecting these diseases is a challenge not only because of the microscopic size of the causative agents but also the time needed for obtaining the results for diagnosis and treatment. Some of the most commonly available techniques for microbial disease diagnosis include immunoassays such as ELISA, RT-PCR, bacterial culture analysis, viral isolation in chicken embryos, and a few other techniques which involve complex steps or time-consuming techniques [26]. In several cases, these techniques have problems with sensitivity or problems with a precise analysis of the bacterial/viral load. Recently, the COVID-19 pandemic showed how important fast diagnosis of such microbial diseases can help to control the spread and severity of the disease [27]. Nanostructures, due to their excellent surface-to-volume ratio and small size, can play an important role in identifying the microbes in a highly efficient manner within the sample received from the infected persons.

Bacterial infections can be treated efficiently if diagnosed early and accurately. Detection of bacteria is easier based on the various bacterial enzymes and by-products which have been widely studied. A device was developed by Vasudevan and colleagues (2020) for quorum sensing of the signalling molecules known as N-Acyl-Homoserine Lactones (AHLs) from gram-negative bacteria *Pseudomonas aeruginosa* to diagnose urinary tract infection (UTI) [28]. The device is based on photoluminescence (PL) and has probes made of ZnO NPs functionalized with cysteamine which acts as a linker (ZnO-Cys). The ZnO NPs were synthesized using a microwave and following calcination, the cysteamine was immobilized on the NPs in ethanol. The biosensor showed a maximum sensitivity of 97% and a linear detection range of 10-120 nM in Artificial Urine Media (AUM) samples during the lag phase of the bacteria. A significant achievement of this biosensor was that it was able to show the PL signature peak of 468 nm even in the presence of interference factors in AUM.

Point-of-care (POC) devices have been in trend in diagnostics due to several factors like portability, simple operation, low time to result, and ease of handling without the need for professional intervention [29]. For improving the performance of such POC devices, the incorporation of ZnO nanostructures in them is being explored. In 2019, a POC device was developed by Xia and colleagues, for virus detection using ZnO nanorods [30]. The microfluidic device was made of PDMS on a SU-8 mold. The PDMS channels were immersed in ZnO growth media after activation with $KMnO_4$ to allow the synthesis of the ZnO nanorods on the surface. Mouse antibodies were introduced to these channels after silanization. The nanorods aid in improving the surface area of the detection system (the PDMS channels in this case) by providing a 3-D surface for attachment of the detection probes (the antibodies). The virus sample can now be introduced to this surface and is captured by the antibodies. Another antibody conjugated with gold nanoparticles (AuNPs) is added to develop a sandwich immunoassay. After adding a silver enhancer to this system, the viral load can be estimated by colorimetric analysis using a smartphone-integrated system (Figure 5). The imaging system is developed with a lens, an LED, and two polarizers, together with the smartphone camera. The mean grayscale value of the images was used for analysis. The device has a detection limit of 2.7×10^4 EID_{50}/mL with naked eyes and 8×10^3 EID_{50}/mL with the smartphone-based imaging system which is one order of magnitude and three times better than the conventional fluorescence-based ELISA, respectively. The time to result from the virus capture process is only 1.5 hours. The device was used for Avian Influenza virus (AIV) analysis in this study, but it has been mentioned to be usable for other viral detections as well.

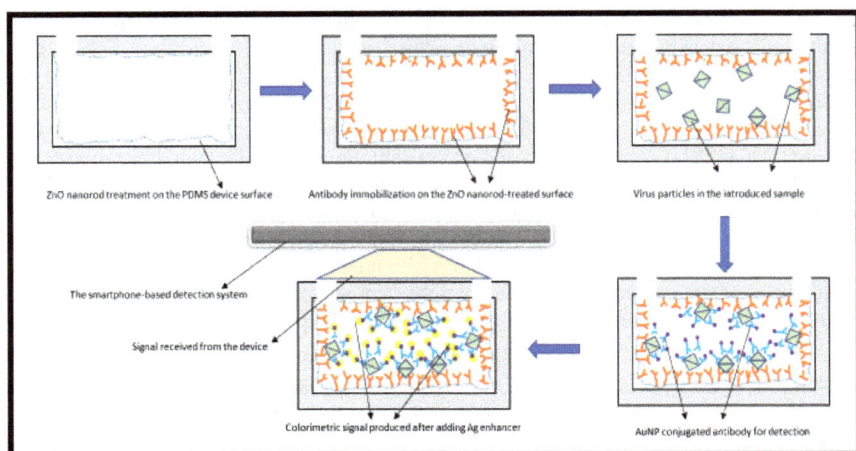

Figure 5: The ZnO-nanorod-based POC system developed by Xia and colleagues for diagnosis of AIV infection.
(Adapted with permission from {Xia et al. (2019). Smartphone-based point-of-care microfluidic platform fabricated with a ZnO nanorod template for colorimetric virus detection. ACS sensors, 4(12), 3298-3307.}. Copyright © {2019} American Chemical Society)

DNA isolation from bacteria or fungi has been a common technique for microbial diagnostics. The conventional commercial DNA extraction kits available currently involve disruption of the cell wall of the microbe, extraction of DNA in sodium dodecyl sulphate (SDS), removal of unnecessary proteins, followed by the precipitation of DNA using isopropanol [31]. In the case of fungal DNA extraction, an added step of lyophilization of mycelia is done before the above-mentioned steps. Qiao and colleagues have explored ZnO NPs as an alternative to the standard lysis buffer used in the isolation kits due to their inhibitory effects on microbes. The ZnO NPs was found to be useful as a lysis buffer for bacterial and eukaryotic cell but had limitations of application in fungal DNA isolation due to their cell wall. However, in 2020, Qiao and colleagues found a combined form of ZnO (ZnO-S-300) which could overcome this limitation in the fungi *Aspergillus*. The resulting assay not only makes the process much less cumbersome but also reduces the use of bulky, expensive equipment. Unlike normal ZnO NPs, the ZnO-S-300 surface is positively charged leading to easy binding of this material to the spores.

Biosensors with ZnO-based nanomaterials have been developed for detecting bacteria and virus infections. One such device was made by Huang and co-workers in 2020 for acid-responsive and bimodal detection of the common pathogenic bacteria *Salmonella* which spreads through contaminated food/water and from animals and can cause severe food poisoning and diseases such as typhoid [32]. In this system, ZnO is used as a capping agent for a conjugated system with mesoporous silica nanoparticles (MSNs) and Curcumin (CUR) [33]. The CUR acts as a dual-signal reporter on acid stimulation with acetic acid (HAc) which is unique among the existing *Salmonella* biosensors. The microfluidics-based system used three inlets for 3 different things: (i) amine-modified MSNs were incubated with CUR to form NPs followed by capping with amino-modified ZnO NPs to form MSN@CUR@ZnO NPs (MCZ NPs) and finally polyclonal antibodies against *Salmonella* were functionalized with tetrazine and trans-cyclooctene, (ii) magnetic NPs with monoclonal antibodies against *Salmonella*, and (iii) the *Salmonella* sample to be detected. After mixing in a Koch fractal structured chamber, a sandwich immunoassay is established. The HAc is then introduced to this system which allows dissolution of CUR which is then collected for fluorescence and absorbance measurements. The biosensor worked for 102 to 107 CFU/mL of the *Salmonella* bacteria and showed results within 1.5 h. The lower limit of detected colonies was 63 CFU/mL and 40 CFU/mL for colorimetric and fluorescent measurements, respectively with high recovery percentages.

Among other uses of ZnO NPs in microbial diagnosis, they have been used for improving diagnostics using techniques such as PCR. In 2020, Upadhyay and colleagues presented a method of using a relatively newer type of ZnO nanomaterial, nano-flowers, for developing a nano-PCR technique to diagnose canine vector-borne disease (CVBD)-causing pathogens [34]. Hydrothermally prepared ZnO nanoflowers were incorporated with the pathogenic samples in a concentration of 1 mM which showed significant enhancement of obtained results. The pathogenic strains of *Babesia canis vogeli* and *Hepatozoon canis* were diagnosed successfully using the nano-PCR method. The adsorption of the ZnO nanoflowers to the isolated DNA samples thereby improving the sensitivity of the PCR and simultaneously helping reduce the turnaround time of the process without affecting the amplification process in any way. Figure 5 shows the intricate structure of the nanoflowers and the confirmation of the nano-PCR diagnosis of *Babesia canis vogeli* and *Hepatozoon canis* at 619 bp and 666 bp respectively in the agarose gel electrophoresis analysis.

ZnO nanomaterials are being investigated to find novel applications in microbial diagnostics. Only a few of such works have been discussed in detail above. Other than the works mentioned in this section, there have been several such research involving ZnO nanomaterials as components in probes or sensor devices for microbial analyte detection. The next section deals with biosensors based on ZnO nanomaterials.

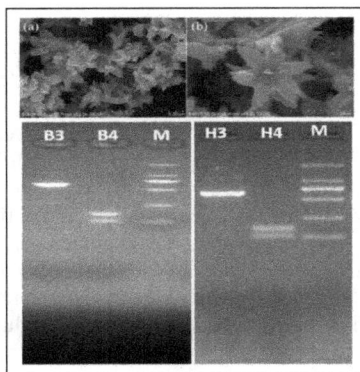

*Figure 6: ZnO nanoflowers observed using scanning electron microscopy ((a) and (b))
(above); Agarose gel electrophoresis of B. canis vogeli (B) and H.canis (H) with ZnO
nanoflowers (B3 and H3), and without them (B4 and H4) with DNA ladder (M) (below).
(Reprinted from: Upadhyay et al. (2020). ZnO Nanoflower-Based NanoPCR as an
Efficient Diagnostic Tool for Quick Diagnosis of Canine Vector-Borne Pathogens.
Pathogens, 9(2), 122.)*

3.3   ZnO-NPs in biochemical biosensors

The ZnO nanomaterials do not only find application in detecting tumours or cancers, and
microbial nucleotides but can also be used for the detection of important chemicals and
biomolecules such as biomarkers and macro or micro-nutrients [35]. This makes ZnO NPs
important as materials of choice for biosensors. Some electrochemical detection-based
biosensors have been developed using ZnO NPs.

Detection of chemicals such as glucose is particularly important for diagnosing and
monitoring the common disorder, Diabetes, which affects a significant part of the
population [36]. Diabetes (Type 1 and Type 2) occur due to the improper functioning or
absence of the hormone insulin which breaks down the glucose in the blood to be converted
into glycogen which is accepted into the cells for metabolic activities. Glucose levels of
3.9mM (70mg/dl) indicates hypoglycaemia (low blood glucose) and levels greater than
10mM (180mg/dl) indicates hyperglycaemia (high blood glucose) both conditions being
potentially lethal for a human if left untreated [37]. Mutuchamy and co-workers (2018)
developed a high-performance electrochemical biosensor with Nitrogen-doped Carbon
sheets (NDCS) embedded with ZnO NPs for estimating blood glucose levels [38].
Interestingly, a green approach was followed for synthesizing the ZnO@NDCS using a
hydrothermal method with zinc powder (for ZnO), aqueous ammonia (for nitrogen), and
peach fruit (for carbon sheets) as precursors. The sensor was fabricated on a glassy carbon

electrode (GCE) with glucose oxidase (GOx). The GCE/ZnO@NDCS/GOx biosensor showed a linear detection range from 0.2 to 12 mM which nicely covers the clinical range of normal blood glucose levels. The system showed the lowest detection limit of 6.3 μM, a sensitivity of 231.7 μA mM^{-1} cm^{-2} and an extremely low time of 3 seconds for amperometric current-based detection from human blood serum. The synthesis method and mechanism of glucose detection are depicted in the image below (Figure 7).

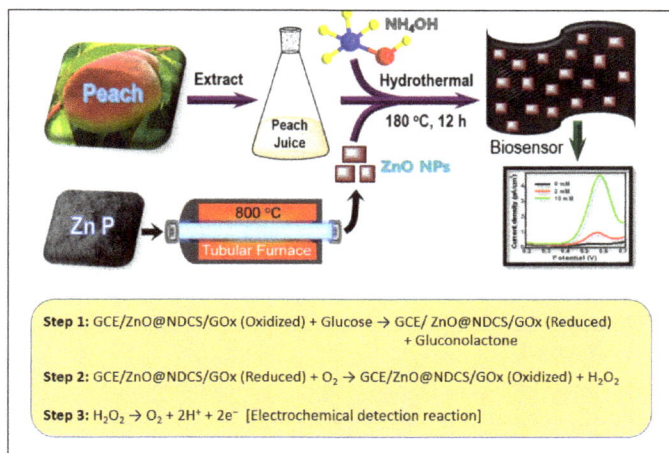

Figure 7: Green synthesis of ZnO NPs embedded- Nitrogen doped carbon nanosheets and the reactions occurring at the modified electrode for glucose detection. (Adapted with permission from {Muthuchamy et al. (2018). High-performance glucose biosensor based on green synthesized zinc oxide nanoparticle embedded nitrogen-doped carbon sheet. J of Electro. chemistry, 816, pp.195-204.). Copyright © {2018} American Chemical Society)

In 2011, Devi and colleagues developed a ZnO NP-based biosensor for detecting the purine base compound, xanthine in fish meat [39]. In previous studies, it has been indicated that xanthine metabolizes in the body to produce uric acid (UA) which can be dangerous if not excreted out of the system correctly. Uric acid accumulation can lead to diseases such as diabetes, gout, fatty liver disease, and even cardiovascular diseases [40]. Certain purine-rich foods such as fish and meat can lead to accumulation of UA triggering such diseases, based on their freshness leading to an increase in xanthine levels. This makes it important for detecting xanthine levels of such foods before determining their consumption period and sending them to market [41]. The biosensor developed by Devi and colleagues involved making a nanocomposite film on a platinum (Pt) electrode composed of ZnO NPs and pyrrole (ZnO/polyPyrrole/Pt) by electro-polymerization followed by physisorption of xanthine oxidase (XOD) to form the working electrode. With Ag/AgCl as the reference electrode and a Pt wire as auxiliary electrode, the xanthine biosensor was constructed which showed linear detection levels for xanthine from 0.8 μM to 40 μM within just 5 seconds. The biosensor was found to lose just 40% of its efficiency over 100 days of use (200 tests) which proves the coating to be quite stable [39].

The very commonly available sterol in the human body, cholesterol, is produced mainly in the liver and is present as low-density and high-density lipoprotein (LDL and HDL). The cholesterol serves as a biomarker for several cardiovascular diseases (CVD), stroke, cardiac arrest, and even can indicate Type II diabetes [42-43]. Recently, Agrawal and colleagues developed a biosensor for detecting cholesterol levels based on Localized Surface Plasmon Resonance (LSPR) unlike the traditionally used amperometric electrochemical sensors [44]. They used Optical Fiber Sensors (OFSs) such as multimode-photosensitive-multimode (MPM) and single-photosensitive-single (SPS) (together known as core-mismatch Fibers) due to several advantages such as robustness, ease of fabrication, low cost, and a broad range of measurements. For immobilization, AuNPs of varied sizes (10 nm and 30 nm) were used for making 3 probes and coated with ZnO NP and combined with the OFSs. The 10 nm probe with MPM showed the best sensitivity of around 0.6898 nm/mM. Linear results were observed for a wide range of 0.1- 10 mM which included the clinical level of around 5.17 mM of Cholesterol in blood serum. A limit of detection as low as 0.6161 mM was observed.

Another important compound in the human body is hydrogen peroxide (H_2O_2) which has found significance as a biomarker for diseases such as Alzheimer's disease (AD), inflammatory diseases, CVD, and cancer [45]. When H_2O_2 levels increase in the body beyond normal levels, the above-mentioned diseases can be triggered. But under normal conditions, H_2O_2 helps in physiological activities such as cell signalling, vascular development, and immune cell activation. In 2018, Sekar and his group developed a biosensor for detecting H_2O_2 levels based on ZnO-PVA nanocomposite synthesized by electrospinning [46]. Bioelectrodes were made with this nanocomposite in conjugation with the enzyme catalase (CAT) which is a heme protein-based enzyme that allows the conversion of H_2O_2 into oxygen and water and enhances the specificity and sensitivity of the biosensor. A thin membrane of chitosan was added further to the electrode surface which provides a smooth and stable morphology to the electrode surface for better sensing. The nanocomposite enhanced electrical conductivity, improved mechanical strength, and better immobilization of CAT. Analysis was done with cyclic voltammetry and the Au/ZnO-PVA/CAT/Chitosan bio-electrode showed linearity for the range of 1 μM to 17 μM with a very low limit of detection of 9.13 nM. A high sensitivity of 210 $\mu A\ \mu M^{-1}\ cm^{-2}$ was observed with a time to result of even lesser than a second.

Biomarker detection has been an especially important field of application where ZnO nanomaterials have been extensively used. Many other groups have researched the development of ZnO-based biosensors for glucose [47-50], cholesterol [51-54], xanthine [55-57] and hydrogen peroxide [58-61]. Some of the other significant studies include the metal-organic framework-derived cobalt-doped ZnO NR gas biosensor for acetone detection as a diabetes biomarker [62]. A similar copper-doped ZnO NP was developed for the CVD biomarker myoglobin [63]. Based on the semiconductor-like properties of ZnO NPs, a field effect transistor system was developed for the CVD biomarker Troponin I for enhanced modulation of current and antibody-based biomarker molecule capture [64]. A self-assembly based fabrication strategy was used to produce ZnO NP- rGO electrodes for dopamine detection [65].

Other than biomarker detections, ZnO NPs can be applied in biosensors for estimating drug concentrations as well which can indicate the dosages and efficacy of administered drugs for instance in chemotherapy. In a very recent study, Karimi-Maleh and co-workers developed a DNA-based biosensor for estimating levels of the chemotherapeutic drug, idarubicin (IDR). The drug, IDR, is used to treat several types of cancers with leukaemia and lymphomas being the common ones. IDR can intercalate into the major groove of the DNA double-helix and binds through the guanine and cytosine bases [66]. In the biosensor device, the ZnO NPs were incorporated with Platinum (Pt) and Palladium (Pd), and these NPs were added to the single-walled carbon nanotubes (SWCNTs) and were obtained by the combination of chemical precipitation and one-pot method, to apply as a modification to GCE [67]. A double-stranded DNA (dS-DNA) from calf thymus was added, using a layer-by-layer deposition on the nanocomposite, as a biological recognition element to improve the sensitivity and efficiency of the system. The signal from the guanine base of the dS-DNA was measured and a range of detection from 1.0 nM to 65 μM was obtained with a low limit of detection of 0.8 nM. A recovery of ~98% to ~105% was observed for this system. The device could be used for both, diagnosis of drug levels and as a support for docking analysis of the drug molecule. A similar application was done for ZnO NPs after doping with Calcium to detect concentrations of an anti-viral drug named Acyclovir [68].

The biosensors involving ZnO NPs are huge in number. A few of the recent and interesting ones were discussed in this part. Several groups around the world are synthesising novel combinatorial materials with ZnO NPs to develop new diagnostic platforms for biosensor development.

3.4   ZnO for theranostic applications

The ZnO nanomaterials have several advantages over other materials one of which is their multi-functionality. They not only find application in diagnostics but also the same materials can fulfil therapeutic actions in the patient's system. A lot of ZnO-based nanomaterials have been used for theranostic applications such as immunotherapy, drug delivery, or cytotoxic destruction for treating diagnosed tumours/cancers [69]. The ZnO NPs show certain advantages like low toxicity and high stability under normal biological conditions. As described in previous sections of this chapter, ZnO NPs are widely applicable in bioimaging due to PL and semiconductor-like properties [70]. They are also quite cost-effective and have an eco-friendly manufacturing process unlike the non-biocompatible cadmium (Cd)-based nanocomposites (NCs) which had highly polluting production processes [71]. The ZnO quantum dots (QDs) have shown significant applications in theranostics due to their quantum-level properties such as broad absorption and narrow emission bands, low self-absorption, tunable emission wavelengths, large Stokes shifts and stable both photochemically and metabolically [72].

On irradiation of ZnO crystals with UV, when in aqueous suspension, electron/hole pair generation takes place which is a characteristic of semiconductors [73]. In ZnO nanomaterials, this leads to photochemical reactions being triggered in the system which leads to the generation of reactive oxygen species (ROS). The valence band holes present on the surface of these materials react with the hydroxyl ions present in water generating hydroxyl radicals (OH•), while oxygen is also reduced to generate the superoxide anion (O_2^-). The ROS production is also enhanced by any pro-inflammatory response of the cell against these nanomaterials. This can be extremely useful for local heat-based ablation of cancer cells as in photodynamic therapy (PDT) where a photosensitive material is excited at a particular wavelength for ROS production inducing localized photodamage in tumour/cancer cells [74]. The figure below (Figure 8) shows how a theranostic NP helps in PDT.

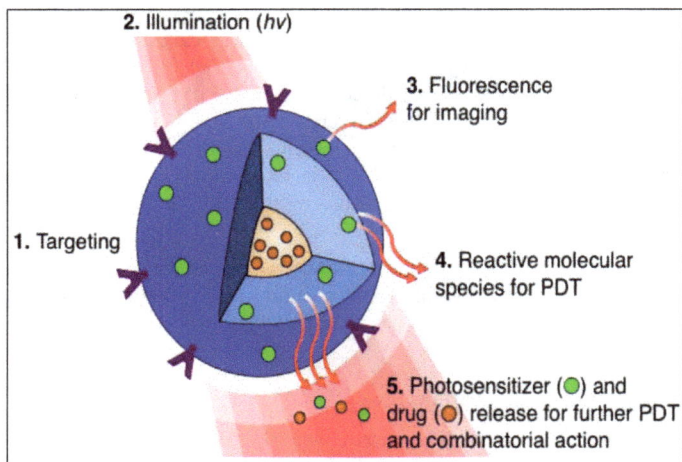

Figure 8: The mechanism of action of nanoparticles as theranostic agents for PDT [101].
(Reprinted from: {Huang and Hasan (2014). The "Nano" world in photodynamic therapy.
Austin J Nanomed Nanotechnol., 2(3).)

A drawback of these heat-based techniques is that the cells acquire resistance to thermal stress quite easily thereby negating the efficiency of the process. Chen and colleagues (in 2014) developed an integrated nanoassembly-based platform using ZnO NPs along with reduced graphene oxide (rGO) and Hyaluronic Acid (HA) [75]. The photoactivity and ROS-producing capacity of ZnO; the graphene acts as an enhancer for the ZnO by preventing the recombination of the electron-hole pairs generated. Graphene also has high optical absorption at the nano-infrared (NIR) range. The nanoassembly allows sequential irradiation with NIR light which acts as a stimulus for the hybrid ZnO-rGO to generate ROS. The HA is combined with deoxycholic acid (DA) to provide colloidal stability to the nanoassembly. The FITC labelled HA-DA breaks down with the help of the hyaluronidase-1 (Hyal-1) enzyme (abundant in tumour microenvironment) to produce high fluorescence signals with the rGO for target bioimaging of cancer cells [76]. The platform allowed highly selective apoptosis induction in the cancer cells by both PDT and Photothermal therapy (PTT) and allowed fluorescence-based detection of the cancer cells.

Another technique that has been commonly used in cancer theranostics is the use of imaging agents for immunotherapies. In immunotherapy, molecules that play a significant role in immunity such as antigens, antibodies, and cytokines (also adjuvants), are delivered to the cancer tissues. These molecules then trigger the host's immune response against these cancerous cells and induce apoptosis [77]. In 2011, Cho and colleagues used ZnO-based NPs for dendritic cells (DCs) to trigger an immune response against tumour cells. The NPs had a core-shell structure with the core made up of superparamagnetic iron oxide (SPIO) and the shell made up of ZnO with ZnO-binding peptide (ZBP) which carries tumour antigens which they demonstrated with CEA in their research paper [78]. The SPIO acts as a high-contrast MRI agent whereas the photonic ZnO allows PL-based detection while simultaneously acting as a carrier for the antigens. The NP-antigen complex was delivered to the DCs which then accepted these complexes without any transfection agents. The loaded DCs triggered a good T-cell response against the tumour cells which was confirmed after *in vivo* analysis in mice where these complex-loaded DCs showed a reduction in tumour growth and greater survivability. For bioimaging, analysis was done using a confocal microscope for *in vitro* samples and with MRI for the *in vivo* detection of the tumours in mice.

ZnO NPs have also been identified as a good drug delivery agent which has been another approach commonly used for cancer theranostics involving ZnO nanomaterials. The nanomaterial-based drug delivery systems (DDSs) serve as strategies for increasing the bioavailability and solubility of drugs, controlling the release of drugs at the specific site to prevent cytotoxicity to healthy cells, and providing protection to the drugs from untimely degradation [79-80]. The ZnO nanomaterials have been found to be good pH-responsive DDSs which allow the formation of linkers between the host and the foreign molecules carrying the drugs and can also generate polymeric micelles which carry the drugs inside which are sensitive to pH [71]. The ZnO NPs can also act as pH-responsive "caps" or "gatekeepers" for other drug-carrying nanomaterials such as MSNs to cover the pores present in them thereby preventing leakage of the carried drug [81]. A similar system has been described in Section 2.2 of this chapter for a *Salmonella* biosensor.

A very innovative multi-functional ZnO-based nanoplatform was developed by Wang and co-workers in 2014 based on the above-mentioned "gatekeeping" role [82]. The system uses lanthanide-doped upconverting nanoparticles (UCNPs) for fabricating the core which was developed by doping with different rare earth metals without the need for any further functional modification. The UCNP core (with NaYF$_4$: 20%Yb^{3+}, 2%Er^{3+}/ NaGdF$_4$: 2%Yb^{3+}) is surrounded by the MSN shell which carries the chemotherapeutic drug Doxorubicin (DOX) and the pores in this MSN shell are sealed by amino-functionalized ZnO QDs. The UCNP core allows trimodal imaging for high-resolution imaging of the cancer tissues with upconversion luminescence (UCL), computed tomography (CT) scan, and MRI, to get more information about the cancer cells as compared to single-modal imaging. The acidic environment at the tumour site triggers the pH-dependent sustained release of the drug from the MSN capped with ZnO which ensures the drug is not released until the platform reaches the tumour site preventing cytotoxicity to the normal cells. The core allows highly enhanced signals for CT and MRI imaging of tumours in mice which are more durable than standard contrast agents. ZnO QDs dissolve on reaching the tumour site which is found by characterisation using TEM and EDX and quenching of green fluorescence of the ZnO QDs (due to PL) on entering the tumour site.

DOX-carrying or adsorbed diagnostic probes are very popular in ZnO QD-based research and have been extensively studied in various forms and different combinations for either *in vitro* or both *in vivo* and *in vitro* theranostic applications [83-86]. These nanosystems were mostly stimulated by pH alterations or by radiation. ZnO NP-based theranostic agents were also developed for other carrying other drugs such as Paclitaxel and even some antibiotics [87-89]. Based on the inherent properties of ZnO nanomaterials, the effect of the delivered drug gets enhanced [90].

Theranostic applications of ZnO-based nanomaterials are not restricted to cancer/ tumour diagnostics and treatment. Several studies have been performed to use the antiseptic properties of ZnO for anti-microbial (anti-bacterial and anti-fungal) therapy post-diagnosis [91]. Some of the common mechanisms for such anti-microbial activity include ROS production, accumulation of NPs inside bacterial cells and liberation of Zn^{2+} ions, and disruption of the bacterial cell wall by electrostatic interactions. Chen and colleagues synthesised a fluorescent nano-probe with BSA-conjugated ZnO QDs (ZnO@BSA) for both diagnosis of bacterial infection and destroying the detected pathogens which in this case are Methicillin-resistant *S. aureas* (MRSA) [92]. The human antimicrobial peptide (UBI$_{29-41}$) and the NIR dye, hydrophilic indocyanine green (ICG) derivative, or MPA are covalently functionalized to ZnO@BSA. This system could properly identify bacterial infection compared to cancer or inflammation-induced infections. They also doped the antibiotic Vancomycin, which works as a last-resort antibiotic against MRSA, on the ZnO@BSA-MPA system. The conjugated system showed better drug activity compared to isolated Vancomycin. Also, Methicillin was conjugated to the system instead of Vancomycin and it was surprisingly found to act upon the resistant MRSA strains by making pores on the cell wall of the bacteria due to the conjugated structure. The conjugated ZnO@BSA molecules enable non-invasive diagnosis of bacterial infection and simultaneously plays the role of DDSs for the antibiotics Vancomycin and Methicillin improving their inhibitory effects. Similarly, ZnO-based nanomaterials have been used for treating bacterial infections *in vitro* and *in vivo* [93-95].

Among other applications of ZnO nanomaterials in theranostics include anti-fungal infection treatment [96], anti-inflammatory therapy [97], wound healing [98-99], and treatment for diseases such as diabetes [100].

Conclusion

This chapter focused on several interesting applications of the ZnO nanomaterials in the field of diagnostics. Research is in progress for improving these materials to diversify their applications and develop novel devices for improving diagnostics in the current scenario where diseases such as cancer and microbial infections are steadily increasing. The upcoming chapters will deal with other applications of ZnO nanomaterials in other chemical and biological fields.

References

[1] Upadhyay, P.K., Jain, V.K., Sharma, K. and Sharma, R., 2020. Synthesis and applications of ZnO nanoparticles in biomedicine. Research Journal of Pharmacy and Technology, 13(4), pp.1636-1644. https://doi.org/10.5958/0974-360X.2020.00297.8

[2] Tuantranont, A., 2013. Applications of nanomaterials in sensors and diagnostics. Springer series on chemical sensors and biosensors, 14. https://doi.org/10.1007/978-3-642-36025-1

[3] Zhang, Y., R Nayak, T., Hong, H. and Cai, W., 2013. Biomedical applications of zinc oxide nanomaterials. Current molecular medicine, 13(10), pp.1633-1645. https://doi.org/10.2174/1566524013666131111130058

[4] Bogutska, K.I., Sklyarov, Y.P. and Prylutskyy, Y.I., 2013. Zinc and zinc nanoparticles: biological role and application in biomedicine. Ukrainica bioorganica acta, 1, pp.9-16.

[5] Tereshchenko, A., Bechelany, M., Viter, R., Khranovskyy, V., Smyntyna, V., Starodub, N. and Yakimova, R., 2016. Optical biosensors based on ZnO nanostructures: advantages and perspectives. A review. Sensors and Actuators B: Chemical, 229, pp.664-677. https://doi.org/10.1016/j.snb.2016.01.099

[6] Barron, A.R., 2015. Physical methods in chemistry and nano science.

[7] Adams, F. and Barbante, C., 2015. Chemical imaging analysis. Elsevier.

[8] Grover, V.P., Tognarelli, J.M., Crossey, M.M., Cox, I.J., Taylor-Robinson, S.D. and McPhail, M.J., 2015. Magnetic resonance imaging: principles and techniques: lessons for clinicians. Journal of clinical and experimental hepatology, 5(3), pp.246-255. https://doi.org/10.1016/j.jceh.2015.08.001

[9] Livieratos, L., 2012. Basic principles of SPECT and PET imaging. In Radionuclide and Hybrid Bone Imaging (pp. 345-359). Springer, Berlin, Heidelberg. https://doi.org/10.1007/978-3-642-02400-9_12

[10] Rezaei, B. and Irannejad, N., 2019. Electrochemical detection techniques in biosensor applications. In Electrochemical Biosensors (pp. 11-43). Elsevier. https://doi.org/10.1016/B978-0-12-816491-4.00002-4

[11] Tang, Y., Zeng, X. and Liang, J., 2010. Surface plasmon resonance: an introduction to a surface spectroscopy technique. Journal of chemical education, 87(7), pp.742-746. https://doi.org/10.1021/ed100186y

[12] Sung, H., Ferlay, J., Siegel, R.L., Laversanne, M., Soerjomataram, I., Jemal, A. and Bray, F., 2021. Global cancer statistics 2020: GLOBOCAN estimates of incidence and mortality worldwide for 36 cancers in 185 countries. CA: a cancer journal for clinicians, 71(3), pp.209-249. https://doi.org/10.3322/caac.21660

[13] Whitaker, K., 2020. Earlier diagnosis: the importance of cancer symptoms. The Lancet Oncology, 21(1), pp.6-8. https://doi.org/10.1016/S1470-2045(19)30658-8

[14] Anjum, S., Hashim, M., Malik, S.A., Khan, M., Lorenzo, J.M., Abbasi, B.H. and Hano, C., 2021. Recent Advances in Zinc Oxide Nanoparticles (ZnO NPs) for Cancer Diagnosis, Target Drug Delivery, and Treatment. Cancers, 13(18), p.4570. https://doi.org/10.3390/cancers13184570

[15] Barui, A.K., Kotcherlakota, R. and Patra, C.R., 2018. Biomedical applications of zinc oxide nanoparticles. In Inorganic frameworks as smart nanomedicines (pp. 239-278). William Andrew Publishing. https://doi.org/10.1016/B978-0-12-813661-4.00006-7

[16] Xu, C., Yang, C., Gu, B. and Fang, S., 2013. Nanostructured ZnO for biosensing applications. Chinese Science Bulletin, 58(21), pp.2563-2566. https://doi.org/10.1007/s11434-013-5714-5

[17] Huang, H., 2018. Matrix metalloproteinase-9 (MMP-9) as a cancer biomarker and MMP-9 biosensors: recent advances. Sensors, 18(10), p.3249. https://doi.org/10.3390/s18103249

[18] Shabani, E., Abdekhodaie, M.J., Mousavi, S.A. and Taghipour, F., 2020. ZnO nanoparticle/nanorod-based label-free electrochemical immunoassay for rapid detection of MMP-9 biomarker. Biochemical Engineering Journal, 164, p.107772. https://doi.org/10.1016/j.bej.2020.107772

[19] Fernández-Baldo, M.A., Ortega, F.G., Pereira, S.V., Bertolino, F.A., Serrano, M.J., Lorente, J.A., Raba, J. and Messina, G.A., 2016. Nanostructured platform integrated into a microfluidic immunosensor coupled to laser-induced fluorescence for the epithelial cancer biomarker determination. Microchemical Journal, 128, pp.18-25. https://doi.org/10.1016/j.microc.2016.03.012

[20] Wang, X., Yu, H., Lu, D., Zhang, J. and Deng, W., 2014. Label free detection of the breast cancer biomarker CA15. 3 using ZnO nanorods coated quartz crystal microbalance. Sensors and Actuators B: Chemical, 195, pp.630-634. https://doi.org/10.1016/j.snb.2014.01.027

[21] Murugan, C., Murugan, N., Sundramoorthy, A.K. and Sundaramurthy, A., 2020. Gradient Triple-Layered ZnS/ZnO/Ta2O5-SiO2 Core-Shell Nanoparticles for Enzyme-Based Electrochemical Detection of Cancer Biomarkers. ACS Applied Nano Materials, 3(8), pp.8461-8471. https://doi.org/10.1021/acsanm.0c01949

[22] Pal, S. and Bhand, S., 2015. Zinc oxide nanoparticle-enhanced ultrasensitive chemiluminescence immunoassay for the carcinoma embryonic antigen. Microchimica Acta, 182(9), pp.1643-1651. https://doi.org/10.1007/s00604-015-1489-5

[23] Liu, Z.Y., Shen, C.L., Lou, Q., Zhao, W.B., Wei, J.Y., Liu, K.K., Zang, J.H., Dong, L. and Shan, C.X., 2020. Efficient chemiluminescent ZnO nanoparticles for cellular imaging. Journal of Luminescence, 221, p.117111. https://doi.org/10.1016/j.jlumin.2020.117111

[24] Hong, H., Wang, F., Zhang, Y., Graves, S.A., Eddine, S.B.Z., Yang, Y., Theuer, C.P., Nickles, R.J., Wang, X. and Cai, W., 2015. Red fluorescent zinc oxide nanoparticle: a novel platform for cancer targeting. ACS applied materials & interfaces, 7(5), pp.3373-3381. https://doi.org/10.1021/am508440j

[25] ISLAM, S., MONDAL, N.I., KARIM, R., CHOWDHURY, M.R.K., RAHMAN, A. and KHAN, H.T., Effects of Communicable Diseases on Life Expectancy in Low-and Lower-Middle-Income Countries.

[26] Sousa, A.M. and Pereira, M.O., 2013. A prospect of current microbial diagnosis methods.

[27] Mallakpour, S., Azadi, E. and Hussain, C.M., 2021. The latest strategies in the fight against the COVID-19 pandemic: the role of metal and metal oxide nanoparticles. New Journal of Chemistry, 45(14), pp.6167-6179. https://doi.org/10.1039/D1NJ00047K

[28] Vasudevan, S., Srinivasan, P., Rayappan, J.B.B. and Solomon, A.P., 2020. A photoluminescence biosensor for the detection of N-acyl homoserine lactone using cysteamine functionalized ZnO nanoparticles for the early diagnosis of urinary tract infections. Journal of Materials Chemistry B, 8(19), pp.4228-4236. https://doi.org/10.1039/C9TB02243K

[29] Bissonnette, L. and Bergeron, M.G., 2015. POC tests in microbial diagnostics: Current status. In Methods in Microbiology (Vol. 42, pp. 87-110). Academic Press. https://doi.org/10.1016/bs.mim.2015.09.003

[30] Xia, Y., Chen, Y., Tang, Y., Cheng, G., Yu, X., He, H., Cao, G., Lu, H., Liu, Z. and Zheng, S.Y., 2019. Smartphone-based point-of-care microfluidic platform fabricated

with a ZnO nanorod template for colorimetric virus detection. ACS sensors, 4(12), pp.3298-3307. https://doi.org/10.1021/acssensors.9b01927

[31] Qiao, Z., Liu, H., Noh, G.S., Koo, B., Zou, Q., Yun, K., Jang, Y.O., Kim, S.H. and Shin, Y., 2020. A simple and rapid fungal DNA isolation assay based on ZnO nanoparticles for the diagnosis of invasive aspergillosis. Micromachines, 11(5), p.515. https://doi.org/10.3390/mi11050515

[32] Boyle, E.C., Bishop, J.L., Grassl, G.A. and Finlay, B.B., 2007. Salmonella: from pathogenesis to therapeutics. Journal of bacteriology, 189(5), pp.1489-1495. https://doi.org/10.1128/JB.01730-06

[33] Huang, F., Guo, R., Xue, L., Cai, G., Wang, S., Li, Y., Liao, M., Wang, M. and Lin, J., 2020. An acid-responsive microfluidic salmonella biosensor using curcumin as signal reporter and ZnO-capped mesoporous silica nanoparticles for signal amplification. Sensors and Actuators B: Chemical, 312, p.127958. https://doi.org/10.1016/j.snb.2020.127958

[34] Upadhyay, A., Yang, H., Zaman, B., Zhang, L., Wu, Y., Wang, J., Zhao, J., Liao, C. and Han, Q., 2020. ZnO Nanoflower-Based NanoPCR as an Efficient Diagnostic Tool for Quick Diagnosis of Canine Vector-Borne Pathogens. Pathogens, 9(2), p.122. https://doi.org/10.3390/pathogens9020122

[35] Shetti, N.P., Bukkitgar, S.D., Reddy, K.R., Reddy, C.V. and Aminabhavi, T.M., 2019. ZnO-based nanostructured electrodes for electrochemical sensors and biosensors in biomedical applications. Biosensors and Bioelectronics, 141, p.111417. https://doi.org/10.1016/j.bios.2019.111417

[36] Mukherjee, D. and Amirian, J., 2022. Effect of Diabetes and other Risk Factors on Bone Health. Interventions in Obesity and Diabetes, 5(5), pp.523-526 https://doi.org/10.31031/IOD.2021.05.000623

[37] Inbasekaran, S., Senthil, R., Ramamurthy, G. and Sastry, T.P., 2014. Biosensor using zinc oxide nanoparticles. International Journal of Innovative Research in Science, Engineering and Technology, 3(1), pp.8601-8606.

[38] Muthuchamy, N., Atchudan, R., Edison, T.N.J.I., Perumal, S. and Lee, Y.R., 2018. High-performance glucose biosensor based on green synthesized zinc oxide nanoparticle embedded nitrogen-doped carbon sheet. Journal of Electroanalytical chemistry, 816, pp.195-204. https://doi.org/10.1016/j.jelechem.2018.03.059

[39] Devi, R., Thakur, M., and Pundir, C.S., 2011. Construction and application of an amperometric xanthine biosensor based on zinc oxide nanoparticles-polypyrrole composite film. Biosensors and Bioelectronics, 26(8), pp.3420-3426. https://doi.org/10.1016/j.bios.2011.01.014

[40] Khan, M.Z.H., Ahommed, M.S. and Daizy, M., 2020. Detection of xanthine in food samples with an electrochemical biosensor based on PEDOT: PSS and functionalized

gold nanoparticles. RSC Advances, 10(59), pp.36147-36154.
https://doi.org/10.1039/D0RA06806C

[41] Devi, R., Yadav, S., Nehra, R., Yadav, S. and Pundir, C.S., 2013. Electrochemical biosensor based on gold coated iron nanoparticles/chitosan composite bound xanthine oxidase for detection of xanthine in fish meat. Journal of Food Engineering, 115(2), pp.207-214. https://doi.org/10.1016/j.jfoodeng.2012.10.014

[42] Narwal, V., Deswal, R., Batra, B., Kalra, V., Hooda, R., Sharma, M. and Rana, J.S., 2019. Cholesterol biosensors: A review. Steroids, 143, pp.6-17. https://doi.org/10.1016/j.steroids.2018.12.003

[43] Alexander, S., Baraneedharan, P., Balasubrahmanyan, S. and Ramaprabhu, S., 2017. Modified graphene based molecular imprinted polymer for electrochemical non-enzymatic cholesterol biosensor. European Polymer Journal, 86, pp.106-116. https://doi.org/10.1016/j.eurpolymj.2016.11.024

[44] Agrawal, N., Zhang, B., Saha, C., Kumar, C., Pu, X. and Kumar, S., 2020. Ultra-sensitive cholesterol sensor using gold and zinc-oxide nanoparticles immobilized core mismatch MPM/SPS probe. Journal of Lightwave Technology, 38(8), pp.2523-2529. https://doi.org/10.1109/JLT.2020.2974818

[45] Eguílaz, M., Dalmasso, P.R., Rubianes, M.D., Gutierrez, F., Rodríguez, M.C., Gallay, P.A., Mujica, M.E.L., Ramírez, M.L., Tettamanti, C.S., Montemerlo, A.E. and Rivas, G.A., 2019. Recent advances in the development of electrochemical hydrogen peroxide carbon nanotube-based (bio) sensors. Current Opinion in Electrochemistry, 14, pp.157-165. https://doi.org/10.1016/j.coelec.2019.02.007

[46] Sekar, N.K., Gumpu, M.B., Ramachandra, B.L., Nesakumar, N., Sankar, P., Babu, K.J., Krishnan, U.M. and Rayappan, J.B.B., 2018. Fabrication of electrochemical biosensor with ZnO-PVA nanocomposite interface for the detection of hydrogen peroxide. Journal of nanoscience and nanotechnology, 18(6), pp.4371-4379. https://doi.org/10.1166/jnn.2018.15259

[47] Wang, J.X., Sun, X.W., Wei, A., Lei, Y., Cai, X.P., Li, C.M. and Dong, Z.L., 2006. Zinc oxide nanocomb biosensor for glucose detection. Applied physics letters, 88(23), p.233106. https://doi.org/10.1063/1.2210078

[48] Ren, X., Chen, D., Meng, X., Tang, F., Hou, X., Han, D. and Zhang, L., 2009. Zinc oxide nanoparticles/glucose oxidase photoelectrochemical system for the fabrication of biosensor. Journal of colloid and interface science, 334(2), pp.183-187. https://doi.org/10.1016/j.jcis.2009.02.043

[49] Dayakar, T., Rao, K.V., Bikshalu, K., Rajendar, V. and Park, S.H., 2017. Novel synthesis and structural analysis of zinc oxide nanoparticles for the non enzymatic glucose biosensor. Materials Science and Engineering: C, 75, pp.1472-1479. https://doi.org/10.1016/j.msec.2017.02.032

[50] DÖNMEZ, S., 2020. Green synthesis of zinc oxide nanoparticles using zingiber officinale root extract and their applications in glucose biosensor. El-Cezeri Journal of Science and Engineering, 7(3), pp.1191-1200.

[51] Khan, R., Kaushik, A., Solanki, P.R., Ansari, A.A., Pandey, M.K. and Malhotra, B.D., 2008. Zinc oxide nanoparticles-chitosan composite film for cholesterol biosensor. Analytica Chimica Acta, 616(2), pp.207-213. https://doi.org/10.1016/j.aca.2008.04.010

[52] Umar, A., Rahman, M.M., Vaseem, M. and Hahn, Y.B., 2009. Ultra-sensitive cholesterol biosensor based on low-temperature grown ZnO nanoparticles. Electrochemistry Communications, 11(1), pp.118-121. https://doi.org/10.1016/j.elecom.2008.10.046

[53] Ahmad, R., Tripathy, N. and Hahn, Y.B., 2012. Wide linear-range detecting high sensitivity cholesterol biosensors based on aspect-ratio controlled ZnO nanorods grown on silver electrodes. Sensors and Actuators B: Chemical, 169, pp.382-386. https://doi.org/10.1016/j.snb.2012.05.027

[54] Hayat, A., Haider, W., Raza, Y. and Marty, J.L., 2015. Colorimetric cholesterol sensor based on peroxidase like activity of zinc oxide nanoparticles incorporated carbon nanotubes. Talanta, 143, pp.157-161. https://doi.org/10.1016/j.talanta.2015.05.051

[55] Zhang, X., Dong, J., Qian, X. and Zhao, C., 2015. One-pot synthesis of an RGO/ZnO nanocomposite on zinc foil and its excellent performance for the nonenzymatic sensing of xanthine. Sensors and Actuators B: Chemical, 221, pp.528-536. https://doi.org/10.1016/j.snb.2015.06.039

[56] Xue, G., Yu, W., Yutong, L., Qiang, Z., Xiuying, L., Yiwei, T. and Jianrong, L., 2019. Construction of a novel xanthine biosensor using zinc oxide (ZnO) and the biotemplate method for detection of fish freshness. Analytical Methods, 11(8), pp.1021-1026. https://doi.org/10.1039/C8AY02554A

[57] Sahyar, B.Y., Kaplan, M., Ozsoz, M., Celik, E. and Otles, S., 2019. Electrochemical xanthine detection by enzymatic method based on Ag doped ZnO nanoparticles by using polypyrrole. Bioelectrochemistry, 130, p.107327. https://doi.org/10.1016/j.bioelechem.2019.107327

[58] Zhu, X., Yuri, I., Gan, X., Suzuki, I. and Li, G., 2007. Electrochemical study of the effect of nano-zinc oxide on microperoxidase and its application to more sensitive hydrogen peroxide biosensor preparation. Biosensors and Bioelectronics, 22(8), pp.1600-1604. https://doi.org/10.1016/j.bios.2006.07.007

[59] Xie, L., Xu, Y. and Cao, X., 2013. Hydrogen peroxide biosensor based on hemoglobin immobilized at graphene, flower-like zinc oxide, and gold nanoparticles nanocomposite modified glassy carbon electrode. Colloids and Surfaces B: Biointerfaces, 107, pp.245-250. https://doi.org/10.1016/j.colsurfb.2013.02.020

[60] Al-Hardan, N.H., Abdul Hamid, M.A., Shamsudin, R., Othman, N.K. and Kar Keng, L., 2016. Amperometric non-enzymatic hydrogen peroxide sensor based on aligned zinc oxide nanorods. Sensors, 16(7), p.1004. https://doi.org/10.3390/s16071004

[61] Uribe, P.A., Ortiz, C.C., Centeno, D.A., Castillo, J.J., Blanco, S.I. and Gutierrez, J.A., 2019. Self-assembled Pt screen printed electrodes with a novel peroxidase Panicum maximum and zinc oxide nanoparticles for H2O2 detection. Colloids and Surfaces A: Physicochemical and Engineering Aspects, 561, pp.18-24. https://doi.org/10.1016/j.colsurfa.2018.10.051

[62] Zhu, S., Xu, L., Yang, S., Zhou, X., Chen, X., Dong, B., Bai, X., Lu, G. and Song, H., 2020. Cobalt-doped ZnO nanoparticles derived from zeolite imidazole frameworks: Synthesis, characterization, and application for the detection of an exhaled diabetes biomarker. Journal of Colloid and Interface Science, 569, pp.358-365. https://doi.org/10.1016/j.jcis.2020.02.081

[63] Haque, M., Fouad, H., Seo, H.K., Alothman, O.Y. and Ansari, Z.A., 2020. Cu-doped ZnO nanoparticles as an electrochemical sensing electrode for cardiac biomarker myoglobin detection. IEEE Sensors Journal, 20(15), pp.8820-8832. https://doi.org/10.1109/JSEN.2020.2982713

[64] Fathil, M.F.M., Arshad, M.M., Ruslinda, A.R., Gopinath, S.C., Nuzaihan, M., Adzhri, R., Hashim, U. and Lam, H.Y., 2017. Substrate-gate coupling in ZnO-FET biosensor for cardiac troponin I detection. Sensors and Actuators B: Chemical, 242, pp.1142-1154. https://doi.org/10.1016/j.snb.2016.09.131

[65] Cao, M., Zheng, L., Gu, Y., Wang, Y., Zhang, H. and Xu, X., 2020. Electrostatic self-assembly to fabricate ZnO nanoparticles/reduced graphene oxide composites for hypersensitivity detection of dopamine. Microchemical Journal, 159, p.105465. https://doi.org/10.1016/j.microc.2020.105465

[66] Charak, S. and Mehrotra, R., 2013. Structural investigation of idarubicin-DNA interaction: Spectroscopic and molecular docking study. International journal of biological macromolecules, 60, pp.213-218. https://doi.org/10.1016/j.ijbiomac.2013.05.027

[67] Karimi-Maleh, H., Khataee, A., Karimi, F., Baghayeri, M., Fu, L., Rouhi, J., Karaman, C., Karaman, O. and Boukherroub, R., 2022. A green and sensitive guanine-based DNA biosensor for idarubicin anticancer monitoring in biological samples: A simple and fast strategy for control of health quality in chemotherapy procedure confirmed by docking investigation. Chemosphere, 291, p.132928. https://doi.org/10.1016/j.chemosphere.2021.132928

[68] Ilager, D., Shetti, N.P., Malladi, R.S., Shetty, N.S., Reddy, K.R. and Aminabhavi, T.M., 2021. Synthesis of Ca-doped ZnO nanoparticles and its application as highly efficient electrochemical sensor for the determination of anti-viral drug, acyclovir. Journal of Molecular Liquids, 322, p.114552. https://doi.org/10.1016/j.molliq.2020.114552

[69] Xiong, H.M., 2013. ZnO nanoparticles applied to bioimaging and drug delivery. Advanced Materials, 25(37), pp.5329-5335. https://doi.org/10.1002/adma.201301732

[70] Urban, B.E., Neogi, P., Senthilkumar, K., Rajpurohit, S.K., Jagadeeshwaran, P., Kim, S., Fujita, Y. and Neogi, A., 2012. Bioimaging using the optimized nonlinear optical properties of ZnO nanoparticles. IEEE Journal of Selected Topics in Quantum Electronics, 18(4), pp.1451-1456. https://doi.org/10.1109/JSTQE.2012.2184793

[71] Eixenberger, J.E., Anders, C.B., Wada, K., Reddy, K.M., Brown, R.J., Moreno-Ramirez, J., Weltner, A.E., Karthik, C., Tenne, D.A., Fologea, D. and Wingett, D.G., 2019. Defect engineering of ZnO nanoparticles for bioimaging applications. ACS applied materials & interfaces, 11(28), pp.24933-24944. https://doi.org/10.1021/acsami.9b01582

[72] De, M., Ghosh, P.S. and Rotello, V.M., 2008. Applications of nanoparticles in biology. Advanced Materials, 20(22), pp.4225-424. https://doi.org/10.1002/adma.200703183

[73] Martínez-Carmona, M., Gun'Ko, Y. and Vallet-Regí, M., 2018. ZnO nanostructures for drug delivery and theranostic applications. Nanomaterials, 8(4), p.268. https://doi.org/10.3390/nano8040268

[74] Dougherty, T.J., Gomer, C.J., Henderson, B.W., Jori, G., Kessel, D., Korbelik, M., Moan, J. and Peng, Q., 1998. Photodynamic therapy. JNCI: Journal of the national cancer institute, 90(12), pp.889-905. https://doi.org/10.1093/jnci/90.12.889

[75] Chen, Z., Li, Z., Wang, J., Ju, E., Zhou, L., Ren, J. and Qu, X., 2014. A multi-synergistic platform for sequential irradiation-activated high-performance apoptotic cancer therapy. Advanced Functional Materials, 24(4), pp.522-529. https://doi.org/10.1002/adfm.201301951

[76] Choi, K.Y., Yoon, H.Y., Kim, J.H., Bae, S.M., Park, R.W., Kang, Y.M., Kim, I.S., Kwon, I.C., Choi, K., Jeong, S.Y. and Kim, K., 2011. Smart nanocarrier based on PEGylated hyaluronic acid for cancer therapy. ACS nano, 5(11), pp.8591-8599. https://doi.org/10.1021/nn202070n

[77] Mellman, I., Coukos, G. and Dranoff, G., 2011. Cancer immunotherapy comes of age. Nature, 480(7378), pp.480-489. https://doi.org/10.1038/nature10673

[78] Cho, N.H., Cheong, T.C., Min, J.H., Wu, J.H., Lee, S.J., Kim, D., Yang, J.S., Kim, S., Kim, Y.K. and Seong, S.Y., 2011. A multifunctional core-shell nanoparticle for dendritic cell-based cancer immunotherapy. Nature nanotechnology, 6(10), pp.675-682. https://doi.org/10.1038/nnano.2011.149

[79] Hu, J., Johnston, K.P. and Williams III, R.O., 2004. Nanoparticle engineering processes for enhancing the dissolution rates of poorly water soluble drugs. Drug development and industrial pharmacy, 30(3), pp.233-245. https://doi.org/10.1081/DDC-120030422

[80] Kou, L., Bhutia, Y.D., Yao, Q., He, Z., Sun, J. and Ganapathy, V., 2018. Transporter-guided delivery of nanoparticles to improve drug permeation across cellular barriers and drug exposure to selective cell types. Frontiers in pharmacology, 9, p.27. https://doi.org/10.3389/fphar.2018.00027

[81] Zhao, Y.L., Li, Z., Kabehie, S., Botros, Y.Y., Stoddart, J.F. and Zink, J.I., 2010. pH-operated nanopistons on the surfaces of mesoporous silica nanoparticles. Journal of the American Chemical Society, 132(37), pp.13016-13025. https://doi.org/10.1021/ja105371u

[82] Wang, Y., Song, S., Liu, J., Liu, D. and Zhang, H., 2015. ZnO-functionalized upconverting nanotheranostic agent: multi-modality imaging-guided chemotherapy with on-demand drug release triggered by pH. Angewandte Chemie International Edition, 54(2), pp.536-540. https://doi.org/10.1002/anie.201409519

[83] Zhao, H., Lv, P., Huo, D., Zhang, C., Ding, Y., Xu, P. and Hu, Y., 2015. Doxorubicin loaded chitosan-ZnO hybrid nanospheres combining cell imaging and cancer therapy. RSC Advances, 5(74), pp.60549-60551. https://doi.org/10.1039/C5RA09587E

[84] Wang, J., Lee, J.S., Kim, D. and Zhu, L., 2017. Exploration of zinc oxide nanoparticles as a multitarget and multifunctional anticancer nanomedicine. ACS applied materials & interfaces, 9(46), pp.39971-39984. https://doi.org/10.1021/acsami.7b11219

[85] Vimala, K., Shanthi, K., Sundarraj, S. and Kannan, S., 2017. Synergistic effect of chemo-photothermal for breast cancer therapy using folic acid (FA) modified zinc oxide nanosheet. Journal of Colloid and Interface Science, 488, pp.92-108. https://doi.org/10.1016/j.jcis.2016.10.067

[86] Qiu, H., Cui, B., Zhao, W., Chen, P., Peng, H. and Wang, Y., 2015. A novel microwave stimulus remote controlled anticancer drug release system based on Fe3O4@ ZnO@ mGd2O3: Eu@P (NIPAm-co-MAA) multifunctional nanocarriers. Journal of Materials Chemistry B, 3(34), pp.6919-6927. https://doi.org/10.1039/C5TB00915D

[87] Puvvada, N., Rajput, S., Kumar, B.N., Sarkar, S., Konar, S., Brunt, K.R., Rao, R.R., Mazumdar, A., Das, S.K., Basu, R. and Fisher, P.B., 2015. Novel ZnO hollow-nanocarriers containing paclitaxel targeting folate-receptors in a malignant pH-microenvironment for effective monitoring and promoting breast tumor regression. Scientific reports, 5(1), pp.1-15. https://doi.org/10.1038/srep11760

[88] Muhammad, F., Wang, A., Guo, M., Zhao, J., Qi, W., Yingjie, G., Gu, J. and Zhu, G., 2013. pH dictates the release of hydrophobic drug cocktail from mesoporous nanoarchitecture. ACS Applied Materials & Interfaces, 5(22), pp.11828-11835. https://doi.org/10.1021/am4035027

[89] Qiu, L., Zhao, Y., Li, B., Wang, Z., Cao, L. and Sun, L., 2017. Triple-stimuli (protease/redox/pH) sensitive porous silica nanocarriers for drug delivery. Sensors and Actuators B: Chemical, 240, pp.1066-1074. https://doi.org/10.1016/j.snb.2016.09.083

[90] Jiang, J., Pi, J. and Cai, J., 2018. The advancing of zinc oxide nanoparticles for biomedical applications. Bioinorganic chemistry and applications, 2018. https://doi.org/10.1155/2018/1062562

[91] Król, A., Pomastowski, P., Rafińska, K., Railean-Plugaru, V. and Buszewski, B., 2017. Zinc oxide nanoparticles: Synthesis, antiseptic activity and toxicity mechanism. Advances in colloid and interface science, 249, pp.37-52. https://doi.org/10.1016/j.cis.2017.07.033

[92] Chen, H., Zhang, M., Li, B., Chen, D., Dong, X., Wang, Y. and Gu, Y., 2015. Versatile antimicrobial peptide-based ZnO quantum dots for in vivo bacteria diagnosis and treatment with high specificity. Biomaterials, 53, pp.532-544. https://doi.org/10.1016/j.biomaterials.2015.02.105

[93] Kadiyala, U., Turali-Emre, E.S., Bahng, J.H., Kotov, N.A. and Vanepps, J.S., 2018. Unexpected insights into antibacterial activity of zinc oxide nanoparticles against methicillin resistant Staphylococcus aureus (MRSA). Nanoscale, 10(10), pp.4927-4939. https://doi.org/10.1039/C7NR08499D

[94] Sarwar, S., Chakraborti, S., Bera, S., Sheikh, I.A., Hoque, K.M. and Chakrabarti, P., 2016. The antimicrobial activity of ZnO nanoparticles against Vibrio cholerae: Variation in response depends on biotype. Nanomedicine: Nanotechnology, Biology and Medicine, 12(6), pp.1499-1509. https://doi.org/10.1016/j.nano.2016.02.006

[95] Alves, M.M., Bouchami, O., Tavares, A., Córdoba, L., Santos, C.F., Miragaia, M. and de Fátima Montemor, M., 2017. New insights into antibiofilm effect of a nanosized ZnO coating against the pathogenic methicillin resistant Staphylococcus aureus. ACS applied materials & interfaces, 9(34), pp.28157-28167. https://doi.org/10.1021/acsami.7b02320

[96] Gunalan, S., Sivaraj, R. and Rajendran, V., 2012. Green synthesized ZnO nanoparticles against bacterial and fungal pathogens. Progress in Natural Science: Materials International, 22(6), pp.693-700. https://doi.org/10.1016/j.pnsc.2012.11.015

[97] Ilves, M., Palomäki, J., Vippola, M., Lehto, M., Savolainen, K., Savinko, T. and Alenius, H., 2014. Topically applied ZnO nanoparticles suppress allergen induced skin inflammation but induce vigorous IgE production in the atopic dermatitis mouse model. Particle and fibre toxicology, 11(1), pp.1-12. https://doi.org/10.1186/s12989-014-0038-4

[98] Sudheesh Kumar, P.T., Lakshmanan, V.K., Anilkumar, T.V., Ramya, C., Reshmi, P., Unnikrishnan, A.G., Nair, S.V. and Jayakumar, R., 2012. Flexible and microporous chitosan hydrogel/nano ZnO composite bandages for wound dressing: in vitro and in

Materials Research Forum LLC
https://doi.org/10.21741/9781644902394-7

vivo evaluation. ACS applied materials & interfaces, 4(5), pp.2618-2629. https://doi.org/10.1021/am300292v

[99] Alavi, M. and Nokhodchi, A., 2020. An overview on antimicrobial and wound healing properties of ZnO nanobiofilms, hydrogels, and bionanocomposites based on cellulose, chitosan, and alginate polymers. Carbohydrate polymers, 227, p.115349. https://doi.org/10.1016/j.carbpol.2019.115349

[100] Kitture, R., Chordiya, K., Gaware, S., Ghosh, S., More, P.A., Kulkarni, P., Chopade, B.A. and Kale, S.N., 2015. ZnO nanoparticles-red sandalwood conjugate: a promising anti-diabetic agent. Journal of nanoscience and nanotechnology, 15(6), pp.4046-4051. https://doi.org/10.1166/jnn.2015.10323

[101] Huang, H.C. and Hasan, T., 2014. The "Nano" world in photodynamic therapy. Austin journal of nanomedicine & nanotechnology, 2(3).

Materials Research Forum LLC
https://doi.org/10.21741/9781644902394-8

Chapter 8

Biomedical Applications of Zinc Oxide Nano-Carriers: An Ingenious Tool

Supandeep Singh Hallan[1,2], Jhaleh Amirian[1,2], Rajinder Kaur[3], Agnese Brangule[1,2]* and Dace Bandere[1,2]*

[1] Riga Stradins University, Department of Pharmaceutical Chemistry, Dzirciema 16, LV-1007 Riga, Latvia

[2] Baltic Biomaterials Centre of Excellence, Headquarters at Riga Technical University, Kalku street 1, LV-1658 Riga, Latvia

[3] State Institute of Nursing and Paramedical Sciences, Badal, Sri Muktsar Sahib, Punjab 152113, India

* Agnese.Brangule@rsu.lv (A.B.), Dace.Bandere@rsu.lv (D.B.)

Abstract

The zinc oxide-based nanoparticles have become the center of interest among the research community, especially in the field of biomedical sciences. They have unique inherent features which help with reduction in biofilm development, anti-bacterial/microbial potential, in addition to transporting active drug molecules to the target site. Further, the concept of green synthesis can also be applied in their fabrication. The effectiveness of these nanomaterials can be improved by transferring them into a gel system. This book chapter focuses on the recent advancements, technical challenges related to surface chemistry, shape and in designing these nanomaterials.

Keywords

Drug Delivery, Metal Oxide Nanomaterials, Zinc Oxide, Bio-Imaging, Bio-Film, Bioluminescence, Nanoparticles, Zinc Oxide Nanoparticles

Contents

Biomedical Applications of Zinc Oxide Nano-Carriers:

1.   Introduction

Nanomaterials have recently sparked greater interest because of their prominent biological characteristics and biomedical applications [1]. Unambiguously, metal oxide nanoparticles (NPs) hold tremendous features including simple production methods with desirable diameter, shape, porosity, and functionalization [2]. Among them, zinc-based NPs are most preferable not only to treat cancer, but also to achieve supplementary goals in the cosmetics, diagnostic, electronic, and textile industries [3-5]. On account of good biocompatibility and biodegradability, zinc oxide (ZnO) NPs have been utilized in various applications including dermal care, tissue regeneration, fluorescence probe design, and drug delivery [6].

It should be noted that zinc (Zn) NPs in oxide form is more biocompatible compared to Zn NPs alone [7]. Apart from the general characteristics including thermal conductivity, refractive index, binding energy, and UV protection, they also hold anti-bacterial, and anti-cancer potential [8]. Additionally, ZnO can be used in anti-microbial, anti-diabetic, anti-inflammatory, and anti-aging applications. Moreover, their utility has also been reported in the process of wound healing [9-12]. ZnO NPs also have the capability to inhibit the growth of *S. mutans, S. aureus, S. sobrinus*, and oral biofilm formation. Moreover, ZnO NPs can be considered a restorative material in the prevention of dental caries [13-15].

In topical applications, ZnO is the main component of the sunscreen/ ointments intended to cure pain and itches because own the microcrystals which have the capability to absorb the light from both UVA and UVB spectral regions. The process of the activation of ZnO by the light induces the anti-bacterial property via facilitating their penetration into bacterial cell wall through diffusion ultimately get accumulated at the cytoplasmic level and interacting with the biomolecules to accomplish cell apoptosis [16]. Further, the oxidative stress induced in this case degrades numerous lipids, carbohydrates, and proteins. The possible reason could be the presence of a positive charge on the Zn NPs and a negative charge on the surface of the biomolecules [17]. Therefore, ZnO NPs can facilitate the transportation of anti-cancer cargo deeper at a cytoplasmic level in order to treat various types of cancer which are usually not feasible with the conventional approaches [18].

The production of ZnO NPs is very straightforward and cost-effective [19]. Moreover, they are available in a wide range of nanostructure forms namely, nanosphere, nanosheet, nanorod, nanobelt, and quantum dots. Interestingly, they can be further categorized based on the drug release pattern due to passive strategies applied or activated by external stimuli by means of pH, microwave, ultrasound, and UV exposure [20]. The classification of ZnO-based smart carriers activated by external stimuli has been presented in Figure 1.

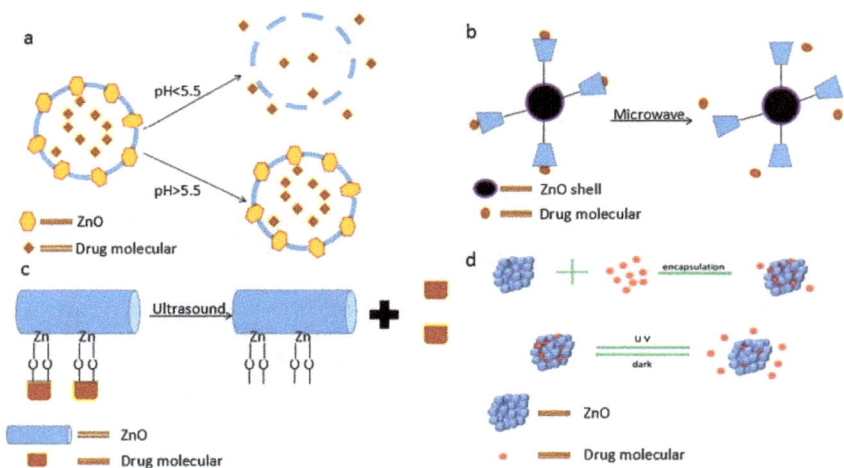

*Figure 1. Classification of the ZnO nanomaterials based on their activation by external stimuli. **a**, pH-response; **b**, Microwave-response; **c**, Ultrasound-response; **d**, UV-response. Reproduced with permission from [20].*

This book chapter will cover the state of art applications od ZnO NPs in biomedical fields, the probability to incorporate them in the gel-based system, and how clinical success can be achieved by considering the critical parameters. The methodology of the review of the literature was carried out using databases Scopus, PubMed, and Web of Sciences to search the relevant publications. Our primary search terms were "Zn NPs"; "ZnO NPs"; "bio-medical applications of zinc-based materials"; "metal oxide nanomaterials" etc.

2. Biomedical applications of ZnO based nanomaterials

The ZnO have multiple bio-medical applications (highlighted in Figure 2) including gene delivery, anti-cancer, anti-bacterial, anti-diabetic, tissue repair, bio-sensing, drug delivery, and immune-therapy [21, 22]. Some of the applications will be discussed in detail in the upcoming sections.

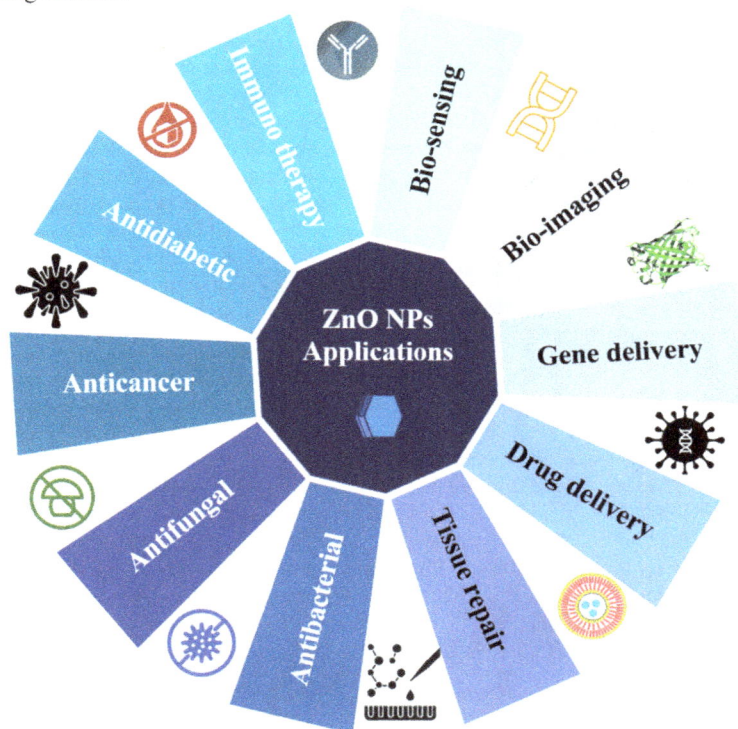

Figure 2. Variety of applications of ZnO NPs in biomedical sciences.

2.1 ZnO in drug delivery and release pattern/ site-specificity

ZnO NPs are a good candidate for drug targeting with higher uptake at the cellular level. Their excellent features include a diameter that is 10-100 nm with high permeability and retention (EPR) effect. The NPs with nano-scale size can leak and travel from the normal vasculature to the cancer cells and finally, undergo easy removal by phagocytosis. They have the potential to achieve higher encapsulation efficiency, and precise control over the drug release [23]. The release of the Zn ion and drug from the NPs can be assessed with different media and pH values. The obtained data from these release studies can describe the drug release pattern in that particular case.

It should be worth underlining that, ZnO NPs have good competence to control the drug release under different pH conditions. The schematic representation has been given in the Figure 3 [24]. The pH sensitive polymer can be employed to coat ZnO NP further stabilize by the surfactant. For instance, the release of quercetin from the ZnO-based nanomaterial has been examined at pH 5.5 and 7.4. Interestingly, the release was very different at these two different pH ranges likely around 70% at the 5.5 and one-fourth part of the initial concentration of the quercetin at pH 7.7. The plausible reason for this trend was the stability of the ZnO-quercetin complex at 7.4. It can be explained further by taking into account that at pH 5.5, the OH group of quercetin remained unionized and could not act as an active ligand in the chelate formation anymore, hence, resulting in the splitting of the chelates. Interestingly, this faster release at acidic pH (actual pH of the cancer cells) can be more effective to combat cancer, otherwise slow or controlled release at pH 7.4 can maintain its activity in the bloodstream for a longer duration without inducing any toxic effects [25]. Therefore, both fast and slow-release have their own importance depending on the objective of the study.

Further, the ZnO nanomaterial guided via external stimuli also might be a promising approach to treating the MDR in cancer. One study based on the pH-responsive release of doxorubicin by ZnO has been reported wherein pH triggered rapid release of the anti-cancer moiety took place upon immediate degradation of the nano-system under acidic conditions ultimately interfered with the cancer cell efflux responsible for Multi-drug resistance (MDR) [26]. In addition, PEGylation can also useful to increase the circulation time of the Zno NPs. The PEGylation was done for the site-specific drug delivery of protoporphyrin IX to combat MDR in Rhabdomyosarcoma via photodynamic therapy. For this, human muscle carcinoma (RD cell line) have been selected to assess apoptosis activity wherein it has been concluded that PEGyalted Zno NPs have shown better activity than bare ZNO in the reversal of the MDR [27].

Moving ahead, the release pattern of the drug and Zn2+ can also be modified by adapting the gel systems approach. Gels hold better control over the release rate based on their expansion or contraction phenomenon with respect to the pH values. Generally, gels expansion at basic pH can be evidenced by a higher release rate as compared to a lower release rate at the acidic pH. The practicable reason could be more the quantity of water entering through widen pores enhances the movements of ions due to higher intraparticle diffusion. This whole event will displace out more Zn ions from the system [28].

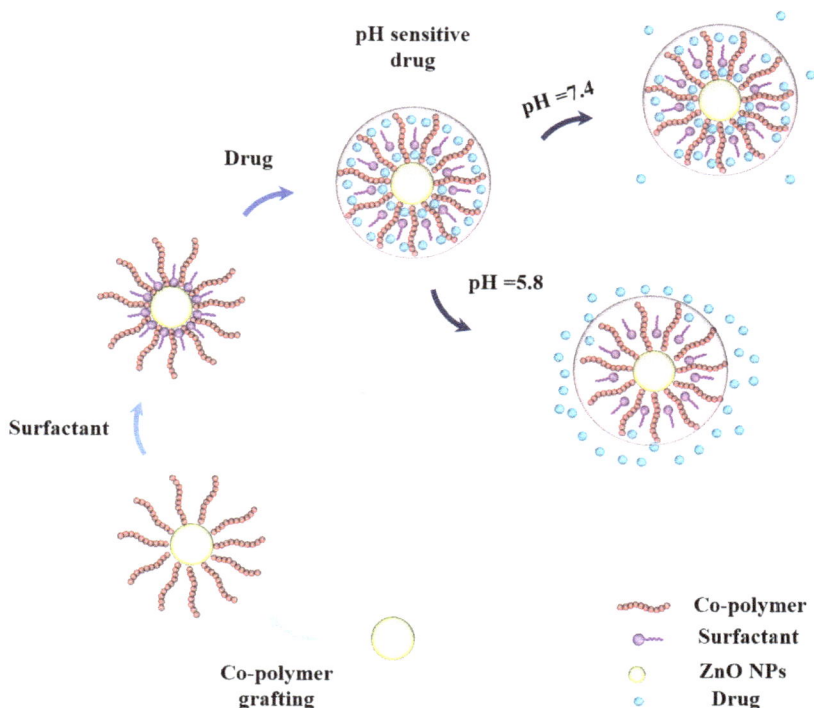

Figure 3. pH responsive ZnO NPs based drug targeting in cancer.

The main complication of the cancer treatment is the resistance being developed by the tumor cells against the chemotherapeutic agents and followed by activation of unavoidable metastasis cascades [29, 30]. Fascinatingly, the ZnO nanomaterial have the ability to kill cancer cells without much affecting the healthy cells [31, 32]. In this manner, the possibility of chitosan-coated ZnO NPs for the delivery of the paclitaxel (a plant-derived alkaloid) in the cancer treatment has been explored wherein, the solubility and bioavailability of the paclitaxel were improved and found to be highly toxic to the MCF-7 breast cancer cell lines with minimal effect on the normal fibroblast cell lines [33]. Similarly, Adipic dihydrazide and heparin have been chosen to modify the surface of the ZnO quantum dots to deliver paclitaxel. The drug release triggered by alteration in pH and toxicity profile on A549 cell lines have been investigated for the treatment of lung cancer. It has been found that lower pH was favorable for the paclitaxel release. Moreover, paclitaxel in bounded form has shown higher anti-cancer activity comparatively lower in the case of paclitaxel in free solution form [34].

Achieving the site-specific drug release in cancer therapy is a major challenge, which can possibly be tackled using other therapeutic alternatives such as gene or photo thermal therapy. This combinational approach can turn into higher anticancer activity with a minimal dose [35, 36]. This type of synergism has been reported in which, ZnO NPs have been coated with the polydopamine followed by the loading of the doxorubicin on the surface due to π- π, hydrophobic and electrostatic interactions [37]. Further, a gene namely, DNAzyme was conjugated to the surface of the multifunctional system. In the above-mentioned approach, the gene is responsible for the gene silencing and doxorubicin for the inhibition of the topoisomerase II, hence synergism was established effectively (Figure 4) [38]. In this way, dependence on anti-cancer drug molecules can also be reduced.

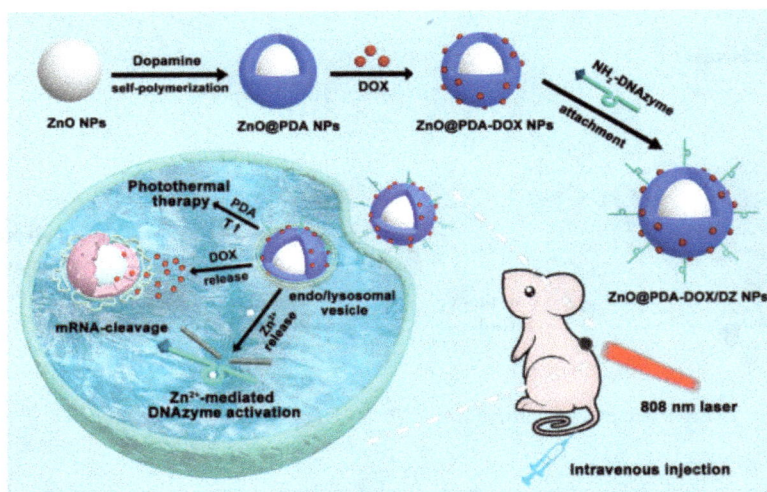

Figure 4. The combined chemo/gene/photothermal therapy in cancer treatment. Reproduced with permission from [27].

Similarly, the delivery of quercetin via ZnO-based smart hexagonal-shaped nano-carriers of the average diameter of 21–39 nm has been accomplished to treat breast cancer. This system has been designed by considering the mechanisms of atomic and molecular-level interactions between quercetin and ZnO NPs with standard-specific surface analytical techniques and density functional theory calculations. The developed ZnO and quercetin complex has shown outstanding anticancer activity assessed on the MCF-7 cell lines compared to quercetin and ZnO NPs alone. It was quite promising in solving the solubility and bioavailability shortcomings associated with quercetin [25]. The role of the ZnO NP in the targeting of various drugs and genes has been mentioned in Table 1.

Table 1. ZnO NP in targeting various drugs and genes.

Active	Diameter (nm)	1% EE, 2%DL	Purpose	Ref.
Quercetin/ ZnO	21/ 39	1 17.4% / 1 82.6%	Synergism with excellent anticancer activity against MCF-7 cancer cells	[25]
ZnO	297.65	1 96.12	Antioxidant activity Animal feed	[39]
Folic Acid drug	N/ A	1 95	Enhanced bioavailability and targeting cancer tumor	[40]
Curcumin	26	1 85	Delivery of hydrophobic drugs with pH triggered anti-cancer activity, moderate antibacterial activity	[41]
Quercetin	31	1 52.08	Activity against normal L929 murine fibroblast cells and A431 human skin carcinoma cell lines	[42]
Anthraquinone	100–120	1 79	Preferential ability to kill HT-29 cancerous cells, UV-blocking, self-cleaning and antibacterial properties	[43]
Naringenin	N/A	1 86.09	Anti-microbial property, cytotoxicity towards A431 human skin carcinoma cells	[44]
Paclitaxel	11.35	N/ A	Low side effects on normal cell line and high cytotoxic effect on breast cancer cell line	[33]
Doxorubicin and DNAzyme	100	2 20	Both anti-cancer effect and gene silencing after endocytosis into cells	[38]

1% EE - Percentage encapsulation efficiency, 2%DL - Percentage drug loading.

2.2   ZnO in antimicrobial/ antibacterial properties

The resistance to antibiotics has become a major concern in healthcare worldwide. It is always a challenging and time taking process to develop and test novel antibiotics and also the bacteria are flexible enough toward all the bactericidal agents. Therefore, one should be focused on the strategies to make existing antibiotics more effective [45, 46]. In this regard, nano-medicine can be an effective approach to tackling MDR and offered unique characteristics including larger surface area, precise drug release, and improved solubility and bioavailability [47, 48]. The likely reason behind that might be a smaller size and larger surface-to-volume ratio enables them to interact with the membranes of the microbes more effectively. The effect is the result of the collective response upstretched from both release of metal ion and generation of the reactive oxygen species (ROS) at the target site [49, 50]. In detail, excluded ROS are the free radicals that can lead to the state of oxidative stress (state of inequivalence in their production and scavenging) hence, death at the cellular level. ZnO nanomaterial are capable to facilitate the free radical formation via releasing free electron/electrons in the presence of UV light, which can produce OH• and H+ in the presence of the water will ultimately give rise to free radical's formation, especially H_2O_2. Later, this H_2O_2 will undergo internalization and damage the bacterial cell, and finally, inhibition of the biofilm formation will take place [51, 52]. In this regard, ZnO NP have been exploited and have undergone conjugation with selected antibiotics. The results from the study support the fact that the use of ZnO nanomaterial conjugation in antibiotic treatment has tremendously improved the cidal activity against both the Gram-positive and Gram-negative bacteria. Significantly, this conjugation has not shown any toxicity against human cell lines [53]. The ZnO NPs are also have remarkable anti-fungal activity shown inhibition of over 95% in the growth of C. albicans [54].

2.3   Topical applications

The dermal penetration of the ZnO NPs have been always conflicting due to the different hypothesis in support. The ZnO NP as an inorganic physical sun blocker with a diameter ranging from 20-100 nm can employ in sunscreen products [55]. They are usually more preferred than organic sun blockers for UV radiation absorption, because, they do not produce any undesirable skin irritation and sensitivity with limited skin penetrations [56, 57].

Some researchers have investigated that ZnO used in commercial ointments can penetrate even deeper inside the skin, their presence has been noticed in blood and urine samples upon their topical administration. Moreover, they also have been traced out in the stratum corneum and upper epidermis of sunburned pork skin [58-60]. On the contrary, some studies have supported by *in vivo* human studies that ZnO NP invaded up to superficial layers of the SC only without diffusing further to viable epidermis when applied onto intact skin [61]. Though they get accumulated in the furrows of the barrier impaired skin [62, 63]

Another less considered factor is concerning the influence of the fluctuations in the climatic conditions after the topical application of the sunscreens. For instance, sunscreen is usually applied in hot weather in order to block harmful UV rays. Wherein, chances of the perspiration are very high. The sweat is composed of lactic acid, urea, and sodium chloride which shifts its pH slightly toward acidic and may have a higher affinity for metal oxide materials [64, 65]. The solubility of the Zn^{2+} generally increases at the low pH value in an aqueous medium [66]. It gives rise to the possibility of the Zn ion penetrating freely as Zn

ZnO and Their Hybrid Nano-Structures   Materials Research Forum LLC
Materials Research Foundations 146 (2023) 234-262   https://doi.org/10.21741/9781644902394-8

salt, evidenced by a study wherein a high concentration of Zn ion form other than that used in sunscreen, might be a new solubilized form [67].

To achieve more accurate targeting, the ligand approach can be adapted for ZnO nanomaterials. Further, the surface modification and drug loading is also possible. To reduces the melanin production and deposition, ZnO quantum dots have been grafted with BQ-788 (an antagonist with a strong affinity to endothelin ETB receptor on the cell sheath of melanocytes) useful to deliver ellagic acid (having poor transdermal delivery) to melanocyte and inhibited the tyrosinase. Here, ZnO quantum dots have increased the transdermal delivery of both ellagic acid and BQ-788 (Figure 5) [68].

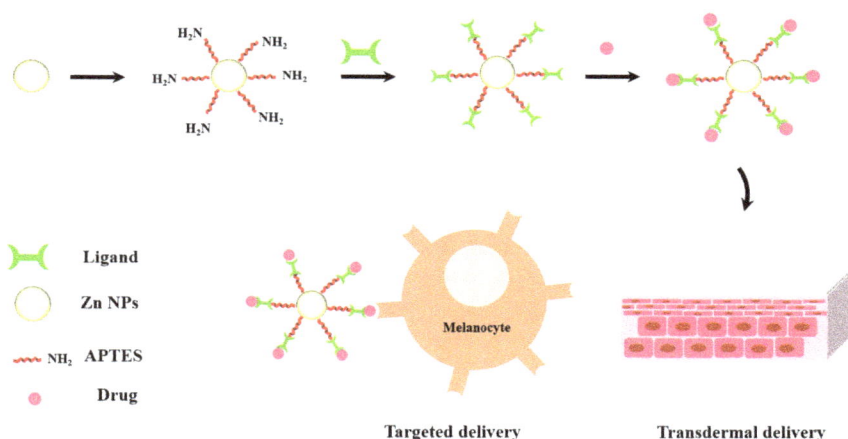

Figure 5. The ligand based drug targeting for transdermal delivery (Redraw [68, p. 7].

It is worth noting that at neutral or acidic pH of the skin, zinc ions released from the ZnO NPs exhibit toxicity in isolated keratinocytes [67]. In order to understand the toxicity exhibited by the ZnO material, quantifying the Zn penetrating across the different skin barriers is a serious flaw. In this regard, Khabir and co-authors have proposed a study wherein stable zinc isotope (^{67}Zn) have been employed in the production of the sunscreen and the quantity of endogenous and exogenous Zn has been examined. The sunscreen was further applied to the excised human skin and employed a couple of techniques including multiphoton microscopy and mass spectroscopy followed by fluorescent microscopy. As a result, both with and without ZnO NP treatment, the zinc content identified in the viable epidermis was shown to be much lower than the HaCaT cytotoxicity threshold. Therefore, the study supports the fact that Zn and Ti dioxide-based sunscreen are FDA approved and can be applied safely on intact skin [69]. It also emphasizes that in comparison to keratinocyte monolayers *in-vitro*, this study shows that excised viable skin is a more relevant model for determining cutaneous toxicity [67].

2.4 ZnO with anti-biofilm property

Biofilms are the dispersion of the microbial colonies immobilized as a matrix composed of various polysaccharides, proteins, and nucleic acids. The biofilm produced by the bacteria first undergoes reversible attachment to biological and non-biological surfaces and later reversible resulting in more aggressive host infection. This infection has become the foremost reason for the morbidity and extremely high health care costs in the European Union [70]. The biofilm formation is being regulated by the quorum-sensing (QS) intracellular communication via molecular chemical signals controls the production of various virulence factors.

The production of QS signaling molecules is regulated by N-acyl homoserine lactone. The film-forming microbes are least sensitive to antibiotics and develop resistance to the immune system of the host. Therefore, this overall event reduces the effectiveness of the antibiotics along with the development of antibiotics' MDR [71].

Thus, it has become very crucial to tackle this issue at the earliest. In this regard, various strategies have been proposed to avoid biofilm formation including applications of antibiotics, bacteriophage, metal coating, etc. which experience various hindrances such as toxicity, high cost, and low efficacy [72].

The ZnO NP are good candidates for the reducing the extent of biofilm formation. The probable reason could be various features being offered by the ZnO at nanoscale production, Firstly, the suitable diameter of ZnO NPs enables them to cross the pores of the biofilm matrix. Secondly, ZnO are available in a variety of shapes (sphere, rod, square, flower) providing an enhanced interactions. Thirdly, the surface modification turns them into more hydrophobic, which is very much effective against bloodstream infections. In addition, the various mechanisms involved are responsible for the antimicrobial activity in the presence or absence of the UV light source [73].

The ZnO nanomaterials can also interfere with the QS mechanism. The above-mentioned statement can be supported by considering the suitable example of the recent study reported based on the production of the ZnO nanospikes. These materials have been tested against chromo-bacterium violaceum (strains 12472 and CVO26) and P. aeruginosa (PAO1) to assess their biofilm inhibitory potential. Further, the inhibition of the QS signaling was evidenced by the production of the violacein in the pathogen even without influencing the overall bacterial growth. As results obtained from a couple of the studies, ZnO nano spikes have drastically reduced the azocasein-degrading protease and elastase activity, and pyocyanin and exopolysaccharide production, hence the biofilm formation interrupted [74].

Moreover, the biofilm at different stages has different responses to the anti-biofilm nanomaterials. ZnO NPs have been exposed to different stages of maturation of biofilm and evaluated for the bacterial adenosine triphosphate (ATP) activity. The finding of this study evidenced that the extent of anti-biofilm activity reduces along with advanced stages of the maturation of biofilm. In other words, resistance develops in the later stages of biofilm formation. Besides this, the presence of extracellular polymeric substances (EPS) also reduces the effectiveness of anti-biofilm nanomaterials [75].

2.5   ZnO in bio imaging

Apart from the therapeutic applications, ZnO have also been explored for diagnostic applications, wherein luminescent potential of these materials has become the center of interest. In general, bio-imaging tools are globally used to study the physio-chemical functioning at the cellular level with the aid of organic dyes and fluorescence proteins. However, these agents are highly prone to degradation, short stokes shift, and photobleaching, hence, these are not appropriate for long durational cell imaging procedures [76, 77].

Cadmium-related species are also applied in the optical and biological fields. The employment of luminescent nano-scale material such as semiconductor quantum dots with cadmium in fluorescence imaging provides high sensitivity with the least invasiveness. Nevertheless, they exhibit some toxicity issues via generation ROS at the cellular level [78]. This has increased the demand for the use of the ZnO nanomaterials as an alternative approach. ZnO NPs possess efficient blue and near-UV emissions, which have green or yellow fluorescence due to oxygen vacancies, allowing them to use in the field of bioimaging [6].

ZnO nano-scale materials have great potential as remarkable semiconductors. In this regard, one study was carried out based on the synthesis of the ZnO/SiO2 core-shell NPs. The silicon shell of the developed system improved both the photoluminescence and aqueous stability of the pure ZnO NPs. At the biological level, the aforementioned NPs have shown very high fluorescence on their attachment to the cells evidenced by visible luminescence. Moreover, the system was highly biocompatible [78].

The quality of the luminescence of ZnO in the UV-blue region can be improved by carrying out the surface modification (capping) and attachment of the biomolecule [79]. ZnO nanomaterial can potentially be employed as a promising choice for cell imaging and pathological research due to its advanced intrinsic fluorescence.

2.6   Green synthesis of ZnO NPs and characterization methods

The eco-friendly production of metallic nanomaterials can be a valuable tool to solve the toxicity, production cost, and solubility issues associated with them. The green synthesis of metallic NPs can be accomplished using extracts from plant, bacterial, algae, and fungi origin. These naturally derived extracts can act as both reducing agents and capping/ stabilizing agents [80]. Additionally, this approach minimizes the involvement of the toxic ingredients because generally derived from natural origins. Some examples of the ZnO-based NPs synthesized by employing extracts against anti-microbial resistance and biofilm inhibition have been discussed in Table 2.

Table 2- Use of ZnO NPs in various green synthesis approaches

Extract	Part used	Diameter (nm)	Purpose	Remarks	Ref.
Cayratia pedata	Leaf	52.24	Immobilization of enzyme	Higher immobilization with cost effective approach	[81]
Myristica fragrans	Fruit	66	Antimicrobial, antileishmanial, antidiabetic, antioxidant, antilarvicidal, and protein kinase inhibitory potential.	Excellent antioxidant and biocompatible nanomaterials	[82]
Cassia fistula and Melia azadarach	leaf	3-68	Antibacterial potential against Escherichia coli (E. coli) and Staphylococcus aureus (S. aureus)	Strong antimicrobial activity against clinical pathogens compared to standard drugs	[83]
Hibiscus subdariffa	Leaf	12–46	Anti-bacterial activity and anti-diabetic activity	Both energy and cost-effective process	[84]
Onion, cabbage, carrot, and tomato	N/A	Below 50	Dye sensitivity solar cell	Improved molecular adsorption onto the surface of ZnO NPs	[85]
Aloe vera/ Cassava starch	Leaf/ colloidal suspension	43.3/ 44.9	Metal trace absorbers in wastewater	Aloe vera have a higher removal efficiency	[86]
Psidium guajava	Leaf	15.8	Antibacterial action	Capped ZnO NP with guava	[87]

3.   Gel system carrying ZnO based materials

There are a variety of dressing materials commercially available composed of natural (chitin, chitosan, gelatin, collagen, alginate, etc) as well as synthetic polymers (poly(urethane), poly (vinyl alcohol), poly (lactic acid), etc) [88, 89]. The toxic product formation and non-hemostatic nature of synthetic polymers are the major challenges whilst the polymers from the natural origin can be beneficial due to their higher biocompatibility, non-toxicity, and non-immunogenicity [89-92]. The hydrogel-based dressing materials have the quality to generate a soothing effect at the wound site by consuming exudate in large volumes ultimately helpful in reducing the infection associated with the wound. The peeling off of the hydrogel is facile due to its humid effect, which is a very important aspect of the wound healing process [93, 94].

However, the grafting approaches face numerous shortcomings in highly burnt patients including infections during dressings. Therefore, it has become highly important to design the dressings that possess biocompatibility and higher mechanical strength with minimal possibility of bacterial infection.

Though, alginate can be an effective candidate even if it lacks a natural degradation phenomenon with the respective enzymes. Interestingly, the ions present majorly calcium salts in the wound exudates, which can help in swelling followed by degradation of the alginate-based hydrogel. The possible explanation for this is that alginate can retain up to 70% water in terms of volume. It is worth mentioning here about chitosan, which can facilitate the degradation of the alginate gel due to their susceptibility toward lysozymal attack [93, 95]. The basic mechanism of gelation via cross linking of ZnO NPs have been described in Figure 6.

In the study focusing on biocompatibility, the tasar variety of silk has been chosen and further coated by the hydrogel. The silk gels compose of a positive charge which can attract negatively charged cells in order to achieve improved adherence [96]. To make the aforementioned gel effective against microbes, ZnO NPs have been deposited on the surface of the silk-based hydrogel by means of high-intensity of ultrasound. Interestingly, the proposed gel has shown better cytocompatibility (adherence and proliferation) of L929 fibroblast cells due to the highly porous nature of the silk gel that allows not only the passage to various nutrients, and growth factors but also the exchange of gases required for the damages tissue regeneration. In this manner, ZnO-based silk hydrogel exhibited higher anti-microbial activity against E coli than simple silk gel [97].

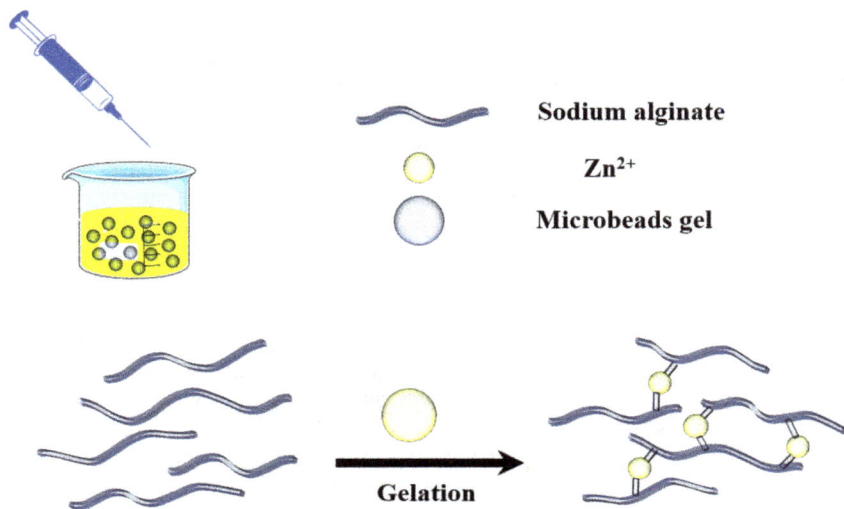

Figure 6. Schematic representation of the cross linking of Zn- ions into gel system.

Apart from the possibility of surface-modified gel with the ZnO NP, an opposite approach which is an encapsulation of the ZnO NPs into the gel has been employed. In order to achieve the photocatalytic activity, ZnO NPs have been loaded in alginate gel. This study evidenced that the ZnO-based gel was stiffer than alginate gel with more advantages including larger pore sizes and higher stability and the system was biocompatible as well [98].

4. Challenges and technical aspects in designing of ZnO NPs

Undoubtedly the utility of the nanomaterial has been raised tremendously, still, the physical, chemical and biological/ safety concerns for their long-term use are under-explored [99]. The aspects related to the colloidal stability should be considered while designing the system based on ZnO NPs. However, some attempts have been made to assess and reduce the toxicity associated with the ZnO nanomaterial utilizing surface modifications, shape, etc.

4.1 Surface chemistry (Lipid coating to avoid agglomeration of ZnO NP)

The production of the Zn NPs is difficult because of the agglomeration which turns them into particles with a larger diameter. These large-sized particles cannot dissolve in media at certain pH values which intrude precise release control of the cargo drug molecules [34,

100].Therefore, a suitable approach should be applied to bypass this agglomeration effect. This issue can be resolved by stabilizing the NPs with surfactants, polymers, and organic polymers. For instance, chitosan can be a good candidate possess a positive charge, is highly biocompatible, and is cost-effective. Moreover, the deposition of the chitosan on the surface of the NPs can also facilitate the Zn NPs interaction with the negatively charged tumor cells [101, 102].

The nanoscale particles present a large surface area means more reactive sites and the ultimately higher surface-to-volume ratio [103, 104]. It should be underlined here that the surface of the newly formed ZnO NPs exposes to the neutral hydroxyl groups which are responsible for the surface charge pattern [105]. For instance, on the exposure of the positively charged outer layer of the NPs to the aqueous medium with higher pH, the chemisorbed proton releases and results in the formation a negatively charged oxygen atoms ($ZnO^{-)}$ surface. This phenomenon can be reversed at lower pH that is from negative to positive surface charged $ZnOH^{+2}$ groups [106].

Similarly, there is a negative charge around the cancer cells due to the presence of the anionic phospholipids, charged proteins, and carbohydrates on their plasma lamella. Therefore, the charge is the main concern to design the system with optimum colloidal stability. This gives rise to the need for the coating which should be biocompatible and stable enough in biological and simulated media. In this regard, lipid coatings have been considered the most reliable tool. One study supports this fact wherein the freshly prepared uncoated ZnO NP enriched with hydroxyl groups, amino-propyl surface modified ZnO, and lipid-coated have been compared. Among three strategies, the highest uptake and colloidal stability have been noticed in the case of lipid-coated ZnO NPs whilst in uncoated, aggregation occurred within the 30 minutes of the exposure to the biological media. Hence, the lipid-based system can move one step ahead toward the clinical success of these inorganic systems [107, 108].

After the surface, shape is an extremely crucial aspect of ZnO nanomaterials that influences the overall performance of the systems. As the shape is very important factor responsible for the interaction between cells and NPs and highly depend upon bio-distribution, clearance and finally, biocompatibility. One *in-vivo* distribution study shows that mesoporous silica NPs in long and short rod shaped have been compared. It has been found that short-rod NPs have a more rapid clearance rate than long-rod NPs. Moreover, Short-rod MSNs were straightforwardly entombed in the liver, while long-rod NPs distributed further in the spleen [109].

The shape tuning is possible in the case of ZnO nanomaterials are available in various shapes namely spheres, squares, rods, flowers, spikes, thorns, etc. The shape of the ZnO NPs is highly dependent on the method of preparation and the type and concentration of the surfactant being employed. A study focusing on the agitation speed of the NPs production. Where the different shapes have been tuned at different rpm enables to make the system more stable thermodynamically [110].

4.2 Toxicity profile over the storage time

A complete mechanism behind the interactions of NPs toward the biological medium needs to be understood to predict the release pattern of ions and the resultant toxicity in the environment. Moreover, the coating process is not only effective to avoid the physical and *in-vitro* agglomeration but also very promising at *in-vivo* level. Concerning this, natural organic matters are the right choice, known to provide colloidal stability and optimize the toxicity of the materials. The natural organic matter namely, humic acid has the potential to reduce the toxicity induced by the zinc ions by doing reduction of free $Zn2+$ being released from NPs containing humic acid complexation [75].

Since ZnO materials have been declared safe by the U.S. Food and Drug Administration and got entitled of "Generally Recognized as Safe" (GRAS) [50]. The toxicity associated with the zinc material has always point of conflict in their clinical use. The two mechanisms are responsible for the cytotoxicity are

(1) Release of $Zn+2$ on intracellular dissolution can give rise to the damage in the mitochondria followed by both interferences of the zinc homeostasis and disequilibrium in the protein activity.

(2) The pro-inflammatory response associated with the ZnO NPs leads to ROS generation. This induced oxidative stress is responsible for cell death even in the normal cells [111, 112]. This cytotoxicity can be managed using a biocompatible coat over the surface of the ZnO NPs. In this concern, polydopamine (autooxidation product of the dopamine)can be chosen as the coating material is a melanin-like polymer that can be deposited spontaneously as a film [113]. With the radical scavenging capability of the polydopamine, ZnO Np can be detoxified, hence, the biocompatibility of the system will be improved as compared to bare ZnO nanomaterials [38].

It is evident on the latest literature that most of the biocompatible studies have been conducted on the freshly prepared zinc oxide NPs and are supportive enough to claim that the ZnO NPs are safe on exposure to the numerous proteins and biological fluids (human saliva, sweat, and broncho alveolar lavage). However, very few studies have been carried out on the toxicity profile by considering the fact of aging. Surprisingly, longer exposure to these biological fluids can result in the conversion of ZnO NPs into metastable $ZnHPO4$ and $Zn3(PO4)2$ [114]. On the contrary, the toxicity associated with ZnO NPs is reduced even after conversion into $Zn5(CO3)2(OH)6$, $Zn(OH)2$, and $Zn2+$ after storage for 40-120 days compared to freshly prepared NPs. Whilst, the reason for the occurrence of the above-stated fact is unclear [115]. However, it has been found that apoptotic genes were less affected by aged NPs, the possible reason might be a reduction in the ability in triggering cell apoptosis along with the aging of NPs [116]. Therefore, more intense investigations should be done on this issue.

5.   Future prospective

As in the field of health sciences, the utilization of nano-scale products has been rising day by day, which subsequently has increased safety concerns. More precisely, the worldwide output of the ZnO NPs has reached up to 3400 tons in recent years [117]. In spite of the fact that some nanomaterials were claimed as biologically inert in the beginning, later studies revealed their extent of toxicity and risk to toxicity at the cellular level [118]. Therefore, toxicity associated with ZnO NPs impedes their utility. As above highlighted strategies being exploited including coating, surface functionalization, and modifying oxidation state to reduce the toxicity are effective up to a certain degree, are unable to eradicate the problems completely and to make the materials safe for human use [116].

Moreover, their tendency towards agglomeration in simulated human plasma is another concern that reduces their overall colloidal stability. Interestingly, the idea of encapsulating the ZnO NPs into lipid-based nano-carriers can be a promising strategy to tailor their surface and resolve the aforementioned issues. Furthermore, the uptake of lipid-coated ZnO NPs by the target cells can also be improved by modifying the surface charge. In this regard, persistent immersion in physiological and cell culture media can be accomplished via making the right selection of the type, composition, and characterization techniques of lipid nano transporter employed to coat the ZnO NP. Additionally, the lipid coating can further avoid the deterioration of ZnO material into toxic by-products. Therefore, they outweigh all these challenges ZnO can attain clinical success.

Conclusions

There's no denying that nanotechnologies have improved people's lives. Selecting the right nanomaterials and minimizing the potential drawbacks are the key elements for the promising future of nanotechnology. To minimize dangers to human health and the environment, intense risk analyses are required to be done before new nano-based products are approved for clinical and commercial use. To determine the long-term viability and safety of their use, a thorough life cycle assessment is required. Conventional methods have limitations that can be overcome by nanotechnology via focusing on a more sustainable approach, which can help to create a more sustainable and safer future products.

References

[1]   M. Jarosz, M. Olbert, G. Wyszogrodzka, K. Młyniec, and T. Librowski, 'Antioxidant and anti-inflammatory effects of zinc. Zinc-dependent NF-κB signaling', *Inflammopharmacol*, vol. 25, no. 1, pp. 11–24, Feb. 2017, https://doi.org/10.1007/s10787-017-0309-4

[2]   M. P. Nikolova and M. S. Chavali, 'Metal Oxide Nanoparticles as Biomedical Materials', *Biomimetics*, vol. 5, no. 2, p. 27, Jun. 2020, https://doi.org/10.3390/biomimetics5020027

[3] F. F. Vidor, T. Meyers, K. Müller, G. I. Wirth, and U. Hilleringmann, 'Inverter circuits on freestanding flexible substrate using ZnO nanoparticles for cost-efficient electronics', *Solid-State Electronics*, vol. 137, pp. 16–21, Nov. 2017, https://doi.org/10.1016/j.sse.2017.07.011

[4] S. Vasantharaj *et al.*, 'Enhanced photocatalytic degradation of water pollutants using bio-green synthesis of zinc oxide nanoparticles (ZnO NPs)', *Journal of Environmental Chemical Engineering*, vol. 9, no. 4, p. 105772, Aug. 2021, https://doi.org/10.1016/j.jece.2021.105772

[5] S. Anjum *et al.*, 'Recent Advances in Zinc Oxide Nanoparticles (ZnO NPs) for Cancer Diagnosis, Target Drug Delivery, and Treatment', *Cancers*, vol. 13, no. 18, p. 4570, Sep. 2021, https://doi.org/10.3390/cancers13184570

[6] J. Jiang, J. Pi, and J. Cai, 'The Advancing of Zinc Oxide Nanoparticles for Biomedical Applications', *Bioinorganic Chemistry and Applications*, vol. 2018, pp. 1–18, Jul. 2018, https://doi.org/10.1155/2018/1062562

[7] W. Jiang, H. Mashayekhi, and B. Xing, 'Bacterial toxicity comparison between nano- and micro-scaled oxide particles', *Environmental Pollution*, vol. 157, no. 5, pp. 1619–1625, May 2009, https://doi.org/10.1016/j.envpol.2008.12.025

[8] Ü. Özgür, D. Hofstetter, and H. Morkoç, 'ZnO Devices and Applications: A Review of Current Status and Future Prospects', *Proc. IEEE*, vol. 98, no. 7, pp. 1255–1268, Jul. 2010, https://doi.org/10.1109/JPROC.2010.2044550

[9] D.-K. Kim *et al.*, 'EVpedia: an integrated database of high-throughput data for systemic analyses of extracellular vesicles', *Journal of Extracellular Vesicles*, vol. 2, no. 1, p. 20384, Jan. 2013, https://doi.org/10.3402/jev.v2i0.20384

[10] Z.-Y. Zhang and H.-M. Xiong, 'Photoluminescent ZnO Nanoparticles and Their Biological Applications', *Materials*, vol. 8, no. 6, pp. 3101–3127, May 2015, https://doi.org/10.3390/ma8063101

[11] P. K. Mishra, H. Mishra, A. Ekielski, S. Talegaonkar, and B. Vaidya, 'Zinc oxide nanoparticles: a promising nanomaterial for biomedical applications', *Drug Discovery Today*, vol. 22, no. 12, pp. 1825–1834, Dec. 2017, https://doi.org/10.1016/j.drudis.2017.08.006

[12] N. Wiesmann, W. Tremel, and J. Brieger, 'Zinc oxide nanoparticles for therapeutic purposes in cancer medicine', *J. Mater. Chem. B*, vol. 8, no. 23, pp. 4973–4989, 2020, https://doi.org/10.1039/D0TB00739K

[13] I. M. Garcia *et al.*, 'Antibacterial response of oral microcosm biofilm to nano-zinc oxide in adhesive resin', *Dental Materials*, vol. 37, no. 3, pp. e182–e193, Mar. 2021, https://doi.org/10.1016/j.dental.2020.11.022

[14] N. L. Brandão, M. B. Portela, L. C. Maia, A. Antônio, V. L. M. e Silva, and E. M. da Silva, 'Model resin composites incorporating ZnO-NP: activity against S. mutans

and physicochemical properties characterization', *J. Appl. Oral Sci.*, vol. 26, no. 0, May 2018, https://doi.org/10.1590/1678-7757-2017-0270

[15] M. Z. I. Nizami, V. W. Xu, I. X. Yin, O. Y. Yu, and C.-H. Chu, 'Metal and Metal Oxide Nanoparticles in Caries Prevention: A Review', *Nanomaterials*, vol. 11, no. 12, p. 3446, Dec. 2021, https://doi.org/10.3390/nano11123446

[16] K. S. Siddiqi, A. ur Rahman, Tajuddin, and A. Husen, 'Properties of Zinc Oxide Nanoparticles and Their Activity Against Microbes', *Nanoscale Res Lett*, vol. 13, no. 1, p. 141, Dec. 2018, https://doi.org/10.1186/s11671-018-2532-3

[17] Biotests and biosensors in ecotoxicological risk assessment of field soils polluted with zinc, lead, and cadmium, *Environ Toxicol Chem*, vol. 24, no. 11, p. 2973, 2005, https://doi.org/10.1897/05-002R1.1

[18] P. Ruenraroengsak *et al.*, 'Frizzled-7-targeted delivery of zinc oxide nanoparticles to drug-resistant breast cancer cells', *Nanoscale*, vol. 11, no. 27, pp. 12858–12870, 2019, https://doi.org/10.1039/C9NR01277J

[19] C. Jayaseelan *et al.*, 'Novel microbial route to synthesize ZnO nanoparticles using Aeromonas hydrophila and their activity against pathogenic bacteria and fungi', *Spectrochimica Acta Part A: Molecular and Biomolecular Spectroscopy*, vol. 90, pp. 78–84, May 2012, https://doi.org/10.1016/j.saa.2012.01.006

[20] X. Huang, X. Zheng, Z. Xu, and C. Yi, 'ZnO-based nanocarriers for drug delivery application: From passive to smart strategies', *International Journal of Pharmaceutics*, vol. 534, no. 1–2, pp. 190–194, Dec. 2017, https://doi.org/10.1016/j.ijpharm.2017.10.008

[21] A. K. Barui, R. Kotcherlakota, and C. R. Patra, 'Biomedical applications of zinc oxide nanoparticles', in *Inorganic Frameworks as Smart Nanomedicines*, Elsevier, 2018, pp. 239–278. https://doi.org/10.1016/B978-0-12-813661-4.00006-7

[22] M. Martínez-Carmona, Y. Gun'ko, and M. Vallet-Regí, 'ZnO Nanostructures for Drug Delivery and Theranostic Applications', *Nanomaterials*, vol. 8, no. 4, p. 268, Apr. 2018, https://doi.org/10.3390/nano8040268

[23] Y. Yao *et al.*, 'Nanoparticle-Based Drug Delivery in Cancer Therapy and Its Role in Overcoming Drug Resistance', *Front. Mol. Biosci.*, vol. 7, p. 193, Aug. 2020, https://doi.org/10.3389/fmolb.2020.00193

[24] R. Dhivya, J. Ranjani, P. K. Bowen, J. Rajendhran, J. Mayandi, and J. Annaraj, 'Biocompatible curcumin loaded PMMA-PEG/ZnO nanocomposite induce apoptosis and cytotoxicity in human gastric cancer cells', *Materials Science and Engineering: C*, vol. 80, pp. 59–68, Nov. 2017, https://doi.org/10.1016/j.msec.2017.05.128

[25] P. Sathishkumar, Z. Li, R. Govindan, R. Jayakumar, C. Wang, and F. Long Gu, 'Zinc oxide-quercetin nanocomposite as a smart nano-drug delivery system:

Molecular-level interaction studies', *Applied Surface Science*, vol. 536, p. 147741, Jan. 2021, https://doi.org/10.1016/j.apsusc.2020.147741

[26] J. Liu *et al.*, 'Zinc Oxide Nanoparticles as Adjuvant To Facilitate Doxorubicin Intracellular Accumulation and Visualize pH-Responsive Release for Overcoming Drug Resistance', *Mol. Pharmaceutics*, vol. 13, no. 5, pp. 1723–1730, May 2016, https://doi.org/10.1021/acs.molpharmaceut.6b00311

[27] M. Fakhar-e-Alam *et al.*, 'ZnO nanoparticles as drug delivery agent for photodynamic therapy', *Laser Phys. Lett.*, vol. 11, no. 2, p. 025601, Feb. 2014, https://doi.org/10.1088/1612-2011/11/2/025601

[28] M. E. Villanueva, M. L. Cuestas, C. J. Pérez, V. Campo Dall' Orto, and G. J. Copello, 'Smart release of antimicrobial ZnO nanoplates from a pH-responsive keratin hydrogel', *Journal of Colloid and Interface Science*, vol. 536, pp. 372–380, Feb. 2019, https://doi.org/10.1016/j.jcis.2018.10.067

[29] D. McManus, 'Gastric Carcinoma', in *Histopathology Reporting*, D. P. Boyle and D. C. Allen, Eds. Cham: Springer International Publishing, 2020, pp. 39–53. https://doi.org/10.1007/978-3-030-27828-1_3

[30] Q. Wei, Y. Qian, J. Yu, and C. C. Wong, 'Metabolic rewiring in the promotion of cancer metastasis: mechanisms and therapeutic implications', *Oncogene*, vol. 39, no. 39, pp. 6139–6156, Sep. 2020, https://doi.org/10.1038/s41388-020-01432-7

[31] A. E. Nel *et al.*, 'Understanding biophysicochemical interactions at the nano–bio interface', *Nature Mater*, vol. 8, no. 7, pp. 543–557, Jul. 2009, https://doi.org/10.1038/nmat2442

[32] F. Muhammad *et al.*, 'Acid degradable ZnO quantum dots as a platform for targeted delivery of an anticancer drug', *J. Mater. Chem.*, vol. 21, no. 35, p. 13406, 2011, https://doi.org/10.1039/c1jm12119g

[33] M. Akbarian, S. Mahjoub, S. M. Elahi, E. Zabihi, and H. Tashakkorian, 'Green synthesis, formulation and biological evaluation of a novel ZnO nanocarrier loaded with paclitaxel as drug delivery system on MCF-7 cell line', *Colloids and Surfaces B: Biointerfaces*, vol. 186, p. 110686, Feb. 2020, https://doi.org/10.1016/j.colsurfb.2019.110686

[34] C. Xie, Y. Zhan, P. Wang, B. Zhang, and Y. Zhang, 'Novel Surface Modification of ZnO QDs for Paclitaxel-Targeted Drug Delivery for Lung Cancer Treatment', *Dose-Response*, vol. 18, no. 2, p. 155932582092673, Apr. 2020, https://doi.org/10.1177/1559325820926739

[35] W. Fan *et al.*, 'A smart upconversion-based mesoporous silica nanotheranostic system for synergetic chemo-/radio-/photodynamic therapy and simultaneous MR/UCL imaging', *Biomaterials*, vol. 35, no. 32, pp. 8992–9002, Oct. 2014, https://doi.org/10.1016/j.biomaterials.2014.07.024

[36] R. L. Setten, J. J. Rossi, and S. Han, 'The current state and future directions of RNAi-based therapeutics', *Nat Rev Drug Discov*, vol. 18, no. 6, pp. 421–446, Jun. 2019, https://doi.org/10.1038/s41573-019-0017-4

[37] W.-Q. Li *et al.*, 'Mitochondria-Targeting Polydopamine Nanoparticles To Deliver Doxorubicin for Overcoming Drug Resistance', *ACS Appl. Mater. Interfaces*, vol. 9, no. 20, pp. 16793–16802, May 2017, https://doi.org/10.1021/acsami.7b01540

[38] M. Liu *et al.*, 'Co-delivery of doxorubicin and DNAzyme using ZnO@polydopamine core-shell nanocomposites for chemo/gene/photothermal therapy', *Acta Biomaterialia*, vol. 110, pp. 242–253, Jul. 2020, https://doi.org/10.1016/j.actbio.2020.04.041

[39] F. Luo *et al.*, 'Synthesis of Zinc Oxide Eudragit FS30D Nanohybrids: Structure, Characterization, and Their Application as an Intestinal Drug Delivery System', *ACS Omega*, vol. 5, no. 20, pp. 11799–11808, May 2020, https://doi.org/10.1021/acsomega.0c01216

[40] E. H. Al-Tememe, F. H. Malk, and R. M. O. Hraishawi, 'Synthesis, characterization and in vitro drug release studies of polymeric drug coated zinc oxide nanoparticles', Al-Samawa, Iraq, 2021, p. 060003. https://doi.org/10.1063/5.0069882

[41] V. J. Sawant and S. R. Bamane, 'PEG-beta-cyclodextrin functionalized zinc oxide nanoparticles show cell imaging with high drug payload and sustained pH responsive delivery of curcumin in to MCF-7 cells', *Journal of Drug Delivery Science and Technology*, vol. 43, pp. 397–408, Feb. 2018, https://doi.org/10.1016/j.jddst.2017.11.010

[42] D. George, P. U. Maheswari, and K. M. M. S. Begum, 'Synergic formulation of onion peel quercetin loaded chitosan-cellulose hydrogel with green zinc oxide nanoparticles towards controlled release, biocompatibility, antimicrobial and anticancer activity', *International Journal of Biological Macromolecules*, vol. 132, pp. 784–794, Jul. 2019, https://doi.org/10.1016/j.ijbiomac.2019.04.008

[43] D. Kundu, C. Hazra, A. Chatterjee, A. Chaudhari, and S. Mishra, 'Extracellular biosynthesis of zinc oxide nanoparticles using Rhodococcus pyridinivorans NT2: Multifunctional textile finishing, biosafety evaluation and in vitro drug delivery in colon carcinoma', *Journal of Photochemistry and Photobiology B: Biology*, vol. 140, pp. 194–204, Nov. 2014, https://doi.org/10.1016/j.jphotobiol.2014.08.001

[44] D. George, P. U. Maheswari, and K. M. M. S. Begum, 'Cysteine conjugated chitosan based green nanohybrid hydrogel embedded with zinc oxide nanoparticles towards enhanced therapeutic potential of naringenin', *Reactive and Functional Polymers*, vol. 148, p. 104480, Mar. 2020, https://doi.org/10.1016/j.reactfunctpolym.2020.104480

[45] A. Mukheem *et al.*, 'Fabrication and Characterization of an Electrospun PHA/Graphene Silver Nanocomposite Scaffold for Antibacterial Applications', *Materials*, vol. 11, no. 9, p. 1673, Sep. 2018, https://doi.org/10.3390/ma11091673

[46] K. Ssekatawa *et al.*, 'Nanotechnological solutions for controlling transmission and emergence of antimicrobial-resistant bacteria, future prospects, and challenges: a systematic review', *J Nanopart Res*, vol. 22, no. 5, p. 117, May 2020, https://doi.org/10.1007/s11051-020-04817-7

[47] L. Wang, C. Hu, and L. Shao, 'The antimicrobial activity of nanoparticles: present situation and prospects for the future', *IJN*, vol. Volume 12, pp. 1227–1249, Feb. 2017, https://doi.org/10.2147/IJN.S121956

[48] A. Anwar *et al.*, 'Synthesis of 4-(dimethylamino)pyridine propylthioacetate coated gold nanoparticles and their antibacterial and photophysical activity', *J Nanobiotechnol*, vol. 16, no. 1, p. 6, Dec. 2018, https://doi.org/10.1186/s12951-017-0332-z

[49] R. K. Dutta, B. P. Nenavathu, M. K. Gangishetty, and A. V. R. Reddy, 'Studies on antibacterial activity of ZnO nanoparticles by ROS induced lipid peroxidation', *Colloids and Surfaces B: Biointerfaces*, vol. 94, pp. 143–150, Jun. 2012, https://doi.org/10.1016/j.colsurfb.2012.01.046

[50] P. J. P. Espitia, N. de F. F. Soares, J. S. dos R. Coimbra, N. J. de Andrade, R. S. Cruz, and E. A. A. Medeiros, 'Zinc Oxide Nanoparticles: Synthesis, Antimicrobial Activity and Food Packaging Applications', *Food Bioprocess Technol*, vol. 5, no. 5, pp. 1447–1464, Jul. 2012, https://doi.org/10.1007/s11947-012-0797-6

[51] M. Fang, J. Chen, X. Xu, P. Yang, and H. Hildebrand, 'Antibacterial activities of inorganic agents on six bacteria associated with oral infections by two susceptibility tests', *International Journal of Antimicrobial Agents*, vol. 27, no. 6, pp. 513–517, Jun. 2006, https://doi.org/10.1016/j.ijantimicag.2006.01.008

[52] L. S. Reddy, M. M. Nisha, M. Joice, and P. N. Shilpa, 'Antimicrobial activity of zinc oxide (ZnO) nanoparticle against *Klebsiella pneumoniae*', *Pharmaceutical Biology*, vol. 52, no. 11, pp. 1388–1397, Nov. 2014, https://doi.org/10.3109/13880209.2014.893001

[53] N. Akbar, Z. Aslam, R. Siddiqui, M. R. Shah, and N. A. Khan, 'Zinc oxide nanoparticles conjugated with clinically-approved medicines as potential antibacterial molecules', *AMB Expr*, vol. 11, no. 1, p. 104, Dec. 2021, https://doi.org/10.1186/s13568-021-01261-1

[54] A. Lipovsky, Y. Nitzan, A. Gedanken, and R. Lubart, 'Antifungal activity of ZnO nanoparticles—the role of ROS mediated cell injury', *Nanotechnology*, vol. 22, no. 10, p. 105101, Mar. 2011, https://doi.org/10.1088/0957-4484/22/10/105101

[55] T. Smijs and Pavel, 'Titanium dioxide and zinc oxide nanoparticles in sunscreens: focus on their safety and effectiveness', *NSA*, p. 95, Oct. 2011, https://doi.org/10.2147/NSA.S19419

[56] G. P. Dransfield, 'Inorganic Sunscreens', *Radiation Protection Dosimetry*, vol. 91, no. 1, pp. 271–273, Sep. 2000, https://doi.org/10.1093/oxfordjournals.rpd.a033216

[57] C. Antoniou, M. G. Kosmadaki, A. J. Stratigos, and A. D. Katsambas, 'Sunscreens--what's important to know', *J Eur Acad Dermatol Venereol*, vol. 22, no. 9, pp. 1110–1118, Sep. 2008, https://doi.org/10.1111/j.1468-3083.2008.02580.x

[58] B. Gulson *et al.*, 'Small Amounts of Zinc from Zinc Oxide Particles in Sunscreens Applied Outdoors Are Absorbed through Human Skin', *Toxicological Sciences*, vol. 118, no. 1, pp. 140–149, Nov. 2010, https://doi.org/10.1093/toxsci/kfq243

[59] B. Gulson *et al.*, 'Comparison of dermal absorption of zinc from different sunscreen formulations and differing UV exposure based on stable isotope tracing', *Science of The Total Environment*, vol. 420, pp. 313–318, Mar. 2012, https://doi.org/10.1016/j.scitotenv.2011.12.046

[60] N. A. Monteiro-Riviere, K. Wiench, R. Landsiedel, S. Schulte, A. O. Inman, and J. E. Riviere, 'Safety Evaluation of Sunscreen Formulations Containing Titanium Dioxide and Zinc Oxide Nanoparticles in UVB Sunburned Skin: An In Vitro and In Vivo Study', *Toxicological Sciences*, vol. 123, no. 1, pp. 264–280, Sep. 2011, https://doi.org/10.1093/toxsci/kfr148

[61] M. Lodén *et al.*, 'Sunscreen use: controversies, challenges and regulatory aspects: Sunscreen use', *British Journal of Dermatology*, vol. 165, no. 2, pp. 255–262, Aug. 2011, https://doi.org/10.1111/j.1365-2133.2011.10298.x

[62] V. R. Leite-Silva *et al.*, 'Human skin penetration and local effects of topical nano zinc oxide after occlusion and barrier impairment', *European Journal of Pharmaceutics and Biopharmaceutics*, vol. 104, pp. 140–147, Jul. 2016, https://doi.org/10.1016/j.ejpb.2016.04.022

[63] V. R. Leite-Silva *et al.*, 'Effect of flexing and massage on *in vivo* human skin penetration and toxicity of zinc oxide nanoparticles', *Nanomedicine*, vol. 11, no. 10, pp. 1193–1205, May 2016, https://doi.org/10.2217/nnm-2016-0010

[64] T. Abe, J. Mayuzumi, N. Kikuchi, and S. Arai, 'Seasonal variations in skin temperature, skin pH, evaporative water loss and skin surface lipid values on human skin.', *Chem. Pharm. Bull.*, vol. 28, no. 2, pp. 387–392, 1980, https://doi.org/10.1248/cpb.28.387

[65] A. B. Stefaniak, M. G. Duling, L. Geer, and M. A. Virji, 'Dissolution of the metal sensitizers Ni, Be, Cr in artificial sweat to improve estimates of dermal bioaccessibility', *Environ. Sci.: Processes Impacts*, vol. 16, no. 2, p. 341, 2014, https://doi.org/10.1039/c3em00570d

[66] F. Pirot, J. Millet, Y. N. Kalia, and Ph. Humbert, 'In vitro Study of Percutaneous Absorption, Cutaneous Bioavailability and Bioequivalence of Zinc and Copper from Five Topical Formulations', *Skin Pharmacol Physiol*, vol. 9, no. 4, pp. 259–269, 1996, https://doi.org/10.1159/000211423

[67] A. M. Holmes, I. Kempson, T. Turnbull, D. Paterson, and M. S. Roberts, 'Penetration of Zinc into Human Skin after Topical Application of Nano Zinc Oxide Used in Commercial Sunscreen Formulations', *ACS Appl. Bio Mater.*, vol. 3, no. 6, pp. 3640–3647, Jun. 2020, https://doi.org/10.1021/acsabm.0c00280

[68] X. Huang *et al.*, 'Transdermal BQ-788/EA@ZnO quantum dots as targeting and smart tyrosinase inhibitors in melanocytes', *Materials Science and Engineering: C*, vol. 102, pp. 45–52, Sep. 2019, https://doi.org/10.1016/j.msec.2019.04.042

[69] Z. Khabir *et al.*, 'Human Epidermal Zinc Concentrations after Topical Application of ZnO Nanoparticles in Sunscreens', *IJMS*, vol. 22, no. 22, p. 12372, Nov. 2021, https://doi.org/10.3390/ijms222212372

[70] E. Tacconelli *et al.*, 'Discovery, research, and development of new antibiotics: the WHO priority list of antibiotic-resistant bacteria and tuberculosis', *The Lancet Infectious Diseases*, vol. 18, no. 3, pp. 318–327, Mar. 2018, https://doi.org/10.1016/S1473-3099(17)30753-3

[71] S. S. Hallan *et al.*, 'Design of Nanosystems for the Delivery of Quorum Sensing Inhibitors: A Preliminary Study', *Molecules*, vol. 25, no. 23, p. 5655, Nov. 2020, https://doi.org/10.3390/molecules25235655

[72] T. Kaur, C. Putatunda, A. Vyas, and G. Kumar, 'Zinc oxide nanoparticles inhibit bacterial biofilm formation via altering cell membrane permeability', *Preparative Biochemistry & Biotechnology*, vol. 51, no. 4, pp. 309–319, Apr. 2021, https://doi.org/10.1080/10826068.2020.1815057

[73] P. P. Mahamuni-Badiger *et al.*, 'Biofilm formation to inhibition: Role of zinc oxide-based nanoparticles', *Materials Science and Engineering: C*, vol. 108, p. 110319, Mar. 2020, https://doi.org/10.1016/j.msec.2019.110319

[74] Mohd. F. Khan *et al.*, 'Anti-quorum Sensing and Anti-biofilm Activity of Zinc Oxide Nanospikes', *ACS Omega*, vol. 5, no. 50, pp. 32203–32215, Dec. 2020, https://doi.org/10.1021/acsomega.0c03634

[75] K. Ouyang, X.-Y. Yu, Y. Zhu, C. Gao, Q. Huang, and P. Cai, 'Effects of humic acid on the interactions between zinc oxide nanoparticles and bacterial biofilms', *Environmental Pollution*, vol. 231, pp. 1104–1111, Dec. 2017, https://doi.org/10.1016/j.envpol.2017.07.003

[76] K. Deshmukh, M. M. Shaik, S. R. Ramanan, and M. Kowshik, 'Self-Activated Fluorescent Hydroxyapatite Nanoparticles: A Promising Agent for Bioimaging and Biolabeling', *ACS Biomater. Sci. Eng.*, vol. 2, no. 8, pp. 1257–1264, Aug. 2016, https://doi.org/10.1021/acsbiomaterials.6b00169

[77] A. R. Kherlopian *et al.*, 'A review of imaging techniques for systems biology', *BMC Syst Biol*, vol. 2, no. 1, p. 74, Dec. 2008, https://doi.org/10.1186/1752-0509-2-74

[78] A. P. S. Prasanna, K. S. Venkataprasanna, B. Pannerselvam, V. Asokan, R. S. Jeniffer, and G. D. Venkatasubbu, 'Multifunctional ZnO/SiO2 Core/Shell Nanoparticles for Bioimaging and Drug Delivery Application', *J Fluoresc*, vol. 30, no. 5, pp. 1075–1083, Sep. 2020, https://doi.org/10.1007/s10895-020-02578-z

[79] K. Senthilkumar *et al.*, 'Preparation of ZnO nanoparticles for bio-imaging applications', *Phys. Status Solidi (b)*, vol. 246, no. 4, pp. 885–888, Apr. 2009, https://doi.org/10.1002/pssb.200880606

[80] H. Agarwal, S. Venkat Kumar, and S. Rajeshkumar, 'A review on green synthesis of zinc oxide nanoparticles – An eco-friendly approach', *Resource-Efficient Technologies*, vol. 3, no. 4, pp. 406–413, Dec. 2017, https://doi.org/10.1016/j.reffit.2017.03.002

[81] A. Jayachandran, A. T.R., and A. S. Nair, 'Green synthesis and characterization of zinc oxide nanoparticles using Cayratia pedata leaf extract', *Biochemistry and Biophysics Reports*, vol. 26, p. 100995, Jul. 2021, https://doi.org/10.1016/j.bbrep.2021.100995

[82] S. Faisal *et al.*, 'Green Synthesis of Zinc Oxide (ZnO) Nanoparticles Using Aqueous Fruit Extracts of *Myristica fragrans* : Their Characterizations and Biological and Environmental Applications', *ACS Omega*, vol. 6, no. 14, pp. 9709–9722, Apr. 2021, https://doi.org/10.1021/acsomega.1c00310

[83] M. Naseer, U. Aslam, B. Khalid, and B. Chen, 'Green route to synthesize Zinc Oxide Nanoparticles using leaf extracts of Cassia fistula and Melia azadarach and their antibacterial potential', *Sci Rep*, vol. 10, no. 1, p. 9055, Dec. 2020, https://doi.org/10.1038/s41598-020-65949-3

[84] N. Bala *et al.*, 'Green synthesis of zinc oxide nanoparticles using Hibiscus subdariffa leaf extract: effect of temperature on synthesis, anti-bacterial activity and anti-diabetic activity', *RSC Adv.*, vol. 5, no. 7, pp. 4993–5003, 2015, https://doi.org/10.1039/C4RA12784F

[85] A. Degefa *et al.*, 'Green Synthesis, Characterization of Zinc Oxide Nanoparticles, and Examination of Properties for Dye-Sensitive Solar Cells Using Various Vegetable Extracts', *Journal of Nanomaterials*, vol. 2021, pp. 1–9, Aug. 2021, https://doi.org/10.1155/2021/3941923

[86] J. de O. Primo *et al.*, 'Synthesis of Zinc Oxide Nanoparticles by Ecofriendly Routes: Adsorbent for Copper Removal From Wastewater', *Front. Chem.*, vol. 8, p. 571790, Nov. 2020, https://doi.org/10.3389/fchem.2020.571790

[87] V. Ramya *et al.*, 'Facile Synthesis and Characterization of Zinc Oxide Nanoparticles Using Psidium guajava leaf Extract and Their Antibacterial

Applications', *Arab J Sci Eng*, vol. 47, no. 1, pp. 909–918, Jan. 2022, https://doi.org/10.1007/s13369-021-05717-1

[88] J. Amirian *et al.*, 'In-situ crosslinked hydrogel based on amidated pectin/oxidized chitosan as potential wound dressing for skin repairing', *Carbohydrate Polymers*, vol. 251, p. 117005, Jan. 2021, https://doi.org/10.1016/j.carbpol.2020.117005

[89] J. Amirian, P. Makkar, G. H. Lee, K. Paul, and B. T. Lee, 'Incorporation of alginate-hyaluronic acid microbeads in injectable calcium phosphate cement for improved bone regeneration', *Materials Letters*, vol. 272, p. 127830, Aug. 2020, https://doi.org/10.1016/j.matlet.2020.127830

[90] M. I. Shekh *et al.*, 'Electrospun ferric ceria nanofibers blended with MWCNTs for high-performance electrochemical detection of uric acid', *Ceramics International*, vol. 46, no. 7, pp. 9050–9064, May 2020, https://doi.org/10.1016/j.ceramint.2019.12.153

[91] J. Amirian, T. Sultana, G. J. Joo, C. Park, and B.-T. Lee, 'In vitro endothelial differentiation evaluation on polycaprolactone-methoxy polyethylene glycol electrospun membrane and fabrication of multilayered small-diameter hybrid vascular graft', *J Biomater Appl*, vol. 34, no. 10, pp. 1395–1408, May 2020, https://doi.org/10.1177/0885328220907775

[92] J. Amirian *et al.*, 'Examination of In vitro and In vivo biocompatibility of alginate-hyaluronic acid microbeads As a promising method in cell delivery for kidney regeneration', *International Journal of Biological Macromolecules*, vol. 105, pp. 143–153, Dec. 2017, https://doi.org/10.1016/j.ijbiomac.2017.07.019

[93] R. Jayakumar, P. Sudheesh Kumar, A. Mohandas, V.-K. Lakshmanan, and R. Biswas, 'Exploration of alginate hydrogel/nano zinc oxide composite bandages for infected wounds', *IJN*, p. 53, Oct. 2015, https://doi.org/10.2147/IJN.S79981

[94] J. Amirian, N. T. B. Linh, Y. K. Min, and B.-T. Lee, 'Bone formation of a porous Gelatin-Pectin-biphasic calcium phosphate composite in presence of BMP-2 and VEGF', *International Journal of Biological Macromolecules*, vol. 76, pp. 10–24, May 2015, https://doi.org/10.1016/j.ijbiomac.2015.02.021

[95] J. Koffler *et al.*, 'Biomimetic 3D-printed scaffolds for spinal cord injury repair', *Nat Med*, vol. 25, no. 2, pp. 263–269, Feb. 2019, https://doi.org/10.1038/s41591-018-0296-z

[96] A. Song, A. A. Rane, and K. L. Christman, 'Antibacterial and cell-adhesive polypeptide and poly(ethylene glycol) hydrogel as a potential scaffold for wound healing', *Acta Biomaterialia*, vol. 8, no. 1, pp. 41–50, Jan. 2012, https://doi.org/10.1016/j.actbio.2011.10.004

[97] S. Majumder *et al.*, 'Zinc Oxide Nanoparticles Functionalized on Hydrogel Grafted Silk Fibroin Fabrics as Efficient Composite Dressing', *Biomolecules*, vol. 10, no. 5, p. 710, May 2020, https://doi.org/10.3390/biom10050710

[98] C. M. Cleetus *et al.*, 'Alginate Hydrogels with Embedded ZnO Nanoparticles for Wound Healing Therapy', *IJN*, vol. Volume 15, pp. 5097–5111, Jul. 2020, https://doi.org/10.2147/IJN.S255937

[99] A. Elsaesser and C. V. Howard, 'Toxicology of nanoparticles', *Advanced Drug Delivery Reviews*, vol. 64, no. 2, pp. 129–137, Feb. 2012, https://doi.org/10.1016/j.addr.2011.09.001

[100]   J. Zhang, D. Wu, M.-F. Li, and J. Feng, 'Multifunctional Mesoporous Silica Nanoparticles Based on Charge-Reversal Plug-Gate Nanovalves and Acid-Decomposable ZnO Quantum Dots for Intracellular Drug Delivery', *ACS Appl. Mater. Interfaces*, vol. 7, no. 48, pp. 26666–26673, Dec. 2015, https://doi.org/10.1021/acsami.5b08460

[101]   A. Javid, S. Ahmadian, A. A. Saboury, S. M. Kalantar, and S. Rezaei-Zarchi, 'Chitosan-Coated Superparamagnetic Iron Oxide Nanoparticles for Doxorubicin Delivery: Synthesis and Anticancer Effect Against Human Ovarian Cancer Cells', *Chem Biol Drug Des*, vol. 82, no. 3, pp. 296–306, Sep. 2013, https://doi.org/10.1111/cbdd.12145

[102]   D. Singh, M. Rashid, S. S. Hallan, N. K. Mehra, A. Prakash, and N. Mishra, 'Pharmacological evaluation of nasal delivery of selegiline hydrochloride-loaded thiolated chitosan nanoparticles for the treatment of depression', *Artificial Cells, Nanomedicine, and Biotechnology*, pp. 1–13, Jun. 2015, https://doi.org/10.3109/21691401.2014.998824

[103]   D. A. Jefferson, 'The surface activity of ultrafine particles', *Philosophical Transactions of the Royal Society of London. Series A: Mathematical, Physical and Engineering Sciences*, vol. 358, no. 1775, pp. 2683–2692, Oct. 2000, https://doi.org/10.1098/rsta.2000.0677

[104]   V. Stone and K. Donaldson, 'Signs of stress', *Nature Nanotech*, vol. 1, no. 1, pp. 23–24, Oct. 2006, https://doi.org/10.1038/nnano.2006.69

[105]   S. Stassi, V. Cauda, C. Ottone, A. Chiodoni, C. F. Pirri, and G. Canavese, 'Flexible piezoelectric energy nanogenerator based on ZnO nanotubes hosted in a polycarbonate membrane', *Nano Energy*, vol. 13, pp. 474–481, Apr. 2015, https://doi.org/10.1016/j.nanoen.2015.03.024

[106]   K. H. Müller *et al.*, 'pH-Dependent Toxicity of High Aspect Ratio ZnO Nanowires in Macrophages Due to Intracellular Dissolution', *ACS Nano*, vol. 4, no. 11, pp. 6767–6779, Nov. 2010, https://doi.org/10.1021/nn101192z

[107]   B. Dumontel *et al.*, 'Enhanced biostability and cellular uptake of zinc oxide nanocrystals shielded with a phospholipid bilayer', *J. Mater. Chem. B*, vol. 5, no. 44, pp. 8799–8813, 2017, https://doi.org/10.1039/C7TB02229H

[108]   D. Cao, X. Shu, D. Zhu, S. Liang, M. Hasan, and S. Gong, 'Lipid-coated ZnO nanoparticles synthesis, characterization and cytotoxicity studies in cancer cell', *Nano*

Convergence, vol. 7, no. 1, p. 14, Dec. 2020, https://doi.org/10.1186/s40580-020-00224-9

[109]   X. Huang *et al.*, 'The Shape Effect of Mesoporous Silica Nanoparticles on Biodistribution, Clearance, and Biocompatibility *in Vivo*', *ACS Nano*, vol. 5, no. 7, pp. 5390–5399, Jul. 2011, https://doi.org/10.1021/nn200365a

[110]   M. F. Khan *et al.*, 'Sol-gel synthesis of thorn-like ZnO nanoparticles endorsing mechanical stirring effect and their antimicrobial activities: Potential role as nano-antibiotics', *Sci Rep*, vol. 6, no. 1, p. 27689, Jun. 2016, https://doi.org/10.1038/srep27689

[111]   G. Bisht and S. Rayamajhi, 'ZnO Nanoparticles: A Promising Anticancer Agent', *Nanobiomedicine*, vol. 3, p. 9, Jan. 2016, https://doi.org/10.5772/63437

[112]   B. Lallo da Silva *et al.*, 'Relationship Between Structure And Antimicrobial Activity Of Zinc Oxide Nanoparticles: An Overview', *IJN*, vol. Volume 14, pp. 9395–9410, Dec. 2019, https://doi.org/10.2147/IJN.S216204

[113]   W. Chen *et al.*, 'Bacteria-Driven Hypoxia Targeting for Combined Biotherapy and Photothermal Therapy', *ACS Nano*, vol. 12, no. 6, pp. 5995–6005, Jun. 2018, https://doi.org/10.1021/acsnano.8b02235

[114]   P. D. Mier and J. J. van den Hurk, 'Lysosomal hydrolases of the epidermis. 2. Ester hydrolases', *Br J Dermatol*, vol. 93, no. 4, pp. 391–398, Oct. 1975, https://doi.org/10.1111/j.1365-2133.1975.tb06512.x

[115]   M. M. Wang *et al.*, 'Mutagenicity of ZnO nanoparticles in mammalian cells: Role of physicochemical transformations under the aging process', *Nanotoxicology*, vol. 9, no. 8, pp. 972–982, Nov. 2015, https://doi.org/10.3109/17435390.2014.992816

[116]   J. Wang *et al.*, 'The Role of Apoptosis Pathway in the Cytotoxicity Induced by Fresh and Aged Zinc Oxide Nanoparticles', *Nanoscale Res Lett*, vol. 16, no. 1, p. 129, Dec. 2021, https://doi.org/10.1186/s11671-021-03587-y

[117]   A. A. Keller and A. Lazareva, 'Predicted Releases of Engineered Nanomaterials: From Global to Regional to Local', *Environ. Sci. Technol. Lett.*, vol. 1, no. 1, pp. 65–70, Jan. 2014, https://doi.org/10.1021/ez400106t

[118]   S. George *et al.*, 'Use of a Rapid Cytotoxicity Screening Approach To Engineer a Safer Zinc Oxide Nanoparticle through Iron Doping', *ACS Nano*, vol. 4, no. 1, pp. 15–29, Jan. 2010, https://doi.org/10.1021/nn901503q

Chapter 9

ZnO Thin Films: Fabrication Routes, and Applications

Minoo Alizadeh Pirposhte[1], Debjita Mukherjee[2†], Azadeh Jafarizadeh Dehaghani[3†], Mojdeh Rahnama Ghahfarokhi[4†], Jhaleh Amirian[5,6*], Agnese Brangule[5,6*], and Dace Bandere[5,6*]

[1]Department of Materials Engineering, Faculty of Advanced materials, Isfahan University of Technology, Isfahan, Iran

[2]College of Medical, Veterinary and Life Sciences, University of Glasgow, University Ave, Glasgow G12 8QQ, United Kingdom

[3]Department of Materials Engineering, Faculty of Materials Processing and Fabrication, Ph.D. Student, Isfahan University of Technology, Isfahan, Iran

[4]Department of Materials Engineering, Faculty of advanced materials, Ph.D. Student, Isfahan University of Technology, Isfahan, Iran

[5] Department of Pharmaceutical Chemistry, Riga Stradiņš University, Dzirciema 16, LV-1007, Riga, Latvia

[6]Baltic Biomaterials Centre of Excellence, Headquarters at Riga Technical University, Kalku street 1, LV-1658 Riga, Latvia

† These authors contributed same

*Prof. Dace Bandere (dace.bandere@rsu.lv) Dr. Agnese Brangule (agnese.brangule@rsu.lv), Jhaleh Amirian (jalehamirian@gmail.com)

Abstract

Thin films have become a hot topic in the field of nanotechnology. Due to their optical and electrical characteristics, thin-film semiconductor oxides are among the semiconductor oxides with the greatest range of applications. The most popular semiconductor component is zinc oxide (ZnO). ZnO, a wide bandgap semiconductor (E_g = 3.37 eV at room temperature), have been widely used in electronic, optoelectronic, and information technology device platforms. Nano-ZnO thin films have a wide range of applications due to their remarkable properties. There are several methods for developing a thin layer of the ZnO nanomaterial. Sputtering, chemical vapor deposition (CVD), molecular beam epitaxy, pulsed laser deposition (PLD), and spray pyrolysis are among these methods. Although sputtering is the most commonly used method for high-quality applications, chemical vapor deposition and spray pyrolysis are also popular due to their low cost and ease of use. This chapter provides a brief overview of the various fabrication routes, characterization techniques, and applications of ZnO thin films, allowing us to investigate the chemical, structural, optical, and electrical properties of ZnO thin films, as well as their various applications.

Keywords

Zinc Oxide (ZnO), Thin Film, Fabrication Techniques, Semiconductor, Routes

Contents

1. Introduction

A 2D nanomaterial technology known as "thin film technology" include "plate-like thin film" has the potential to create efficient and adaptable equipment at the micro- or nanoscale. Microelectronic devices, high energy density harvesters, high sensitivity low power sensors, anti-reflective and protective coatings on modern equipment, transparent electrodes, etc. are just a few of the applications and advantages that thin films have. As a consequence, various methodologies are being developed to generate thin films of good quality [1, 2].

A class of substances known as metal oxides is crucial to several disciplines, including physics, chemistry, and materials science [1]. ZnO, one of the metal oxides, has the potential to be a semiconductor oxide. Additionally, many types of ZnO nanostructures, including nano-ZnO thin films, have also been the focus of numerous research investigations and are being used in business and technology.

High demand in research is beacuse of its high and direct energy gap (3.37 eV) at room temperature, large excitation binding energy (60 meV), high electron mobility, high power stability, inherent piezoelectric properties, environmental compatibility, and relatively easy construction technology [1-3]. The II -VI compound semiconductor exhibits significant electrical, optical, and magnetic properties. Accordingly, this II - VI semiconductor has been widely applied in various fields including electronics, spintronics, sensors, solar cells, piezoelectric nanogenerators, light-emitting diodes, lasers, memory devices, acousto-optic devices, UV detectors, gas sensors, and transparent conductive oxides [4, 5].

ZnO thin film synthesis has gained prominence as a study topic in recent years. Control of the layer morphology must be maintained from a microscopically uniform and flat surface to a nanoscale tailored surface in order to successfully complete these varied applications [6].

1.1 ZnO thin films

The technique of coating a surface with an exceptionally thin layer of material is known as thin film deposition (a few nanometers to 100 micrometres thick, or on the scale of a few atoms thick). However, a significant stage in many applications is the controlled fabrication of thin films. Using thin films, a wide range of innovative and adaptable devices can be generated. Thin films have shown to be important for the development of microelectronic devices, reliable sensors, protective layers on a variety of goods, and antireflective coatings on solar cells, among other uses. Consequently, as the demand for cutting-edge technology grows, efficient methods for thin-film deposition are being developed [7, 8].

ZnO is a versatile oxide that exhibits a unique combination of multifunctional qualities in addition to being a wide bandgap II-VI compound semiconductor and a piezoelectric material [9]. ZnO can be found in a variety of various crystal forms. The three crystal structures of ZnO identified in the literature are wurtzite, rocksalt, and zincblende [2-4]. The latter is hexagonal, while the first two are cubic. The thermodynamically stable form of ZnO is in the Wurtzite crystal structure at room temperature. The lattice constants are a=b=3.25, c=5.2, and c/a=1.60, which is not far from the ideal value for a hexagonal cell, c/a=1.633 [7, 10, 11]. Due to its electro-optical features, high electrochemical stability, big band gap, natural abundance, and low toxicity, ZnO is becoming one of the most widely utilised transparent conductive oxides. The properties of ZnO thin films are dependent on the non-stoichiometry of the films, which is caused by oxygen vacancies and interstitial zinc.

1.2   Historical background

Without a question, semiconductor materials have altered the world in ways that no one could have predicted. Alessandro Volta coined the word "semiconducting" in early 1782.[1]. Since then, the semiconductor industry has grown rapidly around the world. c When sensitive patches on ZnO crystals were tested with thin copper wires as a detector for radio sets, researchers started looking into ZnO materials for semiconductor capabilities in the early 1920s [12]. In 1935, Fritsch et al. synthesized polycrystalline ZnO films by evaporation method, which observed the semiconductor behaviour of undoped ZnO films [13].

It has been shown that a number of deposition processes may deposit thin films of device quality as a result of recent significant efforts in research and development dedicated to the generation and characterization of thin ZnO films. When high-quality single-crystal and epitaxial layers were made accessible, this material underwent a new trend that made it possible to fabricate electro-optical and optical equipment.

ZnO thin films are gaining significance for applications in many disciplines due to its broad energy gap, substantial piezoelectric, electro-optical, and nonlinear optics coefficients, high electrical conductivity, optical transparency, and structural defects. Furthermore, the material has a variety of applications in the field of nanotechnology because it is simultaneously biosafe and biocompatible. A variety of nanostructures can now be generated employing innovative construction methodologies. Additionally, due to its neutral chemical behaviour, it can be used in a wide range of scientific and industrial applications.[2, 4, 14]. However, discovering methods to produce ZnO thin films at scales that are both affordable and efficient, as well as developing these films with enhanced characteristics, remain substantial difficulties in the fields of production and research and development (R&D)[7].

Following that, in 1950, the bulk ZnO crystal was investigated as a typical II-VI compound semiconductor. A significant portion of research at this time was devoted to the study of the electrical and optical properties of ZnO (such as n-type conductivity, absorption spectrum, and electroluminescence decay parameters) [15, 16].

It is important to note that thin films have grabbed the bulk of the scientific community's attention since 1960, initially due to the utilization of ZnO's piezoelectric potential in surface acoustic wave devices [17, 18]. Moreover, there was a considerable amount of growth in the ZnO research in the mid 1960s, with a significant amount of work and research being carried out in both theory and practice. Laudise et al. reported that large macroscopically sound ZnO crystals suitable for preliminary transducer and other experiments have been grown hydrothermally [19].

Significant work on the creation of ZnO devices with a lower complexity was accomplished in the 1970s [20, 21]. Many techniques have been examined for fabrication of high-quality ZnO thin films in the 1980s, including metal-organic chemical vapor deposition (MOCVD), spray pyrolysis, and radio-frequency (RF) magnetron sputtering [22].

In the 1990s, researchers analyzed further into fabrication of high-purity bulk ZnO wafers, and in 1998 they succeeded adopting the transition vapor phase technique. Researchers tended to concentrate more on the mechanisms of ZnO development and the factors that influence its applicability for smart technologies in the early 2000s.

The development of ZnO thin films for acoustical and optical devices has thus been the subject of several experiments on a range of substrates (glass, sapphire, and diamond). In the early studies, thin ZnO films were formed using techniques including chemical vapor deposition and magnetron sputtering. However, the nature of the ZnO thin films was primarily polycrystalline. By using radio frequency magnetron sputtering and other growth techniques that allowed for refined control of the deposition process, such as molecular beam epitaxy (MBE), pulsed laser deposition (PLD), metal-organic chemical vapor deposition (MOCVD), and hydride or halide vapor-phase epitaxy, subsequent research tends to result in the production of high quality ZnO single crystal films (HVPE)[2, 4].

1.3   Scope of this chapter

We will present a good rundown of ZnO thin films in this chapter. The special characteristics, various fabrication techniques, and uses of ZnO thin films will all be covered in the section that follows.

2.   Structural features

According to our discussion, a thin layer consisting of this material has attracted a lot of attention in recent years, and researchers have conducted extensive research in this area. Generally, in crystal form, atoms or ions follow a periodic order. The crystal structures found in oxide compounds are numerous and, in some cases, complex and are found as single microstructures in compounds that contain two and three oxides. The general structures of monoxide (Mo), where M represents the metal cation, have been observed in the form of wurtzite and sodium chloride (NaCl) [23]. In general, Most of the group II–VI binary compound semiconductors crystallize in either cubic zinc blende or hexagonal wurtzite structure where each anion is surrounded by four cations at the corners of a tetrahedron, and vice versa. As a compound II-VI, ZnO exists as three forms: wurtzite, zinc bland, and rock salt. The wurtzite lattice of ZnO has been shown in Figure 1 [2, 4].

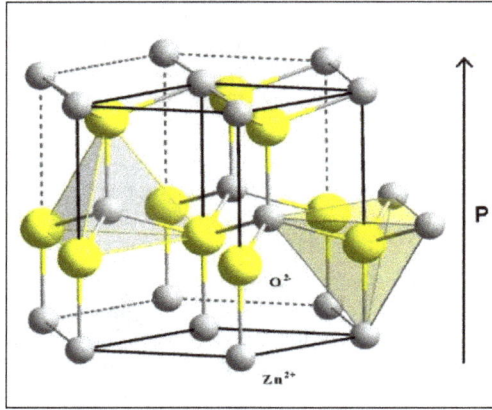

Figure 1: The wurtzite crystal lattice of ZnO: small circles represent zinc atoms, whereas large circles illustrate oxygen atoms [4].

Only by emerging on cubic substrates can the structure of zinc blende ZnO be established. A hexagonal wurtzite structure with the lattice constants a = b = 3,249, c = 5, 204, and a/b = 1.633 is thermodynamically stable at normal temperature, in contrast to the rocksalt structure, which is found under exceptionally high pressures [10 GPa]. Two sub-lattices (Zn^{2+} and O^{2-}) that are layered underneath of one another were taken into consideration to characterize this structure. This signifies that each O ion surrounds a Zn ion in a tetrahedral arrangement, and vice versa. The chosen structure for optoelectronic applications is a wurtzite structure [2, 24]. Table 1 indicates the properties of wurtzite lattice structure of ZnO.

Table 1. the properties of bulk wurtzite lattice structure of ZnO [4].

Property	Value
Lattice parameter	a= b=3.250 Å c= 5.206 Å u= 0.348 c/a= 1.593-1.6035
Density	5.606 gm/cm^3
Melting point	2248 K
Stable crystal structure	wurtzite
Dielectric constant	8.66
Refractive index	2.008, 2.029
Band gap (E$_g$)	3.37 eV (direct)
Exciton binding energy	60 meV
Electron/ Hole effective mass	0.24 m$_0$ /0.59 m$_0$
Hole mobility (300) K	5-50 cm^3/Vs
Electron mobility (300) K	100-200 cm^2/Vs
Heat capacity	9.6 cal/molK
Born effective charge	2.10
Heat of crystallization	62 KJ/mol
Interinsic carrier concentration	<106 cm^{-3}
Bond length	1.977 um
Thermal conductivity	0.6, 1-1.2
Breakdown voltage (10^6 V cm^{-1})	5.0
Saturation velocity (10^{7}cms^{-1})	3.0
Ionicity	62%

3. Synthesis methods

Thin film technology played a crucial role in controlling the properties of ZnO thin films, since the same material deposited using different techniques usually had different physical characteristics. The reason for this is that the electrical and optical properties of the films depended substantially on the structure, morphology, and type of impurities within them. In this sense, films grown using any particular technique may have different properties due to the variation in various deposition parameters, and therefore, the properties can be tailored by varying the deposition parameters. The development of ZnO films with improved optical, electronic, or structural properties is expected to lead to improved conversion efficiencies in the future, while the development of processes for large areas at high growth rates are critical to meeting production cost targets [25].

Various methods have been used to grow ZnO thin films, including chemical vapor deposition (CVD), sputtering, pulsed laser deposition (PLD), molecular beam epitaxy (MBE), spray pyrolysis and so forth, which have improved its electrical and optical properties; These applications are discussed in this section. The first three methods, CVD, sputtering and PLD, lead to the best ZnO films when it comes to high conductivity and transparency. Both of these methods, CVD and particularly the sputtering technique, can also be applied to large areas, making them the most advanced methods for depositing ZnO in the industry [25, 26].

3.1   Commonly used methods for synthesis

The two primary categories for producing ZnO thin films are physical methods and chemical approaches, as was already mentioned. Each of these two strategies has advantages and disadvantages depending on which is picked. Depending on the different uses, a strategic goal might be set.

3.1.1 Sputtering

The sputtering technique is one of the adaptable ways to deposit ZnO thin films. It is seen that the samples have greater uniformity and regulated composition when this method (magnetron sputtering) is compared to chemical methods (such as sol-gel and chemical-vapor deposition methods). Additionally, we can better regulate the film's thickness with this technique. This method's simplicity, affordability, and low operating temperature are further benefits [17, 27]. This technology operates by applying voltage between the cathode and the anode, where the anode operates as the substrate holder and the cathode as the target holder, to establish a gas plasma. It experiences severe ion bombardment before having particles expelled from the cathode. Particles are deposited on a substrate after diffusing away from the cathode surface. Sputtering often takes place between 10^{-2} and 10^{-3} Torr of pressure. A high-frequency generator is linked between the electrodes for non-conducting samples and insulators, and a DC voltage is delivered between the anode and cathode for conducting targets (DC sputtering) (RF Sputtering), as shown in Figure 2 of the schematic [28, 29]. Magnetron sputtering is suitable for applications where a high deposition rate or a low substrate temperature are needed. Magnetron sputtering uses a magnetic field to enhance the sputtering rate, which increases the rate of deposition as well [28] at each optimum conditions.

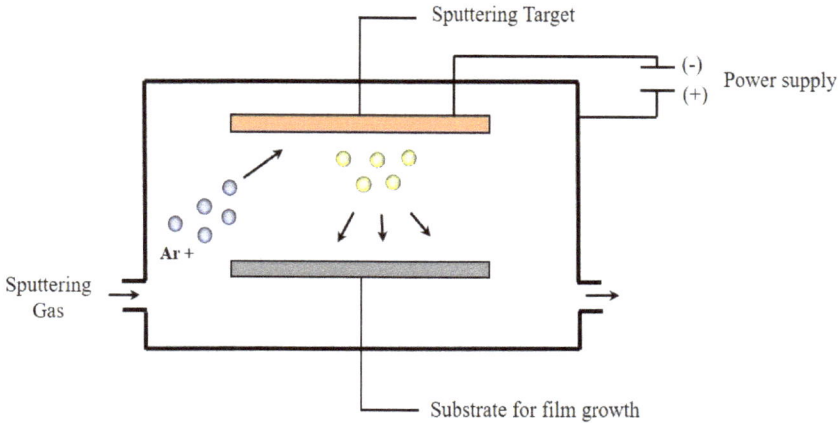

Figure 2. A basic schematic representation of the sputtering process [29].

At a specified substrate temperature, these films are synthesized by magnetron sputtering from high purity ZnO targets. In growth medium with $O_2/Ar+O_2$ gas ratios ranging from 0 to 1 at pressures of 10^{-3} to 10^{-2} Torr, growth occurs predominantly.

O_2 is used as the reactive gas and Ar is employed as a gas to speed up the sputtering process in the approach that has been stated. Another option is to generate ZnO by Zn target in an $Ar+O_2$ combination using a DC sputtering method. Additionally, the amount of sputtering yield rate from the ZnO target may be controlled by adjusting the magnetron sputtering power applied to the plasma. In order to stabilise the system and eliminate impurities from the target surface, this experiment requires the target to be presputtered for 5 to 15 minutes before deposition begins, The structural and optical characteristics of ZnO films produced by RF MSD in an O_2/Ar gas mixture are only marginally affected by the growing temperature in the diapason range of 200 to 300 °C. It has been demonstrated that using pure argon plasma produces non-stoichiometric porosity ZnO sheets with a reflectivity index of approximately 1.65. The ZnO films produced by RF MSD in an oxygen-rich gas combination are stoichiometric solid polycrystals with reflectivity indices and energy band gaps of around 2.0–2.1 eV and 3.12–3.2 eV, respectively [17].

Direct current (DC), radio frequency (RF), and high-power impulse magnetron sputtering (HiPIMS) are only a few of the different types of magnetron sputtering techniques. Radio frequency magnetron sputtering (RF MS)((13.56 MHz, 27 MHz) enables the sputtering of both conductive and nonconductive materials. When non-conducting ZnO is being deposited, a metallic Zn target is typically sputtered in an oxygen environment [30]. Partial pressure is a crucial process parameter in each scenario, and the desired power density is typically no greater than 10 W/cm^{-2}. To manage the stoichiometry of the deposited films, one can regulate the amount of oxygen doping used during the RF process. By sputtering the conductive target, Zn oxide coatings can also be produced using DC MS. This typically necessitates using targets composed of metallic Zn material. Similar to RF MS, the intended power density is limited to 10 W/cm^{-2}. Recently, high-power impulse magnetron sputtering has received a lot of attention. With this technique, the target is sputtered with current pulses that are extremely high and brief—their length is on the order of a dozen or more microseconds. The modulation frequency for current waveforms is several kHz. Though the sputtering efficiency is somewhat lower, it is important to note that the average power released in the target is comparable to that of conventional DC and RF MS procedures. Conversely, due to phenomena associated with the start and pulse suppression of the HiPIMS approach, mechanical properties of deposited films, such as hardness or optical qualities, can be improved [31]. In a recent report, Sm doped ZnO thin films were fabricated by RF sputtering on glass and silicon substrate [32]. AFM technique was utilized to measure the surface roughness and grain size distribution of the films. In the work, the surafce roughness was found to improve with the Sm intercalation. Figure 3 shows the 3D AFM images, grain size distribution, and surface roughness images of the Sm:ZnO thin films. The grain size distribution was found to increase from 73.20 nm to 82.57 nm, while the surface roughness was found to decrease from 6 nm to 2.35 nm with Sm inclusion.

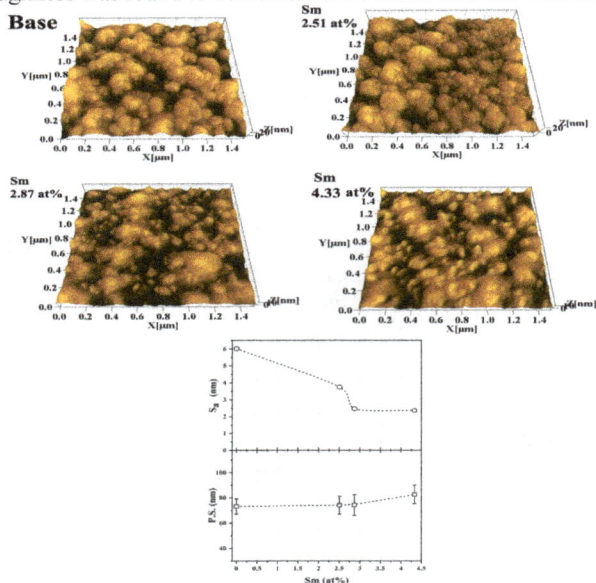

Figure 3: 3D AFM images, grain size distribution and surface roughness of Sm:ZnO thin films[32].

3.1.2 Chemical vapor deposition (CVD)

CVD, one of several techniques for fabricating thin films of semiconductor materials, is particularly alluring due to both its high quality and its capacity for mass manufacturing. CVD or spray pyrolysis preparation processes are comparable to those used for chemical bath deposition (CBD). It is necessary to dissolve and evaporate the zinc precursor in a solution. Precursor evaporation can be accomplished in a number of methods. The zinc precursor first needs to be simple to evaporate or to put it another way, it must be a volatile precursor. The zinc ingot used as a precursor can then evaporate at high temperatures. The final step involves employing equipment like an atomizer or an ultrasonic transducer to atomize the precursor or turn it into a mist. The zinc precursor solution has been transformed into a mist by an ultrasonic device and transported by inert gas into the heat furnace chamber, where a substrate has been positioned as the target for creating a thin layer of ZnO. Some of the zinc precursor particles react with an oxygen-containing molecule in the heater chamber after the precursor mist or vapor has been delivered there and arrived, growing the molecule. The substrate is then covered with this molecule. This molecule breaks down into zinc as a result of a high temperature or sufficient energy, which subsequently interacts with oxygen to create ZnO. The solvent leaves the compartment after evaporating. Likewise, a thin film of ZnO is formed in conjunction with the carrier gas. When zinc precursor enters the chamber, it may adhere to the substrate and then engage in a reaction or bonding with other materials. This alternative reaction processanother oxygen-containing molecule. The zinc precursor molecule and oxygen-containing molecule react with the ZnO due to the high temperature and sufficient energy to decompose, releasing some degraded solvents that are carried away by gas before the ZnO molecule finally transforms into ZnO thin film.

Several CVD-type methods are available [29]:

1.APCVD: (Atmospheric Pressure Chemical Vapor Deposition) deposition under atmospheric pressure;

2.LPCVD: (Low-Pressure Chemical Vapor Deposition) low-pressure deposition;

3.MOCVD: (Metal Organic CVD) the use of organometallic precursors;

4.PACVD: (Plasma Assisted Chemical Vapor Deposition) with the assistance of a plasma. As shown in Figure 4, in this method, one or more gaseous species react with a solid substrate (surface). The metallic oxides are commonly created in this process by vaporising the organo-metallic compounds. Condensate from a vapor is transmitted to a substrate surface, where a heterogeneous process typically breaks it down. Different types of breakdown occur depending on the species that transport the volatiles. Decomposition conditions should be such that the reaction only occurs on the substrate surface or close to the surface, and not in the gaseous state, to avoid the creation of powdery deposits that could cause haziness in the films during deposition [2, 4]. An important aspect of the CVD technique is that the deposited ZnO is the result of chemical reactions between substrates and vapor phase precursors. The vapor is subsequently transported to the growth zone by the carrier gas. The reactions take place in a reactor where the necessary temperature profile is generated in the direction of the gas flow. Another variation of APCVD is AACVD which is knowon as aerosol assisted CVD technique. AACVD has some advantages (non-volatile precursors, easily handeled, wide range) over APCVD and is in current trend for the fabrication of ZnO thin films [33]. Recently, Al doped ZnO thin films were fabricated using AACVD. Different optoelectric properties were optimized by merely varying the precursors. Table 2 summarizes the modified oprtoelectric properties for different precursos.

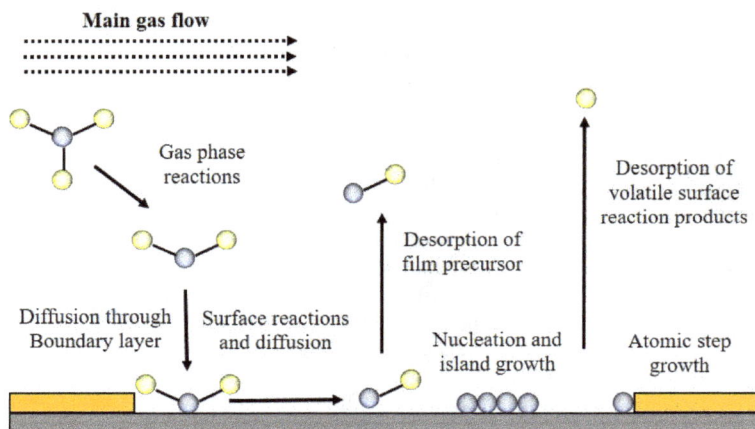

Figure 4. The main steps of deposition by the CVD method.

Table2: Modified optoelectronics properties of Al: ZnO thin films fabricated using aerosol assisted CVD technique [33].

Solvent	$T_{\lambda 400-700}$/%	E_g/eV	$\rho \times 10^{-2}/\Omega$ cm	$n \times 10^{19}/cm^{-3}$	$\mu/cm^2 V^{-1} s^{-1}$
MeOH	83	3.25	0.5	14.0	9.0
MeOH and toluene	78	3.25	27.2	7.1	0.3
MeOH and THF	80	3.25	44.2	5.3	0.3
MeOH and n-hexane	77	3.25	N/A	N/A	N/A
MeOH and cyclohexane	75	3.26	N/A	N/A	N/A
MeOH and ethyl acetate	67	3.24	0.2	9.7	13.3

ZnO and Their Hybrid Nano-Structures   Materials Research Forum LLC
Materials Research Foundations 146 (2023) 263-293   https://doi.org/10.21741/9781644902394-9

3.1.3 Pulsed laser deposition (PLD)

Three stages compensate PLD, and each is repeated thousands of times all throughout deposition process

1. The Vaporization of the target material,

2. The transport of the vapor plume;

3. The growth of a film on a substrate.

Laser pulses are seldom effective in removing material layer-by-layer and in a clean manner. Instead, the target surfaces are chemically and physically deformed [34].

By using this technique, the substance to be deposited, i.e. the target, is ablated by means of laser pulses leading to the formation of plasma. When the plasma is exposed to the background gases supplied into the vacuum chamber, it interacts with the gases. Eventually, it condenses on the surface of the object to be coated with the material, i.e., the substrate, and nucleates to get deposited as a thin film. PLD provides the advantage of being simpler, faster, and cheaper than other technologies; high quality films with the desired crystalline structure can be produced with this technology. However, there are a number of factors involved which determine the quality of the film, such as pressure optimization, target composition and shape, substrate temperature and laser pulse characteristics [7].

A schematic of the typical PLD system is shown in Figure 5. This method has a number of advantages, including [4, 7, 34, 35]:

1. The ability to create particles using high energy sources,
2. Possibility of controlling the film thickness down to even one atomic layer by adjusting the number of pulses,
3. The resulting ZnO films exhibit an extremely smooth surface, compared with films formed by other physical vapor deposition processes,
4. The growth of high-quality films at low substrate temperatures (200 to 800 °C) and creating a thin film with the desired crystalline structure, which occurs because of the high kinetic energies of ejected atoms and ionized particles (>1 eV),
5. The similarity between the target film and the films grown,
6. A simple experiment setup,
7. Easy to produce multilayer films from different materials by successive erosion of different targets,
8. The possibility of growing ZnO films at low temperatures on flexible substrates, in order to produce flexible flat-panel displays [36].

Generally, for the growth of ZnO thin films by PLD method, a combination of UV excimer lasers KrF: =248 nm and ArF: =193 nm along with a pulsed Nd: yttrium aluminum garnet YAG laser at =355 nm is employed for the ablation of the ZnO target in an oxygen medium. In some cases, Cu-vapor laser emitting at =510–578 nm was also used for the same purpose. The common target is a cylindrical ZnO tablet made from compressed ZnO powder that is made from compressed ZnO powder. It has been demonstrated recently that ZnO single crystals can be used to grow high quality ZnO thin films. In general, the properties of the grown ZnO films are affected by the substrate temperature, the ambient oxygen pressure, and the laser intensity [4].

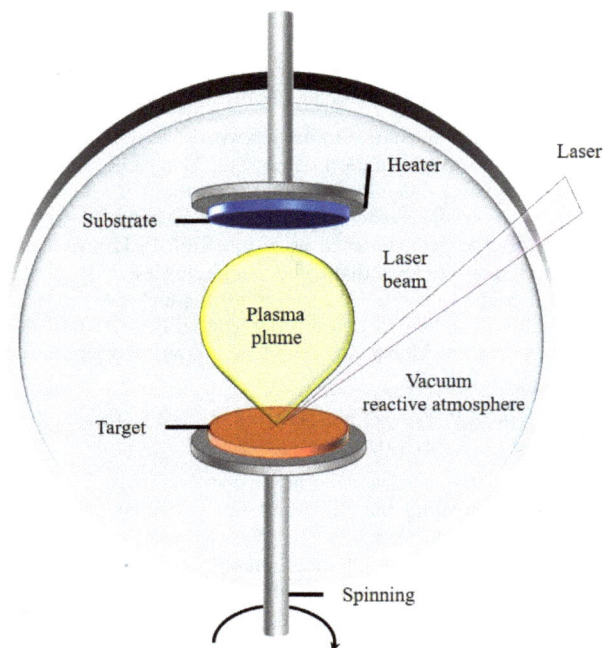

Figure 5: Schematic diagram of a PLD system [2, 4].

3.1.3 Spray pyrolysis (SP)

The SP method, which is one of the chemical methods for the growth of thin films, is widely used for the deposition of a wide variety of thin films. It involves spraying a (usually) aqueous solution containing soluble salts of the desired compound onto a heated substrate. The SP process utilizes endothermic (pyrolytic) decomposition of a metallic compound which is dissolved in a liquid mixture and applied to a heated substrate. By using this method, the doping process is relatively simple and the percentage of doping in the sample is only affected by changing the concentration of the dopant in the solution. This method has the following advantages:

1. The suitability for the deposition of thin films with large areas and good reproducibility at high speeds,

2. The low consumption of materials,

3. The low cost of the material used,

4. There is no need of great quality target and vacuum at various stages.

Also, there are some disadvantages to this approach [29, 37]:

1. It is not suitable for the deposition of thin films of very small thickness,

2. Restrictions on the choice of substrates due to high process temperatures (the effects of substrate temperature on the structural, electrical, and optical properties of ZnO thin films prepared using this method have been investigated),

3. The material used to coat the surface should be soluble,

4. Edge effects (a greater thickness around the edges of the substrate),

5. The difficulty of cleaning the laborious device.

Number of reserachers have utilized SP method for the deposition of pure and doped ZnO thin films [38-40]. In the recent reported work, Srivastha et. al. have fabricated ZnO thin films on glass substrate and studied the effect of substrate temperature on the morphology, defect chemistry and photoconductivity of the films [41]. The results revealed that maximum photosensitivity was observed for thin films deposited at 450 ˚C.

3.1.4 Molecular beam epitaxy (MBE)

The MBE technology facilitates the creation of thin films with exceptional crystalline quality and extremely low roughness at low temperatures and under extremely high vacuum (10^{-10} Torr). The reaction of atomic or molecule fluxes on a single crystal substrate that has attained the proper temperature is the basis for the method's functioning mechanism [29]. In order to generate beams of atoms or molecules with an angular dispersion, the solid source materials are put into evaporation cells. To increase growth homogeneity, the substrate is heated to the necessary temperature and continually rotated as needed.

The formation of high-quality transparent and conductive ZnO thin films was carried out by controlling the growth of the atomic layer at different temperatures on various substrates (including glass, sapphire, and polyethylene tetraphthalate) [42]. The source materials typically used in the growth of ZnO thin films by MBE are Zn metal and O_2. A high purity Zn metal is evaporated from an effusion cell, where the cell temperature can be changed to investigate the effect of the Zn flux on the growth rate and material properties. Morover, by directing an oxygen radical beam on to the film surface, a high oxidation efficiency can be achieved [4].

In general, molecular beam epitaxy has the following advantages [4, 29]:

1. Precise control of the deposition parameters,

2. *In situ* diagnostic capabilities, including possible controls and *in situ* analysis

3. A low growth rate, which allows doping at the atomic level,

4. Very precise control of the thicknesses of the thin layers,

5. The ability to control every step of the process in an automated manner,

6. No boundary layer.

4.   Properties

The following section provides an overview of the structural, optical, and electrical properties of ZnO thin films, including some of the following:

1. High transparency in the visible and near infrared spectral regions,

2. Generating heavily doped sheets with a high free electron density but a low resistance

3. Good contact with the active semiconductors (absorber layers),

4. The possibility of preparing the transparent conductive oxide (TCO) layers on large areas (>1 m^2) using deposition methods such as magnetron sputtering or metal-organic chemical vapor deposition (MOCVD),

5. A possibility for preparing high-quality ZnO films at low substrate temperatures,

6. Ability to prepare tailored surfaces with the desired light scattering properties for light trapping, which is particularly important for thin film solar cells made of Si.

7. Low material costs, nontoxicity, and abundance in the earth's crust.

4.1   Physical properties

The structural attributes of these thin films will be thoroughly explained in this section. Some of the primary characteristics of ZnO are shown in Table 3, as can be seen. Physical parameters may vary with the roughness of the films. Roughness of the films is an critical issue which hinders the utilization of films in device applications. In the last few decades, roughness of the ZnO films have been minimized by number of routes like varying the substrate, substrate temperature, precursors and many other experimental conditions. In general, all thin films fabricated under vacuum suffer from stress (thermal + intrinsic). Where thermal stress originates form the different thermal extinction coefficient (α) of substrate and coating materials. For example, Si substrate exhibit α of $2.5 \times 10^{-6}/°C$. Hexagonal ZnO crystal exhibits α_{11} and α_{33} of 6.05 and $3.53 \times 10^{-6}/°C$ respectively which are quite greater than as reported for Si [43]. This lead to the creation of tensile stress in

ZnO thin films when substrate comes to temperature from higher temperature. Secondly, intrinsic stress results from the crystallographic flaws created during thin film fabrication. Another reason for the comprehensive stress is due to reflection of oxygen and argon atoms from target. The developed stress is responsible for the change in elastic constants and thereby can change many other physical and other properties of ZnO thin films.

Table 3. Physical Parameters of ZnO [44]

Properties	Parameters (Value) at 300 K
Crystalline structure	Wurtzite
Lattice parameters	a=b=3.249 Å , c=5.2042 Å
Refractive index	2.008- 2.029
band gap energy	3.37 eV, direct
Melting point	1975 °C
Density	5.606 g/cm³
Electron effective mass	0.28 m₀
Exciton binding energy	60meV

4.2   Electrical properties

It is challenging to test the electrical characteristics of these oxides due to the fact that the quality of the many ZnO samples available varies greatly. The concentration of charge carriers varies depending on the layer quality, but is typically approximately 10^{16} cm^{-3}. P-type atoms have a maximum of 10^{19} cm^{-3} holes and an upper limit of 10^{20} cm^{-3} electrons for N-type atoms. This high p-type conductivity value, nevertheless, is debatable and hasn't been demonstrated in real-world applications.

Additionally, at 300 K, the bond excitation energy of ZnO, an II-VI semiconductor, is equal to 60 meV, making ZnO an ideal material for opto-electric devices. In other words, a significant exciton binding energy and a wide and direct energy gap (3.44 eV at 4 K and 3.37 eV at normal temperature) show that effective excitonic emission in ZnO can continue at room temperature. ZnO is a suitable material for optoelectronic devices based on excitonic phenomena because excitons often have stronger oscillators than direct electron transfers to holes in direct gap semiconductors [5, 45]. Undoped ZnO is typically regarded as an n-type semiconductor. It is significant to note that ZnO's electrical resistivity can be enhanced through doping, the introduction of more zinc atoms into the interstitial space, or the establishment of oxygen vacancies. The material's electrical resistance is decreased by the presence of zinc interstitials and oxygen vacancies, which function as electron donors. A shallow donor defect called interstitial hydrogen may also be the source of the free electrons [29]. It is crucial to remember that the preparation conditions, particularly heat treatment, have an impact on the electrical resistivity of ZnO thin films. As previously mentioned, it is simple to produce n-type ZnO by replacing zinc atoms with Group III elements (Al, Ga, In, and B) and Group IV elements (Si, Ge, and Zr), or by replacing oxygen atoms with a Group VII element (F, Cl) for p-type ZnO doping with Group I elements (Li, Na and K). It should be noted, nevertheless, that type p is more challenging to obtain than type n.[29, 46].

4.3 Optical properties

In recent times, there has been a significant amount of research conducted on the optical processes and properties of ZnO, including its transmission, absorption, optical gap, and photoluminescence. This property is a direct consequence of the material's wide band gap (about 3.37 eV at room temperature), ultraviolet luminescence (around 380 nm), and good radiative recombination. Additionally, this material's high exciton radiation at room temperature and above, which is 2.4 times as powerful as the thermal energy generated (K B T=25 meV), is facilitated by the exciton binding energy (60 meV) of this substance [14]. These thin films can be utilised as a barrier against ultraviolet radiation as well as transparent in the visible and near ultraviolet-visible wavelength ranges because of their direct and wide band gaps and low permeability in the ultraviolet region. But since excitonic recombination is caused by the high exciton binding energy at ambient temperature, ZnO thin films could one day compete with other materials to produce UV and blue light generating devices. Consequently, epitaxial and polycrystalline ZnO thin films have shown optically pumped lasing. [29, 47, 48]. Several ZnO characteristics are listed in Table 4.

Table 4. Some optical properties of ZnO[4, 49, 50]

Properties	Values
The average transmittance (%)	86-92
Refractive index	2.008–2.029
Absorption coefficient (cm^{-1})	$\sim 10^4$

Optical properties of ZnO thin films are influenced by the following factors [29]:

1. The thickness
2. The optical gap
3. The surface roughness

4. The doping
5. The crystalline quality of the film
6. The deposition method

5.   Potential applications of ZnO thin films

During the last decade, ZnO material has been studied extensively and employed in micro-scale and nano-scale devices. As stated previously, ZnO is a very versatile substance which can be applied in many micro and nano systems. A combination of its wide band gap, piezoelectric, electrical, and optical properties make ZnO thin films unique and crucial for a wide range of technological applications today and in the future. With its various properties, ZnO is a promising candidate for applications in spintronics, medical applications such as sun screen lotions and ointments, solar cells, UV light emitting diodes, sensors, surface acoustic devices, transparent electronics such as transparent thin film transistors and transparent conductive electrodes [10].

5.1   Future potential applications for thin films

ZnO thin films are the potential source of diverse application in numerious areas of technology. Gas sensors are one of the most widespread applications for ZnO thin films. Under temperature-controlled conditions, this thin film can detect a broad array of gases. The fabrication of ZnO gas sensors utilizes a wide range of materials, including thin films and thick films. Some metal ions that can be doped with ZnO are indeed being considered in order to enhance the performance and gas sensing capabilities of the thin film of ZnO. Some metal ions that can be doped with ZnO are being taken into account in order to enhance the efficiency of the thin film of ZnO and its gas sensor technologies. In this regard, ZnO thin films doped with aluminum oxide are effective at detecting tri-ethanolamine, and ZnO thin films doped with indium, copper, and iron are effective at detecting ethanol [51].

Also, transparent semiconducting ZnO thin films have attracted considerable interest due to their excellent properties, such as high infrared reflectance, excellent chemical stability, UV absorption, and high transparency in the visible region, which makes ZnO suitably suitable for many optics and electronics applications, including solar cells, flat displays, and chemical sensors. Further, ZnO is non-toxic and exhibits a strong tendency to grow in a c-axis direction. The versatile nature of ZnO makes it suitable for using in a veriatey of solid state electronic devices [52]. Prakash et al. Prepared ZnO:Tb^{3+} embedded polystyrene nanocomposite thin film. ZnO:Tb^{3+} composite films were found to have a homogeneous distribution of nanophosphor nanoparticles, and the photoluminance characteristics improved when the nanophosphor content was raised. In addition, the photoluminance investigation of the ZnO:Tb^{3+}/PS films that was dependent on temperature indicated that the photoluminance intensity dropped with rising temperature owing to thermal quenching. These results showed that the ZnO:Tb^{3+} nanophosphor inserted in the ZnO:Tb^{3+}/PS nanocomposite film kept their distinctive luminous capabilities without a noticeable spectrum shift, emphasizing their useful applications in solid-state lighting [53].

Use of ZnO's piezoelectric properties in mass-sensitive sensors that focus primarily on the acoustic wave phenomenon is another application. They function similarly to bulk acoustic wave devices and can be used for chemical and biological monitoring. The device resonates at its resonance frequency, which is regulated by the amount of external mass associated to the resonator. This signifies that the frequency changes and the mass that is loaded on the device can be accurately determined when there is a mass-loading effect on the bulk acoustic wave resonator [54]. Furthermore, ZnO is also considered a good choice for UV detectors due to its high on/off current ratio, fast response and recovery, large photo-response current, and wavelength selectivity [55]. Aside from sensor applications, ZnO is also highly sought after for thin film transistors, light emitting diodes, gas sensing, solar cells, surface acoustic wave receivers, and energy harvesting devices. There are several applications associated with each of the properties of ZnO thin films.

ZnO thin film gas sensing behaviour is heavily influenced by O_2/Ar ratios, with higher values indicating greater sensitivity to diverse gases due to reduced carrier density. Without an external bias, ZnO films modify their resistance in two distinct ways depending on the oxygen partial pressure of the surrounding gas and the film's temperature. Bulk dispersion cannot be ignored, even at the reduced temperatures at which these thin film sensors operate. p-ZnO thin films doped with AlN and AlAs are synthesized using RF magnetron sputtering for the purpose of gas sensing. By applying a relatively low potential of 1V, the manufactured p-ZnO films were subjected to various ppm of NH_3 at room temperature, 100 °C, and 150 °C. Compared to Al mono-doped n-ZnO film, both films demonstrated superior responsiveness to NH_3 at ambient temperature. The AlAs co-doped p-ZnO thin film outperforms both films in terms of responsiveness and response/recovery time by a wide margin throughout a wide temperature range. This highlights the fact that AlAs co-doped p-ZnO thin films may be employed in gas sensing applications with performance improvement [56]. Chang et al. investigated the influence of film thickness on sensor response of ZTF with thicknesses ranging from 65 nm to 390 nm produced by RF reactive sputtering on a SiO_2/Si wafer. The film thickness of 65 nm yielded the highest sensibility and the quickest reaction [57, 58]. Batra et al. [59] fabricated Al:ZnO thin film for the development of effective cholesterol biosensor. The produced ChOx/Al:ZnO/Pt/glass bioelectrode has a good sensitivity (173 μAmM^{-1} cm^{-2}) and linear performance in a range of cholesterol concentrations (0.6-12.9 mM) with sufficient certainty. Immobilized ChOx has strong attraction for cholesterol, as evidenced by a reduced Michaelis Menten constant (2.53 mM). These findings provide strong evidence that RF-sputtered Al:ZnO thin films can serve as an efficient substrate for the immobilization of biomolecules such as ChOx in the detection of cholesterol. Bott et al. [60] used a steady voltage source to measure the conductivity shift of polycrystalline ZnO films and compacted discs in air mixes of CO, CH_4, and H_2 from 300 to 500 degrees Celsius. The conductance of polycrystalline oxide films was shown to be more sensitive to CH_4 than that of single crystals [61].

Photovoltaic cells, often known as solar cells, are used to transform sunlight into electricity. The most well-known solar cells contain a massive area p-n junction as their bare minimum component. Whenever exposed to light, the electron-poor p-type semiconductor layer of a p-n junction, such as ZTF, is attracted to the electron-rich n-type semiconductor layer. The highest optical transmittances of 93% can be found in ZnO:Al films, making them ideal for use in solar cells [62, 63]. Ali et al synthesized ZnO thin film by thermal evaporation technique. There is research into the (current-voltage) properties of a ZnO/p-Si heterostructure. There was found to be a 50mA short circuit current, 0.45 V open circuit voltage, 36% effectiveness, and a fill factor of 6%. Findings indicate that ZnO/p-Si heterostructures may be useful in solar cells applications [64]. Using RFMS, coatings of fluorine- and aluminum-co-doped ZnO (FAZO) were formed on glass substrates. The optical transmittance spectra of FAZO films deposited a are displayed in Figure 6(a) in the range between 300 and 1200 nm. Figure 6(b) shows the optical property of prepared thin film and with increase the temperature absorption edges are changes. FAZO films have a lower transmission edge than ITO and FTO films, but their greater transmittance from 400 to 1200 nm makes them appropriate as transparent electrodes for high-efficiency solar cells. Figure 6(c) is a diagram of the structure of the perovskite solar cells with FAZO, FTO, and ITO electrodes. Apart from the clear electrode, the devices were made in the same way, and the J–V curves for each one is presented in Figure 6(b). Perovskite solar cells made with the FAZO film were more efficient at turning light into electricity than reference cells made with F-doped SnO_2 and Sn-doped In_2O_3, showing that it can be used in high-efficiency solar cells [65]. Agrawal et al. employed responsive DC magnetron sputter with variable oxygen flow rates to produce a ZnO thin film of 600 nm thickness on a glass surface, and then annealed the film subsequently. Film was deposited in 21% oxygen environments are transparent at 85% and exhibit a resistance of 15×10^3 cm after being annealed at 300 degree Celsius [66]. Aranovich et al. [67], studied the electrical & photovoltaic characteristics of a hetero-junction formed by spray pyrolyzing ZnO sheets over single crystal p-type CdTe. When the cells are exposed to sunlight, the open-circuit voltage is 0.54 V and the short-circuit current is 19.5 mA/cm^2. This means that the cells are 8.8% efficient. Effectiveness of a p-n heterojunction $CuInS_2$/ZnO solar cell is affected by its deposition parameters, resistivity, and annealing temperature.

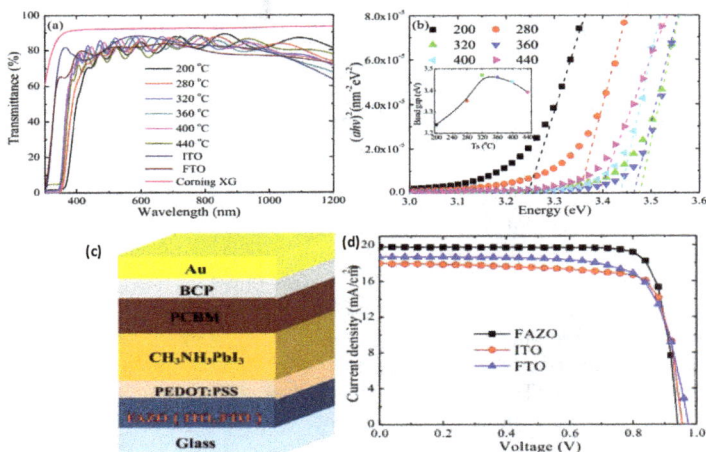

Figure 6. (a) Optical transmittance spectra of FAZO films with different temperature, (b) optical bandgap of the FAZO films (c) Schematic of the structure of FAZO, FTO, or ITO films, and (d) current density–voltage characteristics [65].

ZnO-based LEDs are divided into two categories homojunction LEDs or heterojunction LEDs. Using the PLD method, Rogers et al.[68] grew a thin layer of ZnO on front of a p-type GaN layer in an oxygen environment, which allowed them to examine nonlinear rectification performance at a ZnO band gap equivalent forward bias voltage of 3.3 V. On a ZnO wafer, Guo et al. [69] produced a ZnO homostructural diode with a p-ZnO: N/n-ZnO junction. Created by combining an N-Al co-doped p-type ZnO film with an Al-doped n-type ZnO layer, ZnO p-n homo-junctions were shown by Lu et al. [70]. Lim et al.[71] show how to make UV-emitting ZnO homo-junction LEDs by developing P-doped p-type ZnO atop Ga-doped n-type ZnO. Das et al., n-type SnO_2:F (U-type Asahi) thin films and widely obtainable ZnO:Ga. ZnO:Ga film smooth surface and few spikes imply good surface structure. In order to be most effective in OLED applications, 170 nm ZnO:Ga films need to have an electrical resistivity of 9.6×10^{-5} Ωcm, a sheet resistance of less than 5.6 Ω/cm, an optical transparent of better than 90%, a smooth surface, and fewer spikes [72]. These applications are shown in the table below (as shown in Table 4).

Table 4. The applications related to the properties of ZnO thin film

Propertise	Advantages associated with this properties	Application	Ref.
1. Direct and wide band gap (3.44 eV at low temperature and 3.37 eV at room temperature)	▪ higher breakdown voltages ▪ ability to sustain large electric fields ▪ lower electronic noise ▪ high-temperature and high-power operation	optoelectronics field (in the blue / UV region) light emitting diodes, laser diodes and photo detectors	[4, 8, 73]
2. Large exciton binding energy (60 meV)	▪ efficient excitonic emission in ZnO can persist at room temperature and higher ▪ the stability against thermal dissociation of excitons	optical devices based on excitonic effects	[5, 14]
3. Large piezoelectric constants	▪ Piezoelectric/ pyroelectric properties created by combining two properties (1. low symmetry of the wurtzite crystal structure, 2. Large electro-mechanical coupling)	sensors, transducers and actuators	[74]
4. Strong luminescence (In the green–white region of the spectrum)	▪ The broad peak of the emission spectrum =495 nm [half-width of 0.4 eV]	fluorescent displays and field emission displays	[75]
5. Strong sensitivity of surface conductivity to the presence of adsorbed species	▪ The high surface sensitivity of the ZnO thin films exposed the various gases due to the conductivity of ZnO thin films	cheap smell sensor	[76]
6. High thermal conductivity	▪ in order to increase the thermal conductivity	additive a substrate for homoepitaxy or heteroepitaxy	[77]

5.2   Nanotechnology's thin-film revolution

Nano-ZnO can display the effects of quantum confinement, which include size reduction and an increase in surface area or the surface to volume ratio. This opens up the possibility for the development of new physical, optical, chemical, and chemical characteristics. This encouraged the researchers to fabricate ZnO as a nanostructure material using a variety of techniques and in a diverse range of shapes, comprising nanoparticles, nanowires, nanorods, and nanotubes, nanoplates, etc. [78].

ZnO nanostructures, in particular ZnO thin films, have been significantly used in a broad range of industries and applications as nanotechnology has advanced. It has been shown that this material's nanostructures can be fabricated efficiently at low temperatures utilizing low-cost and low-temperature methodologies [24]. ZnO has a high excitation binding energy and a significant band gap. Additionally, because to its advantageous piezoelectric properties, it has the potential to generate nanoscale electromechanical coupling devices, such as sensors and actuators [55].

For instance, due to its biocompatibility and lack of toxicity to living tissues, ZnO is of tremendous interest in biological applications. Numerous studies have been conducted on this material's nanostructures in a variety of medical specialties, from diagnosis to treatment. Because of their stability, high electrochemical activity, quick electron transfer, biocompatibility, and simplicity of manufacture, ZnO-based biosensors are extensively used [79]. In light of this, ZnO nanostructures are currently the subject of intensive research for the creation of future spintronic, potho-electronic, and medicinal devices.

Conclusion

ZnO thin films have already been briefly covered in this chapter. ZnO and its thin film have been the focus of several research studies due to their significance in industry and technology. Additionally, piezoelectric, semiconducting, and photoconducting qualities can be found in ZnO thin films. They have a high dielectric constant, a wide forbidden energy gap, and a high nonlinear optical coefficient in addition to having high electrical and optical transparency. High piezoelectric, acoustic-optical, and electro-optical coefficients are provided by ZnO films. ZnO films are employed in the creation of surface acoustic wave devices, numerous electro-optical devices, integrated optics, gas and biological sensors, pressure sensors, and photovoltaic solar cells due to the aforementioned qualities as well as their inexpensive cost. ZnO thin films, however, can be synthesized using a variety of deposition techniques. Sputtering, CVD, and SP processes, on the other hand, have been exploited more extensively due to their simplicity and low cost for high-quality applications. Additionally, some of the structural, electrical, and optical characteristics of ZnO films were investigated.

ZnO and Their Hybrid Nano-Structures
Materials Research Foundations 146 (2023) 263-293

Materials Research Forum LLC
https://doi.org/10.21741/9781644902394-9

Aknowlgement

This project has received funding from the European Union's Horizon 2020 research and innovation programme under the grant agreement No 857287.

Reference

[1] G. Busch, Early history of the physics and chemistry of semiconductors-from doubts to fact in a hundred years, European Journal of Physics, 10 (1989) 254. https://doi.org/10.1088/0143-0807/10/4/002

[2] H. Morkoç, Ü. Özgür, ZnO: fundamentals, materials and device technology, John Wiley & Sons, 2008. https://doi.org/10.1002/9783527623945

[3] P. Dhiman, A. Kumar, M. Shekh, G. Sharma, G. Rana, D.-V.N. Vo, N. AlMasoud, M. Naushad, Z.A. Alothman, Robust magnetic ZnO-Fe2O3 Z-scheme heterojunctions with in-built metal-redox for high performance photo-degradation of sulfamethoxazole and electrochemical dopamine detection, Environmental Research, 197 (2021) 111074. https://doi.org/10.1016/j.envres.2021.111074

[4] Ü. Özgür, Y.I. Alivov, C. Liu, A. Teke, M. Reshchikov, S. Doğan, V. Avrutin, S.-J. Cho, Morkoç, A comprehensive review of ZnO materials and devices, Journal of applied physics, 98 (2005) 11. https://doi.org/10.1063/1.1992666

[5] L. Intilla, Study of ZnO properties applied to thin film transistors, in, UCL (University College London), 2016.

[6] M. Khiari, M. Gilliot, M. Lejeune, F. Lazar, A. Hadjadj, Preparation of Very Thin ZnO Films by Liquid Deposition Process: Review of Key Processing Parameters, Coatings, 12 (2022) 65. https://doi.org/10.3390/coatings12010065

[7] B.K. Das, Growth of ZnO Thin Films on Silicon and Glass Substrate by Pulsed Laser Deposition a Thesis, Journal of Physics & Astronomy, 9 (2021) 1-12. https://doi.org/10.2139/ssrn.3840984

[8] C. Jagadish, S.J. Pearton, ZnO bulk, thin films and nanostructures: processing, properties, and applications, Elsevier, 2011.

[9] P. Dhiman, J. Chand, S. Verma, Sarveena, M. Singh, Ni, Fe Co-doped ZnO nanoparticles synthesized by solution combustion method, AIP Conference Proceedings, 1591 (2014) 1443-1445. https://doi.org/10.1063/1.4872990

[10] K.N. Tapily, Synthesis of ALD ZnO and thin film materials optimization for UV photodetector applications, Old Dominion University, 2011.

[11] P. Dhiman, M. Naushad, K.M. Batoo, A. Kumar, G. Sharma, A.A. Ghfar, G. Kumar, M. Singh, Nano FexZn1−xO as a tuneable and efficient photocatalyst for solar powered degradation of bisphenol A from aqueous environment, Journal of Cleaner Production, 165 (2017) 1542-1556. https://doi.org/10.1016/j.jclepro.2017.07.245

[12] A.R. Hutson, Hall effect studies of doped ZnO single crystals, Physical review, 108 (1957) 222. https://doi.org/10.1103/PhysRev.108.222

[13] O. Fritsch, Elektrisches und optisches Verhalten von Halbleitern. X Elektrische Messungen an Zinkoxyd, Annalen der Physik, 414 (1935) 375-401. https://doi.org/10.1002/andp.19354140406

[14] D.C. Look, Recent advances in ZnO materials and devices, Materials Science and Engineering: B, 80 (2001) 383-387. https://doi.org/10.1016/S0921-5107(00)00604-8

[15] G. Heiland, E. Mollwo, F. Stöckmann, Electronic processes in ZnO, in: Solid state physics, Elsevier, 1959, pp. 191-323. https://doi.org/10.1016/S0081-1947(08)60481-6

[16] J. Haynes, Experimental observation of the excitonic molecule, Physical Review Letters, 17 (1966) 860. https://doi.org/10.1103/PhysRevLett.17.860

[17] F. Hickernell, DC triode sputtered ZnO surface elastic wave transducers, Journal of Applied Physics, 44 (1973) 1061-1071. https://doi.org/10.1063/1.1662307

[18] N. Foster, G. Rozgonyi, ZnO film transducers, Applied Physics Letters, 8 (1966) 221-223. https://doi.org/10.1063/1.1754565

[19] R. Laudise, E. Kolb, A. Caporaso, Hydrothermal growth of large sound crystals of ZnO, Journal of the American Ceramic Society, 47 (1964) 9-12. https://doi.org/10.1111/j.1151-2916.1964.tb14632.x

[20] P.R. Emtage, The physics of ZnO varistors, Journal of Applied Physics, 48 (1977) 4372-4384. https://doi.org/10.1063/1.323391

[21] M. Inada, Crystal phases of nonohmic ZnO ceramics, Japanese Journal of Applied Physics, 17 (1978) 1. https://doi.org/10.1143/JJAP.17.1

[22] S. Tiku, C. Lau, K. Lakin, Chemical vapor deposition of ZnO epitaxial films on sapphire, Applied physics letters, 36 (1980) 318-320. https://doi.org/10.1063/1.91477

[23] R.J. Tilley, Crystals and crystal structures, John Wiley & Sons, 2020.

[24] A. Djurišić, A.M.C. Ng, X. Chen, ZnO nanostructures for optoelectronics: Material properties and device applications, Progress in quantum electronics, 34 (2010) 191-259. https://doi.org/10.1016/j.pquantelec.2010.04.001

[25] K. Ellmer, A. Klein, B. Rech, Transparent conductive ZnO: basics and applications in thin film solar cells, (2007). https://doi.org/10.1007/978-3-540-73612-7

[26] M. Habibi, M. Khaledi Sardashti, Preparation and proposed mechanism of ZnO Nanostructure Thin Film on Glass with Highest c-axis Orientation, International Journal of Nanoscience and Nanotechnology, 4 (2008) 13-16. https://doi.org/10.1155/2008/356765

[27] A. Kumar, P. Dhiman, M. Singh, Effect of Fe-doping on the structural, optical and magnetic properties of ZnO thin films prepared by RF magnetron sputtering, Ceramics International, 42 (2016) 7918-7923. https://doi.org/10.1016/j.ceramint.2016.01.136

[28] D. WANG, Fabrication and Characterization of ZnO Related Materials Thin Films for Optical Device Application, (2012).

[29] A. Noua, R. Guemini, Preparation and characterization of thin films nanostructures based on ZnO and other oxides, (2019).

[30] M. Masłyk, M. Borysiewicz, M. Wzorek, T. Wojciechowski, M. Kwoka, E. Kamińska, Influence of absolute argon and oxygen flow values at a constant ratio on the growth of Zn/ZnO nanostructures obtained by DC reactive magnetron sputtering, Applied Surface Science, 389 (2016) 287-293. https://doi.org/10.1016/j.apsusc.2016.07.098

[31] S. Konstantinidis, A. Hemberg, J. Dauchot, M. Hecq, Deposition of ZnO layers by high-power impulse magnetron sputtering, Journal of Vacuum Science & Technology B: Microelectronics and Nanometer Structures Processing, Measurement, and Phenomena, 25 (2007) L19-L21. https://doi.org/10.1116/1.2735968

[32] M.S. Aida, M.S. Abdel-wahab, A.H. Hammad, Impact of samarium on the structural and physical properties of sputtered ZnO thin films, Optik, 250 (2022) 168322. https://doi.org/10.1016/j.ijleo.2021.168322

[33] D.B. Potter, I.P. Parkin, C.J. Carmalt, The effect of solvent on Al-doped ZnO thin films deposited via aerosol assisted CVD, RSC advances, 8 (2018) 33164-33173. https://doi.org/10.1039/C8RA06417B

[34] D. Geohegan, D. Chrisey, G. Hubler, Pulsed laser deposition of thin films, Chrisey and GK Hubler (eds), Wiely, New York, (1994) 59-69.

[35] V. Craciun, S. Amirhaghi, D. Craciun, J. Elders, J.G. Gardeniers, I.W. Boyd, Effects of laser wavelength and fluence on the growth of ZnO thin films by pulsed laser deposition, Applied surface science, 86 (1995) 99-106. https://doi.org/10.1016/0169-4332(94)00405-6

[36] S. Heitsch, C. Bundesmann, G. Wagner, G. Zimmermann, A. Rahm, H. Hochmuth, G. Benndorf, H. Schmidt, M. Schubert, M. Lorenz, Low temperature photoluminescence and infrared dielectric functions of pulsed laser deposited ZnO thin films on silicon, Thin Solid Films, 496 (2006) 234-239. https://doi.org/10.1016/j.tsf.2005.08.305

[37] S. Studenikin, N. Golego, M. Cocivera, Optical and electrical properties of undoped ZnO films grown by spray pyrolysis of zinc nitrate solution, Journal of Applied Physics, 83 (1998) 2104-2111. https://doi.org/10.1063/1.366944

[38] N. Lehraki, M.S. Aida, S. Abed, N. Attaf, A. Attaf, M. Poulain, ZnO thin films deposition by spray pyrolysis: Influence of precursor solution properties, Current Applied Physics, 12 (2012) 1283-1287. https://doi.org/10.1016/j.cap.2012.03.012

[39] A.N. Ech-Chergui, A.S. Kadari, M.M. Khan, A. Popad, Y. Khane, M.h. Guezzoul, C. Leostean, D. Silipas, L. Barbu-Tudoran, Z. Abdelhalim, F. Bennabi, K. Driss-

Khodja, B. Amrani, Spray pyrolysis-assisted fabrication of Eu-doped ZnO thin films for antibacterial activities under visible light irradiation, Chemical Papers, (2022). https://doi.org/10.1007/s11696-022-02543-z

[40] C. Manoharan, G. Pavithra, S. Dhanapandian, P. Dhamodharan, Effect of In doping on the properties and antibacterial activity of ZnO films prepared by spray pyrolysis, Spectrochimica Acta Part A: Molecular and Biomolecular Spectroscopy, 149 (2015) 793-799. https://doi.org/10.1016/j.saa.2015.05.019

[41] M. Srivathsa, P. Kumar, B.V. Rajendra, Ultraviolet photoconductivity and photoluminescence properties of spray pyrolyzed ZnO nanostructure: Effect of deposition temperature, Optical Materials, 131 (2022) 112726. https://doi.org/10.1016/j.optmat.2022.112726

[42] A. Ott, R. Chang, Atomic layer-controlled growth of transparent conducting ZnO on plastic substrates, Materials Chemistry and Physics, 58 (1999) 132-138. https://doi.org/10.1016/S0254-0584(98)00264-8

[43] W. Water, S.-Y. Chu, Physical and structural properties of ZnO sputtered films, Materials Letters, 55 (2002) 67-72. https://doi.org/10.1016/S0167-577X(01)00621-8

[44] S.J. Pearton, D.P. Norton, K. Ip, Y.W. Heo, T. Steiner, Recent progress in processing and properties of ZnO, Superlattices and Microstructures, 34 (2003) 3-32. https://doi.org/10.1016/S0749-6036(03)00093-4

[45] D.C. Look, B. Claflin, Y.I. Alivov, S.-J. Park, The future of ZnO light emitters, physica status solidi (a), 201 (2004) 2203-2212. https://doi.org/10.1002/pssa.200404803

[46] M. Zaharescu, S. Mihaiu, A. Toader, I. Atkinson, J. Calderon-Moreno, M. Anastasescu, M. Nicolescu, M. Duta, M. Gartner, K. Vojisavljevic, ZnO based transparent conductive oxide films with controlled type of conduction, Thin Solid Films, 571 (2014) 727-734. https://doi.org/10.1016/j.tsf.2014.02.090

[47] D. Bagnall, Y. Chen, Z. Zhu, T. Yao, S. Koyama, M.Y. Shen, T. Goto, Optically pumped lasing of ZnO at room temperature, Applied physics letters, 70 (1997) 2230-2232. https://doi.org/10.1063/1.118824

[48] A. Ohtomo, M. Kawasaki, Y. Sakurai, Y. Yoshida, H. Koinuma, P. Yu, Z. Tang, G.K. Wong, Y. Segawa, Room temperature ultraviolet laser emission from ZnO nanocrystal thin films grown by laser MBE, Materials Science and Engineering: B, 54 (1998) 24-28. https://doi.org/10.1016/S0921-5107(98)00120-2

[49] Mursal, Irhamni, Bukhari, Z. Jalil, Structural and Optical Properties of ZnO based Thin Films Deposited by Sol-Gel Spin Coating Method, Journal of Physics: Conference Series, 1116 (2018) 032020. https://doi.org/10.1088/1742-6596/1116/3/032020

[50] E.Y. Muslih, B. Munir, Fabrication of ZnO Thin Film through Chemical Preparations, London: IntechOpen Limited 2018. https://doi.org/10.5772/intechopen.74985

[51] S.A. Hooker, Nanotechnology advantages applied to gas sensor development, in: The nanoparticles 2002 conference proceedings, Business Communications Co., Inc., 2002, pp. 1-7.

[52] S.P. Ghosh, Synthesis & Characterizations of ZnO Thin Films and Nanostructures by Modified Aqueous Chemical Growth Method for Sensor Applications, in, 2015.

[53] J. Prakash, V. Kumar, L. Erasmus, M. Duvenhage, G. Sathiyan, S. Bellucci, S. Sun, H.C. Swart, Phosphor polymer nanocomposite: ZnO: Tb3+ embedded polystyrene nanocomposite thin films for solid-state lighting applications, ACS Applied Nano Materials, 1 (2018) 977-988. https://doi.org/10.1021/acsanm.7b00387

[54] M. Depaz, Processing and characterization of ZnO thin films, (2007).

[55] C.-Y. Lu, S.-P. Chang, S.-J. Chang, T.-J. Hsueh, C.-L. Hsu, Y.-Z. Chiou, I.-C. Chen, A lateral ZnO nanowire UV photodetector prepared on a ZnO: Ga/glass template, Semiconductor science and technology, 24 (2009) 075005. https://doi.org/10.1088/0268-1242/24/7/075005

[56] L.N. Balakrishnan, S. Gowrishankar, N. Gopalakrishnan, ${\rm NH}_{3}$ Sensing by p-ZnO Thin Films, IEEE Sensors Journal, 13 (2013) 2055-2060. https://doi.org/10.1109/JSEN.2013.2244592

[57] S.Y. Nam, Y.S. Choi, J.H. Lee, S.J. Park, J.Y. Lee, D.S. Lee, ZnO/p-GaN Heterostructure for Solar Cells and the Effect of ZnGa2O4 Interlayer on Their Performance, Journal of Nanoscience and Nanotechnology, 13 (2013) 448-451. https://doi.org/10.1166/jnn.2013.6943

[58] W. Liu, M. Wang, C. Xu, S. Chen, X. Fu, One-pot synthesis of ZnO2/ZnO composite with enhanced photocatalytic performance for organic dye removal, Journal of Nanoscience and Nanotechnology, 13 (2013) 657-665. https://doi.org/10.1166/jnn.2013.7091

[59] N. Batra, M. Tomar, V. Gupta, Al: ZnO thin film: An efficient matrix for cholesterol detection, Journal of Applied Physics, 112 (2012) 114701. https://doi.org/10.1063/1.4768450

[60] B. Bott, T. Jones, B. Mann, The detection and measurement of CO using ZnO single crystals, Sensors and Actuators, 5 (1984) 65-73. https://doi.org/10.1016/0250-6874(84)87007-9

[61] A. Jones, T. Jones, B. Mann, J. Firth, The effect of the physical form of the oxide on the conductivity changes produced by CH4, CO and H2O on ZnO, Sensors and Actuators, 5 (1984) 75-88. https://doi.org/10.1016/0250-6874(84)87008-0

[62] S.n. Flickyngerova, V. Tvarozek, P. Gaspierik, ZnO--A Unique Material for Advanced Photovoltaic Solar Cells, Journal of electrical engineering, 61 (2010) 291. https://doi.org/10.2478/v10187-010-0043-2

[63] M. Berginski, J. Hüpkes, M. Schulte, G. Schöpe, H. Stiebig, B. Rech, M. Wuttig, The effect of front ZnO: Al surface texture and optical transparency on efficient light trapping in silicon thin-film solar cells, Journal of applied physics, 101 (2007) 074903. https://doi.org/10.1063/1.2715554

[64] R.S. Ali, K.S. Sharba, A.M. Jabbar, S.S. Chiad, K.H. Abass, N.F. Habubi, Characterization of ZnO thin film/p-Si fabricated by vacuum evaporation method for solar cell applications, NeuroQuantology, 18 (2020) 26. https://doi.org/10.14704/nq.2020.18.1.NQ20103

[65] X. Ji, J. Song, T. Wu, Y. Tian, B. Han, X. Liu, H. Wang, Y. Gui, Y. Ding, Y. Wang, Fabrication of high-performance F and Al co-doped ZnO transparent conductive films for use in perovskite solar cells, Solar Energy Materials and Solar Cells, 190 (2019) 6-11. https://doi.org/10.1016/j.solmat.2018.10.009

[66] S. Agrawal, R. Rane, S. Mukherjee, ZnO thin film deposition for TCO application in solar cell, in: Conference Papers in Science, Hindawi, 2013. https://doi.org/10.1155/2013/718692

[67] J.A. Aranovich, D. Golmayo, A.L. Fahrenbruch, R.H. Bube, Photovoltaic properties of ZnO/CdTe heterojunctions prepared by spray pyrolysis, Journal of Applied Physics, 51 (1980) 4260-4268. https://doi.org/10.1063/1.328243

[68] D. Rogers, F. Hosseini Teherani, A. Yasan, K. Minder, P. Kung, M. Razeghi, Electroluminescence at 375 nm from a Zn O/ Ga N: Mg/ c-Al 2 O 3 heterojunction light emitting diode, Applied physics letters, 88 (2006) 141918. https://doi.org/10.1063/1.2195009

[69] X.-L. Guo, J.-H. Choi, H. Tabata, T. Kawai, Fabrication and optoelectronic properties of a transparent ZnO homostructural light-emitting diode, Japanese Journal of Applied Physics, 40 (2001) L177. https://doi.org/10.1143/JJAP.40.L177

[70] J. Lu, Z. Ye, G. Yuan, Y. Zeng, F. Zhuge, L. Zhu, B. Zhao, S. Zhang, Electrical characterization of ZnO-based homojunctions, Applied physics letters, 89 (2006) 053501. https://doi.org/10.1063/1.2245221

[71] J.H. Lim, C.K. Kang, K.K. Kim, I.K. Park, D.K. Hwang, S.J. Park, UV electroluminescence emission from ZnO light-emitting diodes grown by high-temperature radiofrequency sputtering, Advanced Materials, 18 (2006) 2720-2724. https://doi.org/10.1002/adma.200502633

[72] H.S. Das, R. Das, P.K. Nandi, S. Biring, S.K. Maity, Influence of Ga-doped transparent conducting ZnO thin film for efficiency enhancement in organic light-emitting diode applications, Applied Physics A, 127 (2021) 1-7. https://doi.org/10.1007/s00339-021-04339-6

[73] N.H. Nickel, E. Terukov, ZnO-A Material for Micro-and Optoelectronic Applications: Proceedings of the NATO Advanced Research Workshop on ZnO as a Material for Micro-and Optoelectronic Applications, held in St. Petersburg, Russia, from 23 to 25 June 2004, Springer Science & Business Media, 2006. https://doi.org/10.1007/1-4020-3475-X

[74] R. Ondo-Ndong, G. Ferblantier, F. Pascal-Delannoy, A. Boyer, A. Foucaran, Electrical properties of ZnO sputtered thin films, Microelectronics journal, 34 (2003) 1087-1092. https://doi.org/10.1016/S0026-2692(03)00198-8

[75] S. Shionoya, W.Y.P. Handbook, Phosphor Research Society, in, CRC Press, 1998.

[76] H. Nanto, H. Sokooshi, T. Usuda, Smell sensor using ZnO thin films prepared by magnetron sputtering, in: TRANSDUCERS'91: 1991 International Conference on Solid-State Sensors and Actuators. Digest of Technical Papers, IEEE, 1991, pp. 596-599.

[77] Ü. Özgür, X. Gu, S. Chevtchenko, J. Spradlin, S.-J. Cho, H. Morkoç, F. Pollak, H. Everitt, B. Nemeth, J. Nause, Thermal conductivity of bulk ZnO after different thermal treatments, Journal of electronic materials, 35 (2006) 550-555. https://doi.org/10.1007/s11664-006-0098-9

[78] Z.L. Wang, ZnO nanostructures: growth, properties and applications, Journal of physics: condensed matter, 16 (2004) R829. https://doi.org/10.1088/0953-8984/16/25/R01

[79] S.A. Kumar, S.M. Chen, Nanostructured ZnO particles in chemically modified electrodes for biosensor applications, Analytical Letters, 41 (2008) 141-158. https://doi.org/10.1080/00032710701792612

Materials Research Forum LLC
https://doi.org/10.21741/9781644902394-10

Chapter 10

Future Applications and Perspective of ZnO

Pawan Kumar[1*], Nikhil Thakur[1], Pankaj Sharma[2], Raman Kumar[3]

[1]School of Physics and Materials Science, Shoolini University, Solan 173229, India

[2]Applied Science Department, National Institute of Technical Teachers Training and Research, Sector 26, Chandigarh 160019, India

[3]Department of Physics, Govt. College Dhaliara, Kangra (HP) 177103, India

* kumar74pawan@gmail.com

Abstract

Zinc Oxide (ZnO) is recognized as an outstanding material for preparation of highly specific electrochemical sensors as well as biosensors because of their attractive characteristics like large specific surface area, powerful adsorption ability, and large catalytic efficiency. As a result, ZnO nanostructures are frequently employed to make effective electrochemical sensors as well as biosensors for detecting several analytes. ZnO is a versatile material that has a wide range of applications. The present chapter emphasizes on the current advancements of ZnO-based nanomaterials in the area of energy conversion & storage as well as biological applications. Supercapacitors, Li-ion batteries, and also biomedical applications have all been given special consideration. Lastly, future applications of ZnO-derived materials in fields of energy as well as biological sciences are thoroughly studied.

Keywords

ZnO, Batteries, Biosensors, Supercapacitors, Biomedical, Biological Applications

Contents

1. Introduction

Nanotechnology and nanoscience are terms used to describe the study and manufacturing of extremely small materials or structures that have applications in a variety of fields like physics, chemistry, as well as materials science. Nanotechnology advancements and their applications in several fields have resulted in the formation of novel approaches in diagnostic purposes [1], environmental control [2], pharmaceutical analysis [3]. and food security [4]. Nanostructures have become useful in analytical areas, resulting in characteristics such as large selectivity, large sensitivity since bulk materials lack various particular physiochemical aspects such as high surface ratio or volume ratio, adsorption, as well as reactivity [5, 6].

In high-efficiency electronics, ECS (Energy Conversion and Storage), and environmental cleanup applications, semiconducting metal oxides have indeed been recognized as a significant nanostructured material [7]. Among several metal oxide materials, zinc oxide (ZnO) has received a lot of interest. In materials science, ZnO is categorized as a n-type semiconductor in group II-VI, having a covalence that falls somewhere between ionic or covalent semiconductors. Excellent thermal and mechanical durability at room temperature, as well as a significant excitation binding energy of 60 meV, have made this material appropriate for use in optoelectronics, laser innovative technologies, and electronics [8-10]. Additionally, because of its piezoelectric and pyroelectric properties, ZnO can be used as a photocatalyst, converter, as well as energy producer in hydrogen generation [11, 12]. Because of its hardness, and piezoelectric constant, this material is even commonly employed in the ceramics sector. Nanostructured ZnO materials have received a lot of interest because of their applicability in photonics, optics, as well as in electronics. Furthermore, the remarkable properties of ZnO nanostructure including bio - compatibility, broad band gap of 3.37 eV, chemical and photochemical durability, and large electron communication which make it an ideal material for the development of effective sensors and biosensors. Biosensors as well as electrochemical sensors are presently in widespread usage for a variety of purposes.

The current advancements of ZnO-based nanomaterials in energy as well as biological applications are discussed in this chapter. ZnO-based materials have received special attention for energy storage devices like Li-ion batteries, supercapacitors, and also biological applications like biosensors as well as biomedical applications.

2. Applications of ZnO

2.1 Supercapacitors

The electrochemical supercapacitor is a type of energy storage system that can be used to replace present energy-related technology. It has been regarded as a promising energy storage system in past few years because of its higher power density, durability, and efficiency compared to current energy-related technologies [13, 14]. Transition metal oxides (TMOs) have been recognized as prospective electrode materials in supercapacitors and many other energy-associated applications in past few years [15, 16]. ZnO, a TMOs with quite a high energy density of approximately 650 A hg^{-1}, is widely acknowledged like a battery active material. ZnO also has a high electrical conductivity of approximately 230 S/cm (siemens per centimeter), which is higher than several other metal oxides. In this context, there are a couple of studies on supercapacitors electrode materials relying on ZnO. The hydrothermally produced zinc oxide electrode materials for supercapacitors were described by Ridhuan and colleagues [17]. Moreover, ZnO's weak electrical conductivity and lower specific surface area continues to be a serious challenge, restricting rate capability for high power efficiency and therefore, restricting its widespread usage in energy storage systems. Composite substances are being placed onto ZnO electrode materials to address this problem. To boost certain capacitances, making a hybrid composite electrode appears to be a viable method. The analyzed improvement is attributed to the composites' synergetic effects of individuals by decreasing particle size, enhancing the number of active sites and improving specific surface area. ZnO-derived composite materials have several advantages, including low cost, environmentally friendly nature, excellent electrochemical reversibility, ease of processing, outstanding specific capacitance, proper cycling durability, higher power density, making them a viable candidate for electrochemical supercapacitors. In this chapter, we discuss latest results of various ZnO-derived composite substances, including ZnO/graphene, ZnO/nickel oxide, ZnO/carbon, ZnO/manganese dioxide, as well as others.

2.1.1 ZnO/graphene

Graphene has drawn a lot of attention in recent decades because of its different two-dimensional nanostructure, outstanding electrical conductivity, thermal conductivity, good electrochemical stability, and remarkable theoretical specific surface area. Nanosheets of graphene, as a result, have emerged as one of the most promising materials for use as electrode supportive materials in supercapacitors as well as in certain different electrochemical energy storage systems. However, inhibiting aggregation as well as restacking is critical for progressing the use of graphene in energy storage devices [18, 19]. RGO (Reduced graphene oxide), in specific, is widely utilized in place of graphene, because of its cheap price as well as great manufacturing capabilities in the development of excellent graphene materials [20]. Furthermore, as a result of the Van der Waals interaction, RGO appears to be particularly susceptible to aggregation and agglomeration issues, resulting in a significant reduction of effective SSA and hence minimizing capacitance. Because of its large-area applicability in supercapacitors, great emphasis is currently being placed on increasing graphene's specific surface area.

Lu and colleagues reported a thin film based on graphene-ZnO produced using an ultrasonic spray pyrolysis process that attained a specific capacitance of approximately 635 Fg^{-1}. Lately, Haldorai et al. [21] observed an improved specific capacitance of approximately 314 Fg^{-1} for graphene-ZnO nanocomposites. Furthermore, due to the existence of spacers, Du et al. [22] was able to effectively manufacture honeycomb-like zinc oxide/reduced graphene oxide composites with enhanced capacitive properties of approximately 231.0 F/g^{-1} at 0.1 Ag^{-1}. Saranya et al. [23] observed graphene-ZnO nanocomposites having a greater specific capacitance of approximately 122 F/g at five millivolt per second that were created using a solvothermal method. Li et al. [24] created a sandwiched hybrid layout of zinc oxide/reduced graphene oxide/zinc oxide materials, achieving a specific capacitance of around 275 Fg^{-1} at five millivolt per second. The hybrid zinc oxide/reduced graphene oxide/zinc oxide assembly exhibits outstanding rate capability as well as impressive long-run cycle life in comparison to reduced graphene oxide. Compared to individual specific capacitance, it is predicted that the combination of graphene with zinc oxide will significantly increase worldwide specific capacitance. Jayachandiran and colleagues [25] observed reduced graphene oxide-ZnO through the use of ultrasonic technique and attained a larger specific capacitance of around 312 Fg^{-1} with enhanced cycling durability of maximum one thousand cycles. Rajeshwari and colleagues [26] observed reduced graphene oxide-ZnO nanocomposites for supercapacitors using a hydrothermal method and achieving a capacitance value of almost 719.2 Fg^{-1} at five millivolt per second. Moreover, enhancing the electrochemical characteristics of graphene-ZnO to fulfill a wide range of supercapacitor application requirements remains a challenge. As a result of the huge number of defects as well as the inclusion of a chemical component in the fabrication technique, the skills of graphene-based supercapacitors are severely limited. However, a much more enhanced synthetic method to fabricating graphene-ZnO nanostructures has still been needed, as graphene always exhibits low electrical conductivity. Many people have been interested in 3D-graphene in past few years since it is suitable for the quick transmission of electrical ions as well as has a high potential for solving the aforementioned challenges [27].

Recently, Li and colleagues [28] used a hydrothermal approach to fabricate 3D-graphene/ZnO nanorods, achieving a specific capacitance of around 554 Fg^{-1} at five millivolt per second, with just 6.6 percent of the available capability attenuated over 2300 cycles. Fig. 1 depicts the detailed investigation of the electrochemical characteristics of 3D-graphene/ZnO composite electrodes. When compared to different composited ZnO-graphene substances, the electrochemical characteristics of the 3D-graphene/ZnO composites are extra balanced, with enhanced specific capacitance as supercapacitor electrodes.

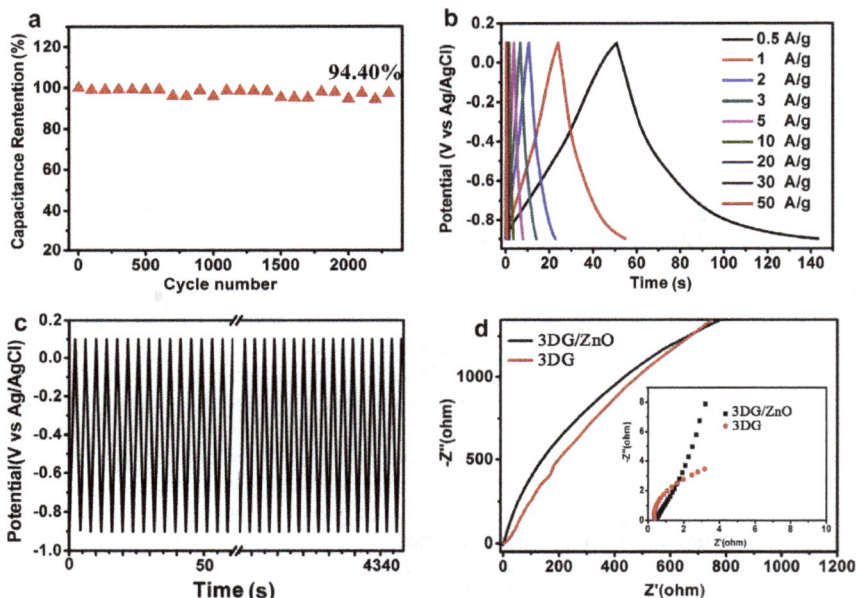

Figure 1: The 3D-graphene/ZnO electrodes' electrochemical performance (a) At a current density of ten Ag^{-1} of two thousand three hundred (2300) cycles, 3D-graphene/ZnO capacitance was observed, (b) The composite electrode's galvanostatic charge/discharge behavior at various current densities, (c) At a current density of ten Ag^{-1}, the composite electrodes' charge/discharge behavior is shown, and (d) Nyquist plot of 3D-graphene electrode as well as 3D-graphene/ZnO composite [28]. Reprinted with permission from Elsevier.

2.1.2 ZnO/Carbon

Many carbon composites, as well as ZnO, can be utilized as supercapacitor electrode materials to overcome the disadvantages of ZnO electrode materials. More research is being conducted to establish ZnO/carbon-derived materials. In this context, Zhang et. al [29] highlighted the ZnO/carbon nanotube composite material, that attained a specific capacitance of approximately 324 Fg^{-1}. Kalpana et al. [30] created a ZnO/CA composite material with a specific capacitance of around 500 Fg^{-1}. Selvakumar et. al [31] used a sol-gel method to make the ZnO/AC composite material, which have a specific capacitance of approximately 160 Fg^{-1} having a stability of five hundred cycles. Sasirekha and colleagues [32] fabricated ZnO/C composite materials for supercapacitors with the help of green chemistry, achieving a specific capacitance of around 820 Fg^{-1} at 1.0 Ag^{-1}. Furthermore, ZnO/C composite materials sustain ninety two percent of capacitance preservation up to four hundred cycles, as well as its symmetric supercapacitors system has a specific capacitance of approximately 92 Fg^{-1} at 2.5 Ag^{-1} as well as an energy density of around 32.78Wh/kg. Li and colleagues [33] described the fabrication of AC/ZnO composites with the help of hydrothermal methods, with a specific capacitance of around 117 Fg^{-1} at 0.5 Ag^{-1}. Using a simple low-temperature water-bath process, Xiao et al. [34] effectively produced CS@ZnO nanocomposites with a specific capacitance of approximately 630.0 F/g at 2.0 A/g.

2.1.3 ZnO/MnO₂

Manganese dioxide is an attractive electrode candidate for supercapacitors due to its higher specific capacitance of approximately 1370.0 Fg^{-1}, various oxidation states of manganese, abundant natural resources, as well as cheap price [35]. However, the weak conductivity of MnO_2 (10^{-5} to 10^{-6} S/cm) must be corrected before it can be employed in supercapacitors. MnO_2 has a low conductivity, which affects electrochemical performance. To address these drawbacks, MnO_2 must be paired with multifunctional materials like ZnO, that offer both an electron transport channel and robust mechanical support [36]. In this regard, Yang et al. [37] used an annealing/hydrogenation process to generate $ZnO-MnO_2$ nanocables with an improved capacitance of approximately 138.70 mF/cm^2 at 1.0 mA/cm^2. Sun and colleagues [38] also succeeded in fabricating a three dimensional $ZnO@MnO_2$ nanocomposites with a specific capacitance of around 31 mF/cm^2 with a core-shell configuration. Radhamani and colleagues [39] investigated the $ZnO@MnO_2$ electrode materials for supercapacitors and found that they had a specific capacitance of around 907 Fg^{-1} at 0.60 Ag^{-1}. In addition, the packed $ZnO@MnO_2$ system has a large power density of around 6.50 kW/kg as well as a large energy density of approximately 17 Wh/kg. Fig. 2 shows that the asymmetric supercapacitors (ASCs) system values outperformed several symmetric as well as asymmetric supercapacitors made of MnO_2.

NiO has been intensively researched for supercapacitors as an electrode material because of its minimum price, large theoretical capacity of approximately 2573 Fg^{-1}, as well as high power density [40, 41].

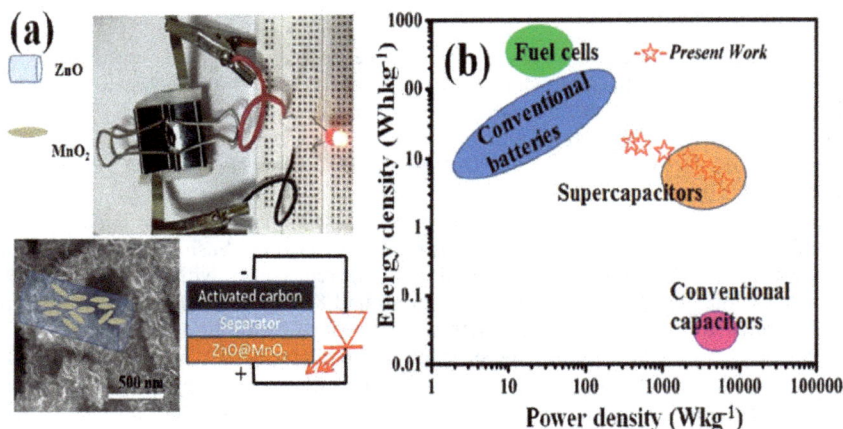

Figure 2: (a) Illumination of light emitting diode by a lab-scale prototype ASC system, and (b) power density vs energy density for AC\ZnO@MnO₂-AC asymmetric supercapacitor in a Ragone plot with conventional capacitors, supercapacitors, conventional batteries as well as fuel cells [39]. Reprinted with the permission from ACS.

Despite this, NiO's low surface area restrict its use in energy-related equipments. In order to solve these problems, three-dimensional nanostructured electrode substances have been extensively studied to reduce electrolyte diffusion distance within pseudocapacitors electrodes [42]. In this context, Pang and colleagues. [43] created porous zinc oxide–nickel oxide nanocomposites and accomplished a specific capacitance of approximately 649 Fg^{-1} with proper cycle stability till four hundred cycles. For electrochemical supercapacitors, table 1 compares the specific capacitance of several types of zinc oxide nanocomposites.

2.2   Lithium-ion batteries (LIBs)

Energy storage systems including batteries as well as supercapacitors are the definitive forms of energy for the future survivors because of their ease of maintenance, extended cycle life, and variable power & energy properties. Owing to rising energy demands in areas like electronics, hybrid vehicles, as well as renewable power generation devices, rechargeable LIBs have received a lot of interest. LIBs stayed the most popular forms of energy for a variety of uses, due to their relatively huge energy density. ZnO has been regarded an impressive potential anode because of its large theoretical capacity of approximately 978 mAh g^{-1}, abundance in nature, minimal toxicity, and cheap price. However, because of the large volume expansion all through the electrochemical lithiation/delithiation procedure, which causes the lack of electrical connections across the anode particles, its practical application is hampered [46, 47].

As a result, multiple attempts have been devoted to prevent capacity degradation and improve extended cycle life, including nano-architectures (one-dimensional, two-dimensional and three-dimensional), diverse surface morphologies, doping with multiple components, and the inclusion of conductive agents. In terms of cathode materials, ZnO may be combined with a variety of lithium storage materials such as $LiMn_2O_4$ (lithium manganese oxide) and $LiFePO_4$ (lithium-ion iron phosphate) [48] to achieve stable as well as high capacities. As a result, the latest work on ZnO for LIB applications is presented below, with diverse structures produced by different processes.

Table 1: Electrochemical properties of ZnO-derived electrodes are compared.

Electrodes	Year	Capacitance	Scan rate per current density	Electrolytes	References
rGO/ZnO	2018	635.00 F/g	5.0 mV/s	1.0 M KOH	[44]
ZnO/C	2018	820.00 F/g	1.0 A/g	1.0 M N_2SO_4	[32]
CS@ZnO core shell	2017	630.00 F/g	2.0 A/g	6.0 M KOH	[34]
3DG/ZnO	2015	554.23 F/g	5.0 mV/s	1.0 M KOH	[28]
ZnO/RGO	2014	314.00 F/g	100.0 mV/s	2.0 M KOH	[21]
MnO_2 nanowire/ZnO	2014	746.70 F/g	2.0 mV/s	1.0 M N_2SO_4	[45]
ZnO/RGO	2014	231.30 F/g	0.10 A/g	6.0 M KOH	[22]
ZnO-NiO composite	2012	649.00 F/g	5.8 A/g	3.0 M KOH	[43]
ZnO/CNT	2009	323.90 F/g	50.0 mV/s	2.0 M KOH	[29]

2.2.1 ZnO/Graphite

Dall'Asta and coworkers [49] created a variety of nanorods (one-dimensional), single- or multilayered nanosheets (two-dimensional), and nano brushes (three-dimensional) as anodes for LIBs. On platinum plated stainless steel substrates, multiple forms of nano-architecture were generated using a simple wet chemical technique. When compared to a number of nano-architecture materials, it was discovered that two-dimensional single-layered nanosheets had a higher reversible capacity. That might be because of the tiny nanoparticles as well as the existence of a mesoporous structure, which maximizes the particular area to promote alloying production and provides ample room for volume adjustments throughout cycling. Quartarone et al. [50] proposed binder-free unique graphite-coated zinc oxide nanosheets for Li-ion battery applications with good capacity and stability. The hybrid ZnO nanosheets was formed on a stainless-steel disc using a hydrothermal technique at sixty degrees Celsius for six hours, then annealed at three hundred fifty degrees Celsius for one hour in air. The thermal evaporation technique was then used to coat ZnO nanosheets with two separate (G_{110} and G_{350}) graphite sheet thicknesses. After hundred cycles, G_{350} thickness graphite sheet coated zinc oxide nanosheets shows better capacity of approximately six hundred mAh g^{-1} having a minimal capacity fading at around one Ag^{-1}. Feng et al. [51] showed varied amounts of graphene inserted into ZnO-graphene composite materials produced by hydrothermal technique (0.2 gram (43.08 wt. percent), 1.0 gram (60.11 wt. percent), and 1.8 gram (71.45 wt. percent)) for Li-ion battery applications. The ZnO-graphene hybrid composites were evaluated at various cut-off voltages ranging from 0.01 volt to 3.0 volt in attempt to better examine the anodic behavior for commercial application of lithium-ion batteries. The results show that graphene added ZnO with 71.45 weight percent has a higher reversible specific capability as well as high-rate performance as compared to many different composites. It also possesses a significant reversible specific capacity of approximately 240 mAh/g at a minimal cut off voltage between 0.01 to 1.0 volts, making it an excellent anode for Li-ion batteries. This outstanding performance was attributed to graphene, which could buffer volume growth during the cycling procedure.

2.2.2 ZnO/Carbon

Using a hydrothermal technique followed by post-heat processing under nitrogen gas atmosphere of zinc citrate ($Zn_3(C_6H_5O_7)_2$) intermediate materials, Xiao and colleagues [52] prepared ZnO nanoparticles enclosed in a three-dimensional hierarchical carbon framework unique material. Raw ZnO as well as ZnO@C were made by calcining $Zn_3(C_6H_5O_7)_2$ at 600°C for four hours under both air and nitrogen gas, in order to analyze the ZnO@CF. The ZnO@CF composite has a higher specific capacity, a better lifespan, and superior rate performance in comparison to the raw ZnO as well as ZnO@C spheres. The enclosed three-dimensional carbon framework can promote efficient electronic connections by providing electron routes, or even the three-dimensional carbon framework's strong and flexible structure helps to buffer the stress as well as volume variations that takes place in electrode materials throughout Li^+ insertion/extraction. Zhao and colleagues [53] used a one-step electrospinning process to intercalate ZnO/CNFs-PA (petroleum asphalt), resulting in unique three-dimensional linked nanofiber films. Films were made by solvent impregnation and following two-step thermal processes with PA amounts of 0.40 grams, 0.60 grams, 1.00 grams, 1.30 grams, and 3.00 grams to analyze the effect of PA inclusion into ZnO/CNFs films. In terms of electrochemical performance, ZnO/CNFs-PA-1.0 outperformed the others. This could be attributed to a unique three-dimensional conductive network structure that effectively permits volume expansion while also facilitating lithium-ion kinetics. Taegyeong et al. [54] described vapor-assisted fabrication of ZnO/N-CF composites with two-step thermal processing. Using a vapor-assisted technique, raw ZnO as well as ZnO/CF samples were created, and by dissolving the zinc oxide from the ZnO/N-CF with twenty-five milliliters of 0.1 molar of HCl solution, etched N-CF was created. In terms of reversible capacity, long cycle life, and rate performance, the ZnO/N-CF composite outperforms the others. Liang et al. [55] described the hydrothermal preparation of nanostructured zinc oxide microspheres as anode materials for Li-ion batteries. To make the carbon-coated zinc oxide microspheres, distinct concentrations of dopamine hydrochloride (20 milligrams, 30 milligrams, and milligrams) were added to the zinc oxide microspheres, which were then thermally treated in N_2. This method has been shown to be a restricted crystallisation strategy for increasing the crystalline nature of carbon coated zinc oxide microspheres. It was discovered that ZnO@C-30 performed better electrochemically as compared to other materials, which might be attributed to a higher degree of crystallisation that could handle volume variations during the cycling procedure. Sun et al. [56] describe a limited growth technique for the production of a new ZnO/carbon composite made up of tiny ZnO nanoparticles whose expansion is restricted in rationally constructed ZnO/NMPCS (N-rich mesoporous carbon nanospheres). The aerogel-assisted co-polymerization technique was used to synthesize NMPCS, which was then carbonized and washed. A hydrothermal approach was utilized, followed by a calcination technique, to develop a new ZnO/NMPCS material. This type of ZnO/NMPCS has various advantages, including an interconnected spherical framework having a huge surface area, the ability to obtain a uniform dispersion of ZnO nanoparticles. Figure 3 shows a schematic diagram for preparing SEM (scanning electron microscope) and TEM (transmission electron microscope) images of NMPCS as well as ZnO/NMPCS composites. Due to the presence of such advantages during electrochemistry, it is available to produce enough Li^+ storage sites while also speeding up the kinetic reaction process, resulting in a large capacity with outstanding cyclability and quick rate capabilities.

2.3 Biomedical applications

ZnO nanoparticles have attracted interest in a number of biomedical areas as a novel kind of less-toxic, minimum price nanomaterial, which include anticancer activity, antibacterial activity, antioxidant activity, antidiabetic activity, and anti-inflammatory activity, along with drug delivery or even bioimaging [57, 58]. ZnO nanoparticles with lower than hundred nanometers are regarded generally biocompatible, which supports their biomedical uses and is a significant property in encouraging biomedical study. The latest advancement on the usage of ZnO nanoparticles in biomedicine is summarized in this section.

2.3.1 Anticancer activity

The use of NPs in drug delivery opens up new possibilities for cancer treatment that are both safe and beneficial. The incapability of anticancer drugs to differentiate between safe & cancerous cells is the biggest hurdle to the rapid advancement of cancer therapy techniques. This is a common source of chemotherapy-related problems and adverse effects. ZnO Nanoparticles are appealing because of their nontoxicity and biodegradability especially in comparison to other nanomaterials. Zn is a crucial trace component that controls the action of several enzymes to preserve homeostasis within the body, hence ZnO nanoparticles have attracted a lot of attention in the cancer drug delivery. Zn is also involved in humeral as well as cellular immunity, that helps to prevent cells from cancer. Zn shortage promotes the beginning and spread of cancer cells through DNA mutation as well as p53 disruption [59]. Especially in comparison to bulk Zn materials, ZnO Nanoparticles have EPR (enhanced permeability & retention) effects towards cancer cells, and they can destroy cancer cells by producing ROS (reactive oxygen species) [60]. Nanomaterial-based nanomedicine in the past few years showed the ability to eliminate these negative effects due to its great biocompatibility, cancer targeting, ease of surface functionalization, as well as drug delivery capability. Adults require Zn^{2+}, and ZnO nanoparticles are believed to be harmless in vivo. Based on these benefits, ZnO nanoparticles can be investigated for treatment of cancer [61, 62]. Table 2 shows ZnO nanoparticles anticancer activity in various cancers.

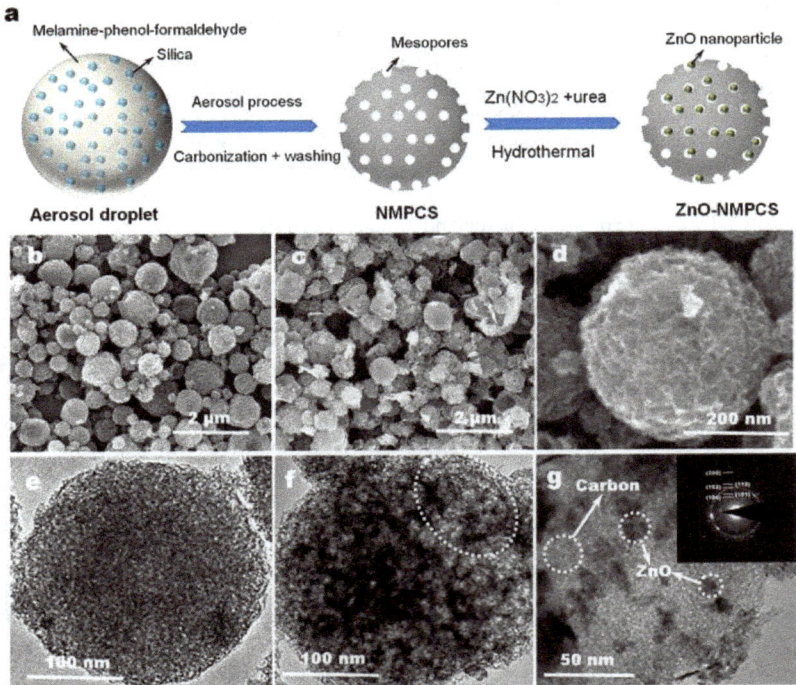

Figure 3: (a) The formation procedure of NMPCS as well as ZnO-NMPCS is depicted schematically, (b) NMPCS SEM image at low magnification, (c) SEM image of ZnO-NMPCS at lower magnification, (d) SEM image of ZnO-NMPCS at higher magnification, (e) An individual NMPCS particle as seen through a TEM, (f) TEM image of an individual ZnO-NMPCS particle, and (g) ZnO-NMPCS high resolution transmission electron microscopy (HRTEM) image; the SAED design for this is shown in the inset [56]. Reprinted with permission from Elsevier.

Table 2: ZnO nanoparticles anticancer activities in various human cancer cell lines.

Cancer's Type	Effects and Mechanism
Hepatocarcinoma	In HepG2 cells, ZnO nanoparticles produced oxidative DNA damage and reactive oxygen species, which resulted in mitochondrial-mediated apoptosis [63]. In SMMC-7721 cells, the Dox-ZnO (Doxorubicin-Zinc Oxide) nanocomplex can be used as a drug delivery method to improve Dox internalization [64].
Lung cancer	Zinc oxide nanoparticles in liposomes not just made the transport carrier pH responsive, but they also had a synergistic chemo-photodynamic anticancer effect [65]. ZnO NP20 as well as Al-ZnO NP20 were found to be toxic to human lungs with an epidermal growth factor receptor (EGFR) mutation, resulting in non-autophagic cell loss [66].
Colon cancer	In Caco-2 cells, ZnO nanoparticles and fatty acids may cause lysosomal instability [67]. Caco-2 cells were cytotoxic after being exposed to ZnO nanoparticles, which resulted in a rise in intracellular zinc ions [68].
Breast cancer	In resistant MCF-7 (Michigan Cancer Foundation-7) cells, a ZnO nanomaterial-based doxorubicin delivery method can bypass the P-glycoprotein rise in drug accumulation [69]. In an acidic pH environment, FA-functionalized PTX (paclitaxel)-ZnO nanocarriers released approximately seventy five percent of the PTX payload inside 6 hours, improving chemotherapy tolerance [70].
Gastric cancer	PMMA-AA/ZnO nanoparticles as well as PMMA-PEG/ZnO nanoparticles were both capable of transporting a considerable proportion of the hydrophobic drug, which had a strong anti-gastric cancer effect [71, 72].

2.3.2 Antibacterial activity

ZnO Nanoparticles can be used as an antibacterial material due to its exceptional features, including a large specific surface area as well as excellent activity to prevent a variety of pathogenic agents. However, until recently, nothing was known about ZnO Nanoparticles' antibacterial activity. Prior findings had shown that major antibacterial toxicity mechanisms of zinc oxide nanoparticles mainly focused on their potential to produce excess ROS synthesis, including superoxide anion, production of H_2O_2, and hydroxyl radicals, as illustrated in Figure 4 [73].

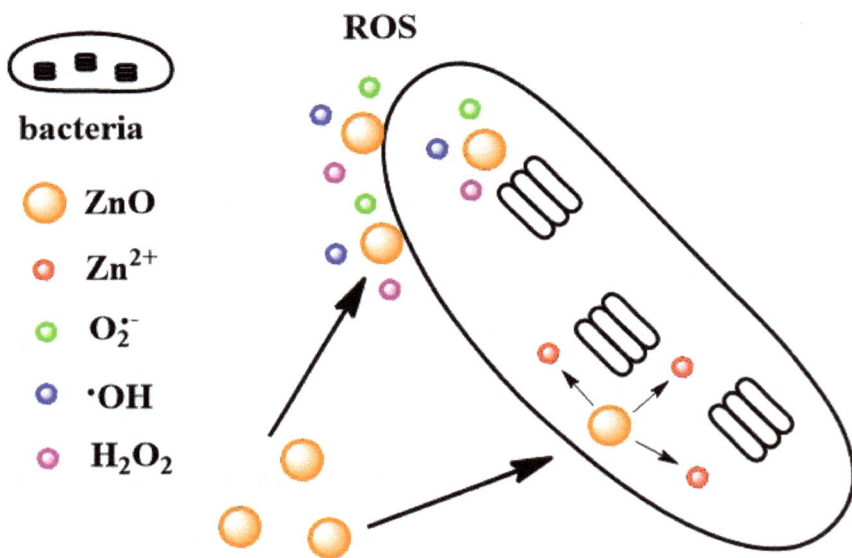

Figure 4: Schematic illustration of antibacterial activity of zinc oxide nanoparticles [73].

The antibacterial activity could be due to the large amount of ZnO nanoparticles in bacterial cells' external membranes or cytoplasm, that would stimulate Zn^{2+} release, causing bacterial cell membrane breakdown, genomic instabilities, as well as, membrane protein degradation ending in bacterial cell death [74-76]. Table 3 shows ZnO Nanoparticles' antibacterial activity against various bacterial species.

Table 3: The antibacterial activity of zinc oxide nanoparticles against a variety of bacterial species.

Name of Material	Material's Size	Bacterial Species	Antibacterial Mechanism
ZnO Nanoparticles	30.0 nanometers	*E. coli*	Damages both the membrane's integrity and the generation of ROS [75].
ZnO Nanoparticles	~80.0 nanometers	*V. cholera*	Production of ROS, enhanced permeabilization as well as DNA damage [77].
ZnO Nanoparticles	~20.0 nanometers	*E. coli* 11634	H_2O_2 [78]
Ag-ZnO composite	64.0 nanometers	*S. aureus*, GFP *E. coli*	The release of Ag^+ as well as Zn^{2+} by ROS [79].
ZnO QDs	4.0 nanometers	*E. coli* MG1655	$Zn2+$ is the primary cause of toxicity [80].
ZnO nanostructures	70-80 nanometers	*S. aureus*, S. *typhimurium*, as well as K. *pneumoniae*	Cell membranes are harmed by reactive oxygen species [81].
Ge-ZnO Nanoparticles	20.0 nanometers	P. *aeruginosa* and E. *faecalis*	Bacterial cell death was caused by the penetration of the cell [82].

Jiang et al. [75] discovered that ZnO nanoparticles have antibacterial properties against *E. coli*. It was discovered that ZnO nanoparticles with an approximate size of around thirty nanometer triggered cell death by immediately touching the membrane's phospholipid bilayer and disrupting its integrity. Reddy et al. [83] made ZnO nanoparticles with a size of approximately thirteen nanometer and tested their antibacterial activity against *E. coli* and *S. aureus*. The findings were described as follows: ZnO nanoparticles totally suppressed *E. coli* progress at concentrations of around 3.40 mM, but suppressed *S. aureus* progress at significantly lower concentrations of less than or equal to 1 mM.

Despite the fact that ZnO in nanoparticle format is a potential antibacterial agent with activity against either Gram-positive or Gram-negative bacteria, the specific antibacterial action of ZnO nanoparticles remains unknown. As a result, researching it in depth provides a great deal of theoretical and practical significance.

2.3.3 Anti-inflammatory activity

Inflammation is the body's complicated biological response to damaging stimuli including irritants, damaged cells, or even germs [84]. The anti-inflammatory properties of ZnO nanoparticles have received a lot of interest since the emergence of nanoparticles.

ZnO nanoparticles' anti-inflammatory activity has been found to be particularly efficient in the treatment of inflammatory illnesses. Nagajyothi et al. [85] reported a simple, affordable, and environmentally friendly ZnO nanoparticles made from the root extraction of P. tenuifolia, and studied the anti-inflammatory properties in Lipopolysaccharide stimulated RAW 264.7 macrophages. ZnO nanoparticles showed strong anti-inflammatory effect by reducing nitric oxide generation and associated protein expressions of iNOS (Inducible nitric oxide synthase), COX-2 (Cyclooxygenase 2), IL-1β, (Interleukin 1 beta), IL-6 (Interleukin 6) and TNF- α (Tumor necrosis factor). Thatoi et al. [86] used aqueous leaf extract of 2 mangrove plants (i.e., Heritiera fomes as well as Sonneratia apetala), to make ZnO nanoparticles beneath photo conditioning, and discovered that ZnO nanoparticles seemed to have a larger anti-inflammatory potential (79 percent) than silver nanoparticles (69.1 percent).

Table 4: ZnO nanoparticles with anti-inflammatory.

Types of nanoparticles	Size	Effects
ZnO Nanoparticles	-	In a mouse model of atopic dermatitis, ZnO nanoparticles had stronger anti-inflammatory activities, reducing pro inflammatory cytokines significantly [87].
ZnO-functionalized textile fibers	-	When patients of atopic dermatitis wore the ZnO textiles continuously for three days in a row, they saw a significant reduction in their atopic dermatitis symptoms, itching, and subjective sleep quality [88].
ZnO Nanoparticles	Spherical: 33.03 nanometers to 73.48 nanometers	ZnO nanoparticles reduced inflammation and suppressed the production of associated proteins in a dose-dependent manner [85].
TNTs/ZnO	-	The proliferation and adherence of macrophages were significantly inhibited by TNTs/ZnO [89].
ZnO Nanoparticles	69.4 nanometers ± 13.0 nanometers	In colitic mice, ZnO nanoparticles reduce proinflammatory cytokines interleukin 1 beta, tumor necrosis factor-α, as well as myeloperoxidase (MPO), and activate Nrf2 signaling [90].
ZnO nanoparticles and Ag nanoparticles	-	ZnO nanoparticles demonstrated a stronger anti-inflammatory potential (79 percent) when compared to Ag Nanoparticles (69.1 percent) [86].

ZnO and Their Hybrid Nano-Structures   Materials Research Forum LLC
Materials Research Foundations 146 (2023) 294-322   https://doi.org/10.21741/9781644902394-10

Table 4 summarizes the findings of anti-inflammatory activity of ZnO nanoparticles. As a result, ZnO Nanoparticles are capable of acting as an anti-inflammatory agent.

2.3.4 Biosensors

ZnO-derived nanostructured materials have demonstrated to be a great option for binding biological molecules in recent times. According to recent study, boosting the surface area to volume proportion of nanostructures for biosensing applications increases their reactivity, allowing them to translocate very effortlessly across cell membranes and effectively bind analyte. For biosensing and biological applications, scientists created an electrochemical sensor relying on ZnO nanostructures as well as its composites. The progress of biosensors has been shown in the use of ZnO as a transductor/transduction material. These ZnO nanostructures feature a unique surface selectivity that makes them ideal for biosensing. ZnO crystallizes in three different types: zinc blende, wurtzite, as well as rock salt; with the hexagonal wurtzite type being the most thermodynamically stable of the three [91]. Particularly in comparison to other metal oxides, ZnO anisotropic structure shows greater electron mobility as well as diffusion coefficient. As a result of this property, ZnO nanostructure is extremely beneficial as an electrochemical biosensor [92]. Nanolaminates are just one of ZnO's optical biosensor achievements. These nanomaterials exchange the qualities of a nanomaterial in that they have built-in properties such as biocompatibility, chemical uniformity, and, non-toxicity. For cysteine detection in various solutions, Kamaci et al. [93] reported a ZnO quantum dots-based fluorescence biosensor exhibiting great selectivity and responsiveness.

A single ZnO Nanowire has been studied to see if it can be used to make effective biosensors. L. M. Zhao et al. [94] suggested an extremely sensitive multipurpose biosensor focused on the SOR (Schottky & Ohmic reversible) structures that can switch between Schottky and Ohmic contacts because of oxygen vacancy diffusion in zinc oxide NMW (nano/micro wire) once pushed by TENG (triboelectric nanogenerator) voltage pulses.

Furthermore, ZnO complex is among the most appropriate material for biosensors. As shown in Figure 5, M. G. Zhao et al. [95] proposed a glucose biosensor made of nickel foam as well as ZnO/BiOI p-n circuit nanorods, that plays a significant part in acting as an enzyme.

Biosensors have proven to be a good choice for ZnO nanowire arrays. H. M. Kim and colleagues [96] developed a largely sensitive plasmonic biosensor using 3D structures made up of optical fiber, ZnO nanowires, and gold nanoparticles.

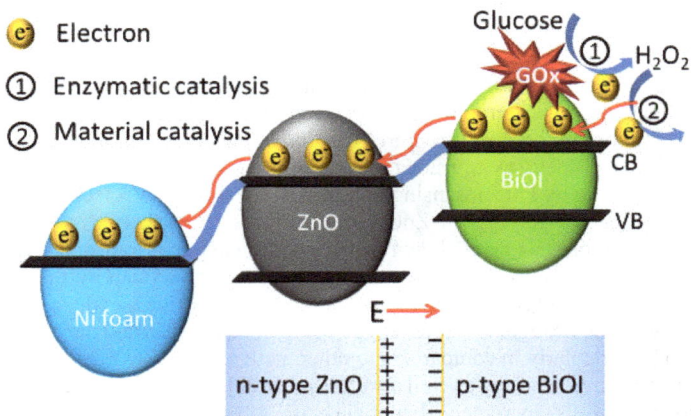

Figure 5: The BiOI/ZnO/Ni foam's effects on glucose sensing at the p-n junction [95]. Reprinted with permission from Elsevier.

As shown in Figure 6, the sensitivity of three-dimensional biosensors to prostate-specific antigen (PSA) was significantly enhanced in comparison to two dimensional biosensors.

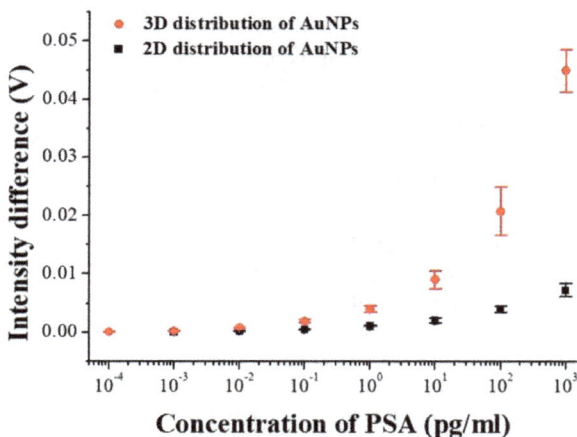

Figure 6: The intensity variations measured by each sensor in response to different PSA concentrations [96]. Reprinted with permission from Scientific Reports.

Materials Research Forum LLC
https://doi.org/10.21741/9781644902394-10

2.3.5 Bioimaging

ZnO material's intrinsic fluorescence characteristic and penetration capacity are often utilized in both in vitro as well as in vivo imaging. Zinc oxide materials' optical properties can be tuned by utilizing distinct dopants like cobalt, copper, and nickel [97]. Under Ultraviolet light, ZnO covered with polymer demonstrated stable luminescence without any toxic effect [98]. Gd (Gadolinium)-doped ZnO quantum dots with a size of six nanometers show maximum intensity at around 550 nanometers, are non-toxic to HeLa cells, and it can be imaged in vitro using MRI (magnetic resonance imaging) and confocal microscopy [99]. In a similar manner, Fe_3O_4-ZnO nanoparticle has been investigated for cancer imaging as well as for therapy [100].

2.3.6 Drug delivery

ZnO QDs have been found to exhibit exceptional luminous qualities in recent decades, in addition to their biocompatibility, minimal price, and low toxicity, have led to their widespread use. Because of these qualities, ZnO has become one of the most popular options for bioimaging as well as for drug delivery applications [101]. In addition to imaging, Nanomaterial made of ZnO can also be employed for drug delivery. The efficiency of ZnO nanomaterials enclosed with doxorubicin was seventy five percent, and it released rapidly at pH 7.4, indicating that it needs to be enhanced further for chemotherapeutic applications. The zinc oxide shell enhances cell internalization and may be comfortably identified using magnetic resonance imaging in mouse lymph nodes [102]. In general, organic dye molecules have enabled the researchers to identify and view a wide range of substances, including amino acids, medicines, nucleotides, either outside or within living cells. They've also been widely utilized to investigate the life chemistry in order to diagnose several diseases [103]. Zinc is more comfortably absorbed in the body when it is in the form of ZnO, which has tiny particle size. As a result, nano-ZnO is widely employed like a food additive. ZnO nanoparticles have received a lot of interest in the medical, biological, as well as clinical fields. ZnO has outstanding clinical and biological uses in comparison to other metal oxides, including anticancer, drug delivery, as well as diabetes treatment [104].

2.3.7 Gene delivery

Gene delivery is aided by the drug delivery system in a variety of applications. The plasmid containing the desired gene contributes in the targeting of a certain tissue [105]. Preventing DNA degradation and increasing cellular uptake efficiency are the two greatest issues in gene therapy. Several studies have shown that ZnO nanoparticles have an encouraging impact in gene delivery. The delivery of pEGFPN1 DNA to A375 living person melanoma cells was investigated using a three-dimensional ZnO nanostructure.

3. Summary and future perspective

The characteristics of nanostructured materials related with ZnO are diverse. The major characteristics include large IEP (isoelectric point) of 9.5, greater specific surface area, toxic free, large catalytic efficiency, broad band gap of 3.2 eV, etc. In terms of functionality and application, each characteristic is outstanding and unique in its own way. Because of its interesting features, various researchers have focused their efforts on the basic features of the creation, modification, and implementation of ZnO-derived nanostructures in energy as well as biological applications. We concentrated and offered a summative overview on ZnO derived nanostructures in this chapter, with a special emphasis on their applications in supercapacitors, lithium-ion batteries, biosensors, and biological areas.

ZnO-based nanostructures have been discovered to have good energy storage properties, indicating that they could be utilized as an electrode material in lithium-ion batteries as well as in supercapacitors. With an estimated energy density of around 650 A hg^{-1}, ZnO is typically considered as an active material in batteries. In addition, ZnO has an electrical conductivity of approximately 230 Scm^{-1} that is greater as compared to many other oxide materials. Under visible light exposure, ZnO-derived materials can also be used as photocatalysts to degrade dye molecules.

Nanostructured materials based on ZnO have demonstrated to be a good choice for connecting biological substances. According to a latest study, enhancing the surface area to volume proportion of nanostructures increases their reactivity for biosensing applications, allowing the system to translocate across cell membranes and effectively bind analyte. For biosensing as well as environmental applications, scientists created an electrochemical sensor centered on ZnO nanostructures. Biological applications for nanostructures based on ZnO typically involve bioimaging, drug delivery, gene delivery etc.

Materials based on ZnO have made significant development with various design processes; it is also suggested that these materials can be included using less harmful chemical sources. Efforts could also be made to create nanocomposites with diverse compositions, and the resulting composites should be subjected to structural, morphological, as well as application evaluation in a variety of energy, environmental, or even in biological areas. ZnO-based materials are feasible and efficient for future applications because to their easy synthesis technique and low price. Despite the fact that there are already several studies that have been conducted to develop ZnO-derived materials for a variety of energy storage applications, more research is needed to enhance long-term stability, large-scale manufacturing, and financial viability for commercial applications.

Despite the fact that nanostructured materials based on ZnO are a fantastic component for biosensors, biomedical, as well as biological applications, but there is still a lot of scope for experimental research to improve their performance and efficiency in an inter-disciplinary field. As a result, more research into the tuneable features of the materials based on ZnO in the fields of energy, the environment, and biology is still expected in the near future.

References

[1] V. Medawar-Aguilar, C.F. Jofre, M.A. Fernández-Baldo, A. Alonso, S. Angel, J. Raba, S.V. Pereira, G.A. Messina, Serological diagnosis of Toxoplasmosis disease using a fluorescent immunosensor with chitosan-ZnO-nanoparticles, Analytical biochemistry, 564 (2019) 116-122. https://doi.org/10.1016/j.ab.2018.10.025

[2] M.R. Willner, P.J. Vikesland, Nanomaterial enabled sensors for environmental contaminants, Journal of nanobiotechnology, 16 (2018) 1-16. https://doi.org/10.1186/s12951-018-0419-1

[3] D. Sharma, C.M. Hussain, Smart nanomaterials in pharmaceutical analysis, Arabian Journal of Chemistry, 13 (2020) 3319-3343. https://doi.org/10.1016/j.arabjc.2018.11.007

[4] V.D. Krishna, K. Wu, D. Su, M.C. Cheeran, J.-P. Wang, A. Perez, Nanotechnology: Review of concepts and potential application of sensing platforms in food safety, Food microbiology, 75 (2018) 47-54. https://doi.org/10.1016/j.fm.2018.01.025

[5] F. Emadi, A. Amini, Y. Ghasemi, A. Gholami, Graphene: recent advances in engineering, medical and biological sciences, and future prospective, Trends in Pharmaceutical Sciences, 4 (2018) 131-138.

[6] K.R. Reddy, H.M. Jeong, Y. Lee, A.V. Raghu, Synthesis of MWCNTs-core/thiophene polymer-sheath composite nanocables by a cationic surfactant-assisted chemical oxidative polymerization and their structural properties, Journal of Polymer Science Part A: Polymer Chemistry, 48 (2010) 1477-1484. https://doi.org/10.1002/pola.23883

[7] Y.-Y. Hu, Z. Liu, K.-W. Nam, O.J. Borkiewicz, J. Cheng, X. Hua, M.T. Dunstan, X. Yu, K.M. Wiaderek, L.-S. Du, Origin of additional capacities in metal oxide lithium-ion battery electrodes, Nature materials, 12 (2013) 1130-1136. https://doi.org/10.1038/nmat3784

[8] K. Rekha, M. Nirmala, M.G. Nair, A. Anukaliani, Structural, optical, photocatalytic and antibacterial activity of zinc oxide and manganese doped zinc oxide nanoparticles, Physica B: Condensed Matter, 405 (2010) 3180-3185. https://doi.org/10.1016/j.physb.2010.04.042

[9] W. Jeong, S. Kim, G. Park, Preparation and characteristic of ZnO thin film with high and low resistivity for an application of solar cell, Thin Solid Films, 506 (2006) 180-183. https://doi.org/10.1016/j.tsf.2005.08.213

[10] D. Vanmaekelbergh, L.K. Van Vugt, ZnO nanowire lasers, Nanoscale, 3 (2011) 2783-2800. https://doi.org/10.1039/c1nr00013f

[11] S. Nair, A. Sasidharan, V. Divya Rani, D. Menon, S. Nair, K. Manzoor, S. Raina, Role of size scale of ZnO nanoparticles and microparticles on toxicity toward bacteria

and osteoblast cancer cells, Journal of Materials Science: Materials in Medicine, 20 (2009) 235-241. https://doi.org/10.1007/s10856-008-3548-5

[12] A.A. Reinert, C. Payne, L. Wang, J. Ciston, Y. Zhu, P.G. Khalifah, Synthesis and characterization of visible light absorbing (GaN) 1-x (ZnO) x semiconductor nanorods, Inorganic chemistry, 52 (2013) 8389-8398. https://doi.org/10.1021/ic400011n

[13] K. Thiagarajan, J. Theerthagiri, R. Senthil, J. Madhavan, Simple and low cost electrode material based on Ca2V2O7/PANI nanoplatelets for supercapacitor applications, Journal of Materials Science: Materials in Electronics, 28 (2017) 17354-17362. https://doi.org/10.1007/s10854-017-7668-x

[14] K. Thiagarajan, J. Theerthagiri, R. Senthil, P. Arunachalam, J. Madhavan, M.A. Ghanem, Synthesis of Ni3V2O8@ graphene oxide nanocomposite as an efficient electrode material for supercapacitor applications, Journal of Solid State Electrochemistry, 22 (2018) 527-536. https://doi.org/10.1007/s10008-017-3788-8

[15] J. Theerthagiri, K. Thiagarajan, B. Senthilkumar, Z. Khan, R.A. Senthil, P. Arunachalam, J. Madhavan, M. Ashokkumar, Synthesis of hierarchical cobalt phosphate nanoflakes and their enhanced electrochemical performances for supercapacitor applications, ChemistrySelect, 2 (2017) 201-210. https://doi.org/10.1002/slct.201601628

[16] M.A. Ghanem, P. Arunachalam, M.S. Amer, A.M. Al-Mayouf, Mesoporous titanium dioxide photoanodes decorated with gold nanoparticles for boosting the photoelectrochemical alkali water oxidation, Materials Chemistry and Physics, 213 (2018) 56-66. https://doi.org/10.1016/j.matchemphys.2018.04.037

[17] N.S. Ridhuan, K. Abdul Razak, Z. Lockman, A. Abdul Aziz, Structural and morphology of ZnO nanorods synthesized using ZnO seeded growth hydrothermal method and its properties as UV sensing, PloS one, 7 (2012) e50405. https://doi.org/10.1371/journal.pone.0050405

[18] P. Tamilarasan, S. Ramaprabhu, Graphene based all-solid-state supercapacitors with ionic liquid incorporated polyacrylonitrile electrolyte, Energy, 51 (2013) 374-381. https://doi.org/10.1016/j.energy.2012.11.037

[19] D. Li, M.B. Müller, S. Gilje, R.B. Kaner, G.G. Wallace, Processable aqueous dispersions of graphene nanosheets, Nature nanotechnology, 3 (2008) 101-105. https://doi.org/10.1038/nnano.2007.451

[20] Y. Sun, Q. Wu, G. Shi, Graphene based new energy materials, Energy & Environmental Science, 4 (2011) 1113-1132. https://doi.org/10.1039/c0ee00683a

[21] Y. Haldorai, W. Voit, J.-J. Shim, Nano ZnO@ reduced graphene oxide composite for high performance supercapacitor: Green synthesis in supercritical fluid, Electrochimica Acta, 120 (2014) 65-72. https://doi.org/10.1016/j.electacta.2013.12.063

[22] G. Du, Y. Li, L. Zhang, X. Wang, P. Liu, Y. Feng, X. Sun, Facile self-assembly of honeycomb ZnO particles decorated reduced graphene oxide, Materials Letters, 128 (2014) 242-244. https://doi.org/10.1016/j.matlet.2014.04.126

[23] M. Saranya, R. Ramachandran, F. Wang, Graphene-zinc oxide (G-ZnO) nanocomposite for electrochemical supercapacitor applications, Journal of Science: Advanced Materials and Devices, 1 (2016) 454-460. https://doi.org/10.1016/j.jsamd.2016.10.001

[24] Z. Li, P. Liu, G. Yun, K. Shi, X. Lv, K. Li, J. Xing, B. Yang, 3D (Three-dimensional) sandwich-structured of ZnO (zinc oxide)/rGO (reduced graphene oxide)/ZnO for high performance supercapacitors, Energy, 69 (2014) 266-271. https://doi.org/10.1016/j.energy.2014.03.003

[25] J. Jayachandiran, J. Yesuraj, M. Arivanandhan, A. Raja, S.A. Suthanthiraraj, R. Jayavel, D. Nedumaran, Synthesis and electrochemical studies of rGO/ZnO nanocomposite for supercapacitor application, Journal of Inorganic and Organometallic Polymers and Materials, 28 (2018) 2046-2055. https://doi.org/10.1007/s10904-018-0873-0

[26] V. Rajeswari, R. Jayavel, A.C. Dhanemozhi, Synthesis and characterization of graphene-zinc oxide nanocomposite electrode material for supercapacitor applications, Materials Today: Proceedings, 4 (2017) 645-652. https://doi.org/10.1016/j.matpr.2017.01.068

[27] Y. Ito, M. Nyce, R. Plivelich, M. Klein, D. Steingart, S. Banerjee, Zinc morphology in zinc-nickel flow assisted batteries and impact on performance, Journal of Power Sources, 196 (2011) 2340-2345. https://doi.org/10.1016/j.jpowsour.2010.09.065

[28] X. Li, Z. Wang, Y. Qiu, Q. Pan, P. Hu, 3D graphene/ZnO nanorods composite networks as supercapacitor electrodes, Journal of Alloys and Compounds, 620 (2015) 31-37. https://doi.org/10.1016/j.jallcom.2014.09.105

[29] Y. Zhang, X. Sun, L. Pan, H. Li, Z. Sun, C. Sun, B.K. Tay, Carbon nanotube-ZnO nanocomposite electrodes for supercapacitors, Solid State Ionics, 180 (2009) 1525-1528. https://doi.org/10.1016/j.ssi.2009.10.001

[30] D. Kalpana, K. Omkumar, S.S. Kumar, N. Renganathan, A novel high power symmetric ZnO/carbon aerogel composite electrode for electrochemical supercapacitor, Electrochimica Acta, 52 (2006) 1309-1315. https://doi.org/10.1016/j.electacta.2006.07.032

[31] M. Selvakumar, D.K. Bhat, A.M. Aggarwal, S.P. Iyer, G. Sravani, Nano ZnO-activated carbon composite electrodes for supercapacitors, Physica B: Condensed Matter, 405 (2010) 2286-2289. https://doi.org/10.1016/j.physb.2010.02.028

[32] C. Sasirekha, S. Arumugam, G. Muralidharan, Green synthesis of ZnO/carbon (ZnO/C) as an electrode material for symmetric supercapacitor devices, Applied Surface Science, 449 (2018) 521-527. https://doi.org/10.1016/j.apsusc.2018.01.172

[33] Y. Li, X. Liu, Activated carbon/ZnO composites prepared using hydrochars as intermediate and their electrochemical performance in supercapacitor, Materials Chemistry and Physics, 148 (2014) 380-386. https://doi.org/10.1016/j.matchemphys.2014.07.058

[34] X. Xiao, B. Han, G. Chen, L. Wang, Y. Wang, Preparation and electrochemical performances of carbon sphere@ ZnO core-shell nanocomposites for supercapacitor applications, Scientific reports, 7 (2017) 1-13. https://doi.org/10.1038/srep40167

[35] M. Huang, F. Li, F. Dong, Y.X. Zhang, L.L. Zhang, MnO 2-based nanostructures for high-performance supercapacitors, Journal of Materials Chemistry A, 3 (2015) 21380-21423. https://doi.org/10.1039/C5TA05523G

[36] W. Zilong, Z. Zhu, J. Qiu, S. Yang, High performance flexible solid-state asymmetric supercapacitors from MnO 2/ZnO core-shell nanorods//specially reduced graphene oxide, Journal of Materials Chemistry C, 2 (2014) 1331-1336. https://doi.org/10.1039/C3TC31476F

[37] P. Yang, X. Xiao, Y. Li, Y. Ding, P. Qiang, X. Tan, W. Mai, Z. Lin, W. Wu, T. Li, Hydrogenated ZnO core-shell nanocables for flexible supercapacitors and self-powered systems, ACS nano, 7 (2013) 2617-2626. https://doi.org/10.1021/nn306044d

[38] X. Sun, Q. Li, Y. Lü, Y. Mao, Three-dimensional ZnO@ MnO 2 core@ shell nanostructures for electrochemical energy storage, Chemical communications, 49 (2013) 4456-4458. https://doi.org/10.1039/c3cc41048j

[39] A. Radhamani, K. Shareef, M.R. Rao, ZnO@ MnO2 core-shell nanofiber cathodes for high performance asymmetric supercapacitors, ACS Applied Materials & Interfaces, 8 (2016) 30531-30542. https://doi.org/10.1021/acsami.6b08082

[40] M. Huang, C. Gu, X. Ge, X. Wang, J. Tu, NiO nanoflakes grown on porous graphene frameworks as advanced electrochemical pseudocapacitor materials, Journal of power sources, 259 (2014) 98-105. https://doi.org/10.1016/j.jpowsour.2014.02.088

[41] P. Arunachalam, M.A. Ghanem, A.M. Al-Mayouf, M. Al-shalwi, O.H. Abd-Elkader, Microwave assisted synthesis and characterization of Ni/NiO nanoparticles as electrocatalyst for methanol oxidation in alkaline solution, Materials Research Express, 4 (2017) 025035. https://doi.org/10.1088/2053-1591/aa5ed8

[42] B.L. Ellis, P. Knauth, T. Djenizian, Three-dimensional self-supported metal oxides for advanced energy storage, Advanced Materials, 26 (2014) 3368-3397. https://doi.org/10.1002/adma.201306126

[43] H. Pang, Y. Ma, G. Li, J. Chen, J. Zhang, H. Zheng, W. Du, Facile synthesis of porous ZnO-NiO composite micropolyhedrons and their application for high power supercapacitor electrode materials, Dalton transactions, 41 (2012) 13284-13291. https://doi.org/10.1039/c2dt31916k

[44] M. Sreejesh, S. Dhanush, F. Rossignol, H. Nagaraja, Microwave assisted synthesis of rGO/ZnO composites for non-enzymatic glucose sensing and supercapacitor applications, Ceramics International, 43 (2017) 4895-4903. https://doi.org/10.1016/j.ceramint.2016.12.140

[45] S. Li, J. Wen, X. Mo, H. Long, H. Wang, J. Wang, G. Fang, Three-dimensional MnO2 nanowire/ZnO nanorod arrays hybrid nanostructure for high-performance and flexible supercapacitor electrode, Journal of Power Sources, 256 (2014) 206-211. https://doi.org/10.1016/j.jpowsour.2014.01.066

[46] S. Sinha, H.V. Ramasamy, D.K. Nandi, P.N. Didwal, J.Y. Cho, C.-J. Park, Y.-S. Lee, S.-H. Kim, J. Heo, Atomic layer deposited zinc oxysulfide anodes in Li-ion batteries: an efficient solution for electrochemical instability and low conductivity, Journal of Materials Chemistry A, 6 (2018) 16515-16528. https://doi.org/10.1039/C8TA04129F

[47] J. Zhang, P. Gu, J. Xu, H. Xue, H. Pang, High performance of electrochemical lithium storage batteries: ZnO-based nanomaterials for lithium-ion and lithium-sulfur batteries, Nanoscale, 8 (2016) 18578-18595. https://doi.org/10.1039/C6NR07207K

[48] Y. Hu, J. Yao, Z. Zhao, M. Zhu, Y. Li, H. Jin, H. Zhao, J. Wang, ZnO-doped LiFePO4 cathode material for lithium-ion battery fabricated by hydrothermal method, Materials Chemistry and Physics, 141 (2013) 835-841. https://doi.org/10.1016/j.matchemphys.2013.06.012

[49] V. Dall'Asta, C. Tealdi, A. Resmini, U.A. Tamburini, P. Mustarelli, E. Quartarone, Influence of the ZnO nanoarchitecture on the electrochemical performances of binder-free anodes for Li storage, Journal of Solid State Chemistry, 247 (2017) 31-38. https://doi.org/10.1016/j.jssc.2016.12.016

[50] E. Quartarone, V. Dall'Asta, A. Resmini, C. Tealdi, I.G. Tredici, U.A. Tamburini, P. Mustarelli, Graphite-coated ZnO nanosheets as high-capacity, highly stable, and binder-free anodes for lithium-ion batteries, Journal of Power Sources, 320 (2016) 314-321. https://doi.org/10.1016/j.jpowsour.2016.04.107

[51] Y. Feng, Y. Zhang, X. Song, Y. Wei, V.S. Battaglia, Facile hydrothermal fabrication of ZnO-graphene hybrid anode materials with excellent lithium storage properties, Sustainable Energy & Fuels, 1 (2017) 767-779. https://doi.org/10.1039/C7SE00102A

[52] C. Xiao, S. Zhang, S. Wang, Y. Xing, R. Lin, X. Wei, W. Wang, ZnO nanoparticles encapsulated in a 3D hierarchical carbon framework as anode for lithium ion battery, Electrochimica Acta, 189 (2016) 245-251. https://doi.org/10.1016/j.electacta.2015.11.045

[53] Q. Zhao, H. Xie, H. Ning, J. Liu, H. Zhang, L. Wang, X. Wang, Y. Zhu, S. Li, M. Wu, Intercalating petroleum asphalt into electrospun ZnO/Carbon nanofibers as enhanced free-standing anode for lithium-ion batteries, Journal of Alloys and Compounds, 737 (2018) 330-336. https://doi.org/10.1016/j.jallcom.2017.12.091

[54] T. Kim, H. Kim, J.-M. Han, J. Kim, ZnO-embedded N-doped porous carbon nanocomposite as a superior anode material for lithium-ion batteries, Electrochimica Acta, 253 (2017) 190-199. https://doi.org/10.1016/j.electacta.2017.09.079

[55] L. Xiao, E. Li, J. Yi, W. Meng, S. Wang, B. Deng, J. Liu, Enhancing the performance of nanostructured ZnO as an anode material for lithium-ion batteries by polydopamine-derived carbon coating and confined crystallization, Journal of Alloys and Compounds, 764 (2018) 545-554. https://doi.org/10.1016/j.jallcom.2018.06.081

[56] F. Sun, J. Gao, H. Wu, X. Liu, L. Wang, X. Pi, Y. Lu, Confined growth of small ZnO nanoparticles in a nitrogen-rich carbon framework: advanced anodes for long-life Li-ion batteries, Carbon, 113 (2017) 46-54. https://doi.org/10.1016/j.carbon.2016.11.039

[57] J.W. Rasmussen, E. Martinez, P. Louka, D.G. Wingett, Zinc oxide nanoparticles for selective destruction of tumor cells and potential for drug delivery applications, Expert opinion on drug delivery, 7 (2010) 1063-1077. https://doi.org/10.1517/17425247.2010.502560

[58] H.M. Xiong, ZnO nanoparticles applied to bioimaging and drug delivery, Advanced Materials, 25 (2013) 5329-5335. https://doi.org/10.1002/adma.201301732

[59] H. Peng, B. Cui, G. Li, Y. Wang, N. Li, Z. Chang, Y. Wang, A multifunctional β-CD-modified Fe3O4@ ZnO: Er3+, Yb3+ nanocarrier for antitumor drug delivery and microwave-triggered drug release, Materials Science and Engineering: C, 46 (2015) 253-263. https://doi.org/10.1016/j.msec.2014.10.022

[60] H. Sharma, K. Kumar, C. Choudhary, P.K. Mishra, B. Vaidya, Development and characterization of metal oxide nanoparticles for the delivery of anticancer drug, Artificial cells, nanomedicine, and biotechnology, 44 (2016) 672-679. https://doi.org/10.3109/21691401.2014.978980

[61] Y. Zhang, T. R Nayak, H. Hong, W. Cai, Biomedical applications of zinc oxide nanomaterials, Current molecular medicine, 13 (2013) 1633-1645. https://doi.org/10.2174/1566524013666131111130058

[62] M. Martínez-Carmona, Y. Gun'Ko, M. Vallet-Regí, ZnO nanostructures for drug delivery and theranostic applications, Nanomaterials, 8 (2018) 268. https://doi.org/10.3390/nano8040268

[63] V. Sharma, D. Anderson, A. Dhawan, Zinc oxide nanoparticles induce oxidative DNA damage and ROS-triggered mitochondria mediated apoptosis in human liver cells (HepG2), Apoptosis, 17 (2012) 852-870. https://doi.org/10.1007/s10495-012-0705-6

[64] Y. Deng, H. Zhang, The synergistic effect and mechanism of doxorubicin-ZnO nanocomplexes as a multimodal agent integrating diverse anticancer therapeutics, International Journal of Nanomedicine, 8 (2013) 1835. https://doi.org/10.2147/IJN.S43657

[65] N. Tripathy, R. Ahmad, H.A. Ko, G. Khang, Y.-B. Hahn, Enhanced anticancer potency using an acid-responsive ZnO-incorporated liposomal drug-delivery system, Nanoscale, 7 (2015) 4088-4096. https://doi.org/10.1039/C4NR06979J

[66] K.-J. Bai, K.-J. Chuang, C.-M. Ma, T.-Y. Chang, H.-C. Chuang, Human lung adenocarcinoma cells with an EGFR mutation are sensitive to non-autophagic cell death induced by zinc oxide and aluminium-doped zinc oxide nanoparticles, The Journal of Toxicological Sciences, 42 (2017) 437-444. https://doi.org/10.2131/jts.42.437

[67] Y. Cao, M. Roursgaard, A. Kermanizadeh, S. Loft, P. Møller, Synergistic effects of zinc oxide nanoparticles and fatty acids on toxicity to caco-2 cells, International Journal of Toxicology, 34 (2015) 67-76. https://doi.org/10.1177/1091581814560032

[68] X. Fang, L. Jiang, Y. Gong, J. Li, L. Liu, Y. Cao, The presence of oleate stabilized ZnO nanoparticles (NPs) and reduced the toxicity of aged NPs to Caco-2 and HepG2 cells, Chemico-Biological Interactions, 278 (2017) 40-47. https://doi.org/10.1016/j.cbi.2017.10.002

[69] J. Liu, X. Ma, S. Jin, X. Xue, C. Zhang, T. Wei, W. Guo, X.-J. Liang, Zinc oxide nanoparticles as adjuvant to facilitate doxorubicin intracellular accumulation and visualize pH-responsive release for overcoming drug resistance, Molecular pharmaceutics, 13 (2016) 1723-1730. https://doi.org/10.1021/acs.molpharmaceut.6b00311

[70] N. Puvvada, S. Rajput, B. Kumar, S. Sarkar, S. Konar, K.R. Brunt, R.R. Rao, A. Mazumdar, S.K. Das, R. Basu, Novel ZnO hollow-nanocarriers containing paclitaxel targeting folate-receptors in a malignant pH-microenvironment for effective monitoring and promoting breast tumor regression, Scientific reports, 5 (2015) 1-15. https://doi.org/10.1038/srep11760

[71] R. Dhivya, J. Ranjani, J. Rajendhran, J. Mayandi, J. Annaraj, Enhancing the anti-gastric cancer activity of curcumin with biocompatible and pH sensitive PMMA-AA/ZnO nanoparticles, Materials Science and Engineering: C, 82 (2018) 182-189. https://doi.org/10.1016/j.msec.2017.08.058

[72] R. Dhivya, J. Ranjani, P.K. Bowen, J. Rajendhran, J. Mayandi, J. Annaraj, Biocompatible curcumin loaded PMMA-PEG/ZnO nanocomposite induce apoptosis and cytotoxicity in human gastric cancer cells, Materials Science and Engineering: C, 80 (2017) 59-68. https://doi.org/10.1016/j.msec.2017.05.128

[73] Z.-Y. Zhang, H.-M. Xiong, Photoluminescent ZnO nanoparticles and their biological applications, Materials, 8 (2015) 3101-3127. https://doi.org/10.3390/ma8063101

[74] L.-E. Shi, Z.-H. Li, W. Zheng, Y.-F. Zhao, Y.-F. Jin, Z.-X. Tang, Synthesis, antibacterial activity, antibacterial mechanism and food applications of ZnO nanoparticles: a review, Food Additives & Contaminants: Part A, 31 (2014) 173-186. https://doi.org/10.1080/19440049.2013.865147

[75] Y. Jiang, L. Zhang, D. Wen, Y. Ding, Role of physical and chemical interactions in the antibacterial behavior of ZnO nanoparticles against E. coli, Materials Science and Engineering: C, 69 (2016) 1361-1366. https://doi.org/10.1016/j.msec.2016.08.044

[76] R. Dutta, B.P. Nenavathu, M.K. Gangishetty, A. Reddy, Antibacterial effect of chronic exposure of low concentration ZnO nanoparticles on E. coli, Journal of Environmental Science and Health, Part A, 48 (2013) 871-878. https://doi.org/10.1080/10934529.2013.761489

[77] S. Sarwar, S. Chakraborti, S. Bera, I.A. Sheikh, K.M. Hoque, P. Chakrabarti, The antimicrobial activity of ZnO nanoparticles against Vibrio cholerae: Variation in response depends on biotype, Nanomedicine: Nanotechnology, Biology and Medicine, 12 (2016) 1499-1509. https://doi.org/10.1016/j.nano.2016.02.006

[78] K. Ghule, A.V. Ghule, B.-J. Chen, Y.-C. Ling, Preparation and characterization of ZnO nanoparticles coated paper and its antibacterial activity study, Green Chemistry, 8 (2006) 1034-1041. https://doi.org/10.1039/b605623g

[79] I. Matai, A. Sachdev, P. Dubey, S.U. Kumar, B. Bhushan, P. Gopinath, Antibacterial activity and mechanism of Ag-ZnO nanocomposite on S. aureus and GFP-expressing antibiotic resistant E. coli, Colloids and Surfaces B: Biointerfaces, 115 (2014) 359-367. https://doi.org/10.1016/j.colsurfb.2013.12.005

[80] X. Bellanger, P. Billard, R. Schneider, L. Balan, C. Merlin, Stability and toxicity of ZnO quantum dots: Interplay between nanoparticles and bacteria, Journal of Hazardous Materials, 283 (2015) 110-116. https://doi.org/10.1016/j.jhazmat.2014.09.017

[81] M. Ramani, S. Ponnusamy, C. Muthamizhchelvan, J. Cullen, S. Krishnamurthy, E. Marsili, Morphology-directed synthesis of ZnO nanostructures and their antibacterial activity, Colloids and Surfaces B: Biointerfaces, 105 (2013) 24-30. https://doi.org/10.1016/j.colsurfb.2012.12.056

[82] M. Divya, B. Vaseeharan, M. Abinaya, S. Vijayakumar, M. Govindarajan, N.S. Alharbi, S. Kadaikunnan, J.M. Khaled, G. Benelli, Biopolymer gelatin-coated zinc oxide nanoparticles showed high antibacterial, antibiofilm and anti-angiogenic activity, Journal of Photochemistry and Photobiology B: Biology, 178 (2018) 211-218. https://doi.org/10.1016/j.jphotobiol.2017.11.008

[83] K.M. Reddy, K. Feris, J. Bell, D.G. Wingett, C. Hanley, A. Punnoose, Selective toxicity of zinc oxide nanoparticles to prokaryotic and eukaryotic systems, Applied physics letters, 90 (2007) 213902. https://doi.org/10.1063/1.2742324

[84] L. Ferrero-Miliani, O. Nielsen, P. Andersen, S. Girardin, Chronic inflammation: importance of NOD2 and NALP3 in interleukin-1β generation, Clinical & Experimental Immunology, 147 (2007) 227-235. https://doi.org/10.1111/j.1365-2249.2006.03261.x

[85] P. Nagajyothi, S.J. Cha, I.J. Yang, T. Sreekanth, K.J. Kim, H.M. Shin, Antioxidant and anti-inflammatory activities of zinc oxide nanoparticles synthesized using Polygala tenuifolia root extract, Journal of Photochemistry and Photobiology B: Biology, 146 (2015) 10-17. https://doi.org/10.1016/j.jphotobiol.2015.02.008

[86] P. Thatoi, R.G. Kerry, S. Gouda, G. Das, K. Pramanik, H. Thatoi, J.K. Patra, Photo-mediated green synthesis of silver and zinc oxide nanoparticles using aqueous extracts of two mangrove plant species, Heritiera fomes and Sonneratia apetala and investigation of their biomedical applications, Journal of Photochemistry and Photobiology B: Biology, 163 (2016) 311-318. https://doi.org/10.1016/j.jphotobiol.2016.07.029

[87] M. Ilves, J. Palomäki, M. Vippola, M. Lehto, K. Savolainen, T. Savinko, H. Alenius, Topically applied ZnO nanoparticles suppress allergen induced skin inflammation but induce vigorous IgE production in the atopic dermatitis mouse model, Particle and fibre toxicology, 11 (2014) 1-12. https://doi.org/10.1186/s12989-014-0038-4

[88] C. Wiegand, U.-C. Hipler, S. Boldt, J. Strehle, U. Wollina, Skin-protective effects of a zinc oxide-functionalized textile and its relevance for atopic dermatitis, Clinical, Cosmetic and Investigational Dermatology, 6 (2013) 115. https://doi.org/10.2147/CCID.S44865

[89] S. Yao, X. Feng, J. Lu, Y. Zheng, X. Wang, A.A. Volinsky, L.-N. Wang, Antibacterial activity and inflammation inhibition of ZnO nanoparticles embedded TiO2 nanotubes, Nanotechnology, 29 (2018) 244003. https://doi.org/10.1088/1361-6528/aabac1

[90] J. Li, H. Chen, B. Wang, C. Cai, X. Yang, Z. Chai, W. Feng, ZnO nanoparticles act as supportive therapy in DSS-induced ulcerative colitis in mice by maintaining gut homeostasis and activating Nrf2 signaling, Scientific Reports, 7 (2017) 1-11. https://doi.org/10.1038/srep43126

[91] V.A. Coleman, C. Jagadish, Zinc oxide bulk, thin films and nanostructures, UK, Elsevier, (2006) 1-5. https://doi.org/10.1016/B978-008044722-3/50001-4

[92] S. Tarish, Y. Xu, Z. Wang, F. Mate, A. Al-Haddad, W. Wang, Y. Lei, Highly efficient biosensors by using well-ordered ZnO/ZnS core/shell nanotube arrays, Nanotechnology, 28 (2017) 405501. https://doi.org/10.1088/1361-6528/aa82b0

[93] U.D. Kamaci, M. Kamaci, Selective and sensitive zno quantum dots based fluorescent biosensor for detection of cysteine, Journal of Fluorescence, 31 (2021) 401-414. https://doi.org/10.1007/s10895-020-02671-3

[94] L. Zhao, H. Li, J. Meng, A.C. Wang, P. Tan, Y. Zou, Z. Yuan, J. Lu, C. Pan, Y. Fan, Reversible conversion between schottky and ohmic contacts for highly sensitive, multifunctional biosensors, Advanced Functional Materials, 30 (2020) 1907999. https://doi.org/10.1002/adfm.201907999

[95] M. Zhao, J. Shang, H. Qu, R. Gao, H. Li, S. Chen, Fabrication of the Ni/ZnO/BiOI foam for the improved electrochemical biosensing performance to glucose, Analytica Chimica Acta, 1095 (2020) 93-98. https://doi.org/10.1016/j.aca.2019.10.033

[96] H.-M. Kim, J.-H. Park, S.-K. Lee, Fiber optic sensor based on ZnO nanowires decorated by Au nanoparticles for improved plasmonic biosensor, Scientific reports, 9 (2019) 1-9. https://doi.org/10.1038/s41598-019-52056-1

[97] F. Xue, J. Liang, H. Han, Synthesis and spectroscopic characterization of water-soluble Mn-doped ZnOxS1− x quantum dots, Spectrochimica Acta Part A: Molecular and Biomolecular Spectroscopy, 83 (2011) 348-352. https://doi.org/10.1016/j.saa.2011.08.045

[98] H.-M. Xiong, Y. Xu, Q.-G. Ren, Y.-Y. Xia, Stable aqueous ZnO@ polymer core−shell nanoparticles with tunable photoluminescence and their application in cell imaging, Journal of the American Chemical Society, 130 (2008) 7522-7523. https://doi.org/10.1021/ja800999u

[99] Y. Liu, K. Ai, Q. Yuan, L. Lu, Fluorescence-enhanced gadolinium-doped zinc oxide quantum dots for magnetic resonance and fluorescence imaging, Biomaterials, 32 (2011) 1185-1192. https://doi.org/10.1016/j.biomaterials.2010.10.022

[100] S.P. Singh, Multifunctional magnetic quantum dots for cancer theranostics, Journal of Biomedical Nanotechnology, 7 (2011) 95-97. https://doi.org/10.1166/jbn.2011.1219

[101] N. Jones, B. Ray, K.T. Ranjit, A.C. Manna, Antibacterial activity of ZnO nanoparticle suspensions on a broad spectrum of microorganisms, FEMS microbiology letters, 279 (2008) 71-76. https://doi.org/10.1111/j.1574-6968.2007.01012.x

[102] N.-H. Cho, T.-C. Cheong, J.H. Min, J.H. Wu, S.J. Lee, D. Kim, J.-S. Yang, S. Kim, Y.K. Kim, S.-Y. Seong, A multifunctional core-shell nanoparticle for dendritic cell-based cancer immunotherapy, Nature nanotechnology, 6 (2011) 675-682. https://doi.org/10.1038/nnano.2011.149

[103] X. Huang, X. Zheng, Z. Xu, C. Yi, ZnO-based nanocarriers for drug delivery application: From passive to smart strategies, International journal of pharmaceutics, 534 (2017) 190-194. https://doi.org/10.1016/j.ijpharm.2017.10.008

[104] S. Kim, S.Y. Lee, H.-J. Cho, Doxorubicin-wrapped zinc oxide nanoclusters for the therapy of colorectal adenocarcinoma, Nanomaterials, 7 (2017) 354. https://doi.org/10.3390/nano7110354

[105] E. Taylor, T.J. Webster, Reducing infections through nanotechnology and nanoparticles, International journal of nanomedicine, 6 (2011) 1463. https://doi.org/10.2147/IJN.S22021

Keyword Index

About the Editors

Dr. Gaurav Sharma

International Research Centre of Nanotechnology for Himalayan Sustainability (IRCNHS), Shoolini University, Solan, 173229, Himachal Pradesh, India

Email: Gaurav.541@shooliniuniversity.com

Dr. Gaurav Sharma research activity started in 2009 at Shoolini University (India) as a master of philosophy student, and then, he continued his research work as PhD student with the preparation and characterization of diverse multifunctional nanomaterials, and their composites, specially focused on potential applications in environmental remediation (as photocatalysts and adsorbents). For four years he worked as assistant professor in the School of chemistry at Shoolini University (India), where he carried out diverse research lines, interrelated to each other based on synthesis and characterization of nanocomposites, hydrogels, bi and trimetallic nanoparticles, ion exchangers, adsorbents and photocatalysts etc. Moreover, he performed and taught different courses as nanochemistry, polymer chemistry, spectroscopy and natural products, among others. On the other hand, he supervised 3 PhD, 5 Master of Philosophy, and more than 25 Master and Bachelors students. He established collaborative research with various professors in countries as Finland, Saudi Arabia, China, Spain and South Africa. In this context, he was invited as visiting research professor from University of KwaZuklu-Natal (South Africa) in 2017 and 2019. In 2017, he joined as postdoctoral fellow at college of materials science and engineering, Shenzhen University. He got project from china postdoctoral science foundation in 2018. The outcome of his research work was depicted in more than 150 publications, in various journals such as Renewable and Sustainable Energy Reviews, Chemical Engineering Journal, Journal of Cleaner Production, Carbohydrate Polymers, ACS Applied Materials and Interfaces, Journal of Hazardous Materials, Applied Catalysis B, and International Journal of Biological Macromolecules etc, 9 book chapters and 7 edited books. He is also serving as Director, International Research Centre of Nanotechnology for Himalayan Sustainability (IRCNHS), Shoolini University, India. He is a Highly Cited Researcher - 2020, 2021 Crossfield (Web of Science); and also Ranked among the World top 2% Scientists (Current year 2019, 2020, &2021 category) as per Stanford.

His h-index is 66, citations: more than 11000 (web of science); Google Scholar: h-index is 68, citations: more than 11000. He is Associate Editor of the International Journal of Environmental Science and Technology (Springer). Editorial Board member of Current Organic Chemistry, Current analytical Chemistry, Materials-MDPI Innovations in Corrosion and Materials Science, Journal of Nanostructure in Chemistry, Nanotechnology for Environmental Engineering(Springer), Letters in Applied

NanoBioScience etc, and Academic Editor of Journal of Nanomaterials, Advances in Polymer Technology.

Dr. Amit Kumar

International Research Centre of Nanotechnology for Himalayan Sustainability (IRCNHS), Shoolini University, Solan, 173229, Himachal Pradesh, India

Email: amitkumar@shooliniuniversity.com

Dr. Amit Kumar started his research career with a PhD in Chemistry at Himachal Pradesh University, Shimla, India in 2008. He was awarded with the prestigious predoctoral scholarship of the Council of Scientific and Industrial Research (Government of India) in 2007. After his PhD, he worked as assistant professor at Shoolini University (India) for 3.5 years, where he was involved in research and teaching at undergraduate, postgraduate and doctoral level (2014-2017). He has experience of supervising three PhD students (all graduated). Furthermore, he has supervised 10 Masters students. He has worked as post-doctoral fellow in the College of Materials Science and Engineering of Shenzhen University (2017-2019). He is involved in field of optical designing of semiconductor heterojunctions as catalysts for environmental remediation and energy production. Currently He is a senior researcher at College of Materials Science and Engineering of Shenzhen University, PR China (01.01.2020 onwards). In addition, he holds position of Co-director, International Research Centre of Nanotechnology for Himalayan Sustainability (IRCNHS), Shoolini University, India. He is also a Visiting Professor at School of technology, Glocal University, India. Moreover, he is a visiting faculty at Department of Chemistry and Physics from University of KwaZulu Natal, South Africa.

He has diverse research collaborations in Instituto de Catálisis y Petroleoquímica (Spain), King Saud University (Saudi Arabia), Chinese Research Academy of Environmental Sciences (China) and Universidade Federal do Rio Grande do Sul (Brazil). He is also ranked among the World top 2% Scientists (Current year 2019, 2020, 2021 category) as per Stanford University rankings, 2020. He is a Highly Cited Researcher – 2021 (Web of Science).

His scientific production encompasses 172 papers (WoS) including Applied Catalysis B: Environmental, Chemical Engineering Journal, ACS Applied Materials and Interfaces, Journal of Hazardous Materials, Journal of Cleaner Production etc. He has received h-index 57 and more than 10000 citations.

Dr.Pooja Dhiman

International Research Centre of Nanotechnology for Himalayan Sustainability (IRCNHS), Shoolini University, Solan, 173229, Himachal Pradesh, India

Email id:poojadhiman@shooliniuniversity.com

Dr. Pooja Dhiman started her research career with a Ph.D. and M.Phil. in Physics from Himachal Pradesh University, Shimla, India. She is currently working as Assistant Professor of Physics at Shoolini University, H.P. She has a total teaching experience of more than 8 years and research experience of 15 years. Currently, she is involved in the field of spintronics, multi-ferroic materials and optical designing of semiconductor heterojunctions as catalysts with high interfacial contact for environmental remediation as persistent and new arising pollutant removal. She is a professional member of IEFRP and also holds fellow membership of ISCA, HSCA. She has been awarded with the best researcher award in 2020, 2021 under NESIN by science father registered under ministry of corporate affairs.

She has published more than 55 publications in journals of repute and has published 7 book chapters and has edited 1 book. Her highest impact factor publication (7.246) is in the journal of cleaner production. She has an h-index of 22 (SCOPUS) and 24 (Google scholar). The total number of citations achieved is 1993 (Google Scholar). She has experience of supervising many research candidates.

www.ingramcontent.com/pod-product-compliance
Lightning Source LLC
Chambersburg PA
CBHW071324210326
41597CB00015B/1338